Cell Interaction - Molecular and Immunological Basis for Disease Management

Edited by Bhawana Singh

Published in London, United Kingdom

IntechOpen

Supporting open minds since 2005

Cell Interaction - Molecular and Immunological Basis for Disease Management
http://dx.doi.org/10.5772/intechopen.78995
Edited by Bhawana Singh

Contributors
Jehane Eid, Maram Abdelghany, Sarah Abdullah Gozai Alghamdi, Isotta Chimenti, Roberto Carnevale, Vittorio Picchio, Vittoria Cammisotto, Francesca Pagano, MªCarmen Duran-Ruiz, Margarita Jimenez-Palomares, Alba Cristobal, Tamiris Azamor, Zilton Vasconcelos, Ana Paula Dinis Ano Bom, Juliana Gil Melgaço, Amanda Torrentes-Carvalho, Rajnikant Dixit, Charu Chauhan, sanjay tevatiya, Seena Kumari, Jyoti Rani, Sharma Punita, Sevil Oskay Halacli, Shyam Babu Prasad, Rahul Kumar, Kadriye Kızılbey, Nelisa Türkoğlu, Fatma Ceren Kırmızıtaş, Jose Luis Muñoz-Carrillo, Juan Francisco Contreras Cordero, Paola Trinidad Villalobos-Gutiérrez, Oscar Gutiérrez-Coronado, Luis Guillermo Ramos-Gracia, Jazmín Monserrat Vargas-Barboza, Siddanakoppalau N. Pramod

First published in London, United Kingdom, 2021 by IntechOpen
IntechOpen is the global imprint of INTECHOPEN LIMITED, registered in England and Wales, registration number: 11086078, 5 Princes Gate Court, London, SW7 2QJ, United Kingdom
Printed in Croatia

British Library Cataloguing-in-Publication Data
A catalogue record for this book is available from the British Library

Additional hard and PDF copies can be obtained from orders@intechopen.com

Cell Interaction - Molecular and Immunological Basis for Disease Management
Edited by Bhawana Singh
p. cm.
Print ISBN 978-1-83968-416-6
Online ISBN 978-1-83968-417-3
eBook (PDF) ISBN 978-1-83968-418-0

For EU product safety concerns:
IN TECH d.o.o., Prolaz Marije Krucifikse Kozulić 3, 51000 Rijeka, Croatia, info@intechopen.com or visit our website at intechopen.com.

Meet the editor

Dr. Bhawana Singh is currently working as a scientist in the Department of Medicine, Institute of Medical Sciences, Banaras Hindu University, India, where she also obtained her Ph.D. working on immuno-genetic aspects of visceral leishmaniasis. She has a broad background in cellular and molecular immunology in parasitic disease with significant epidemiological and field experience working with human subjects. During her Ph.D., Dr. Singh received training at QIMR Berghofer Medical Research Institute, Brisbane, Australia, and presented her work on several international platforms. Her research has been published in peer-reviewed international journals of repute. She obtained post-doctoral experience at Wexner Medical Center, The Ohio State University, Columbus, USA. There, she worked on host-directed immunotherapy for relieving immunosuppression in chronic diseases using immunomodulators.

Contents

Preface

There has been tremendous progress in the field of immunology in recent years, which could be attributed to the recent adoption of high-throughput technological approaches for genomics, transcriptomics, and proteomics. New techniques have revolutionized biomedical science in terms of innovative platforms for diagnostics, therapeutics, and prophylactics. These technological advances and their role in accelerating basic and clinical research sparked the idea for this book, which brings together relevant research of the methodology and applications for gaining better insight into human diseases.

Cells sense and respond to a variety of signals in terms of the physiological functions of living organisms. In the context of immune cells, signalling events are crucial drivers in recruiting the cells to the site of infection and interaction with other cells (cell-cell interaction) that leads to the release of soluble factors such as cytokines. These soluble factors travel through the bloodstream and mediate their effects on neighbouring or distant cells, which eventually determines the outcome of disease via a series of complex cellular interactions. Cellular interactions involved in the recognition of pathogens and their molecules begins the cell signalling cascade by ligation of cell surface receptors, activation of transcription factors and downstream components that drives the outcome of infection. Overall, signal transduction cassettes include the membrane receptors, effector signalling and regulatory proteins that coordinate to detect, amplify, and orchestrate the external signals of cellular responses. During infection, the main immunological players, innate and adaptive immune cells, are key in combatting the infection. As per the conventional cell signalling process, binding of the antigen generates an intracellular signalling cascade, however, there are a diversity of antigenic receptors that trigger many signals for orchestrating immunological outcomes.

We have focussed on understanding the innate and adaptive arms of immunity that recognize foreign protein segments based on a programmed series of immune cell synapses between T cells and antigen-presenting cells (APCs). These interactions are crucial for immediate and long-term responsiveness of T cells to major histocompatibility complex (MHC)–peptide complexes depending on the thresholds that are tuned at the cell synapses and determine immune responses. The fine-tuning of these thresholds remains vital for the host, as an overreaction to self-proteins can be the cause of autoimmune dysfunctions, whereas underreaction to pathogens can equally be detrimental in terms of the development of susceptibilities to a variety of infections and tumours. Apart from immunological synapses, the importance of cytoskeleton dynamics, cell asymmetry, and membrane patterning for the setting threshold for immune cell activation also play an important role in immune cell signalling and regulation of immune response.

Currently, much attention has been paid to understanding the role of molecular events in immunological processes. Our current understanding is based on an array of model systems and with the advent of recent technological platforms, it is likely that many surprises will be revealed.

Keeping these facts in mind and expanding our knowledge about immune cell signalling and its role in disease management, this book has been organized into two sections. It begins with the tools and recent use of model systems for understanding disease pathology, which will eventually help in designing strategies for disease management, which has been covered in the second section. Briefly, the chapter outlines are as follows:

Chapter 1 reviews the flow cytometric approach for primary immunodeficiency diseases. It examines different techniques for immunophenotyping molecules for better understanding of immunological interactions. Chapter 2 briefly discusses the use of organoid models for cell-cell interactions and their benefits for regenerative medicine and designing personalized therapies. Chapter 3 discusses recent innovative models for studying lung diseases. It includes a comprehensive outlook on organ-on-chip and organoids for a better understanding of lung pathologies and designing strategies for the management of lung fibrosis. Chapter 4 examines nanoparticle-based immunotherapy for managing cancer. Chapter 5 comprehensively reviews immunological signalling for the development of allergies. The chapter briefly covers eosinophils, basophils, immunoglobulins, Th2 responses, and histamines as the crucial determinants for the development of allergies. Chapter 6 provides up-to-date information on the molecular understanding of SARS-CoV2 and other coronavirus ancestors. The chapter gives an outlook about the molecular tools for dissecting the virus and that help in targeting specific molecular structures for designing vaccines and managing the epidemic outbreak. Chapter 7 provides a comprehensive outlook on the molecular dynamics of vector and malarial parasite interactions, with the aim to provide insight into host-parasite cellular interactions to abrogate disease transmission. Chapter 8 reviews and summarizes the recent advances in understanding dendritic cells for pathogenic infections. It covers different aspects of immunological interactions orchestrated via dendritic cells for regulating immunological outcomes. Chapter 9 covers recent information and gaps in the understanding of modulation in innate immune responses against Zika virus infection during pregnancy. It provides a brief discussion on recent technological advances that can overcome current knowledge gaps. Finally, Chapter 10 addresses the role of toll-like receptors (TLRs) in cancer progression and how these receptors can be targeted for the management of a disease.

The two sections of this volume include the study of different molecular phenomena that eventually drive infection outcomes as well as manipulation of this phenomena for effective treatment of disease symptoms. Different signalling mediators drive the production of messenger molecules that mediate their action, leading to the elicitation/suppression of immune responses. The book presents a balanced approach in line with the previous edition, which also explained various molecular and immunological tools for better understanding of cellular interactions. I hope this edition provides students and researchers comprehensive information on cellular interactions in terms of disease management aspects and their utility in bench side as well as bedside applications.

Bhawana Singh
Department of Medicine,
Institute of Medical Sciences,
Banaras Hindu University,
Varanasi, India

Section 1

Tools for Studying Cell Interactions

Chapter 1

Flow Cytometric Approach in the Diagnosis of Primary Immunodeficiencies

Sevil Oskay Halacli

Abstract

Primary Immunodeficiencies (PIDs) compose of a large spectrum of diseases characterized by abrogated or dysregulated functions of innate and adaptive immune system components that cause susceptibility to recurrent infections, autoimmunity, neoplasia/malignancy and dysfunction of organs and skeletal system. PIDs are also evaluated as molecular diseases due to the mutations in one or more genes. That affects transcripts and protein expressions as well as their functions. Today, 430 different genes are known to have various functional effects which are related to 403 different PIDs. Analyzing the effects of the mutations on relevant protein expression and function is significant to diagnose and the follow-up of the PIDs. Application of flow cytometry for analyzing protein expression levels and functions in immune cells as well as investigating the cellular functions tender a rapid, quantitative and reliable approach to identify and to prove the genetic background of PIDs. Therefore, the use of flow cytometry aids to have a large spectrum of data from gene to function and from function to clinical relevance in the first-step and differantial diagnosis of PIDs.

Keywords: Primary immunodeficiency, flow cytometry, molecular diagnosis, immunophenotyping, PBMC culture, functional assays, intracellular staining, PI3K pathway analysis- flow, CFSE cell proliferation

1. Introduction

Primary immunodeficiencies (PIDs) are rare and heterogenous genetic diseases of the immune system. According to updated IUIS (International Union of Immunological Societies) classification in 2019, there is a large spectrum of PIDs including 403 different diseases caused by mutations in 430 genes categorized 10 different subclasses with these topics: Severe combined immunodeficiencies (SCIDs), combined immunodeficiencies (CIDs) less profound than SCID, CIDs with associated or syndromic features and predominantly antibody deficiencies including common variable immunodeficiency (CVID), immune dysregulation, phagocyte system defects, innate immune defects, auto-inflammation, complement deficiencies, bone marrow abnormalities and phenocopies of PIDs. Each disease has unique laboratory and clinical manifestations. Decreased or increased immune cell counts, unbalanced immune cell plasticity, decreased or increased immunoglobulin levels and complement factors, dysregulated functions of immune cells due to abrogated intracellular molecular functions cause developing clinical manifestations of PIDs [1]. Use of flow cytometry in these laboratory investigations is a significant approach that offers a quantitative, reliable and rapid results. Evaluation of these laboratory findings helps to clinicians for proper diagnose of PIDs [2, 3].

2. Analysis of inflammatory and regulatory cell profiles in PIDs

Immune dysregulation with autoimmunity is observed in many PIDs such as LRBA, CTLA4, STAT3 GOF, PIK3CD deficiencies as well as IPEX syndrome caused by loss or dysfunctional FOXP3 expression [4–18]. Disrupted T helper cell plasticity is pointed out as a prominent feature of the autoimmunity in PIDs. Deregulated numbers and functions of Treg cells are observed in most of the patients with IPEX or IPEX-like (such as in patients with LRBA deficiency) [6, 7, 19–21]. Decreased Treg cell numbers or loss of Treg cell functions are related to severe form of autoimmunities in PIDs. In contrast, deregulated inflammatory cell numbers/ratios and the inflammatory cytokines produced by inflammatory cells are observed as autoimmune manifestations of PIDs such as LRBA and STAT3 LOF deficiencies. In LRBA deficiency, increased number of circulating T folicular helper (Tfh) is associated with autoimmune manifestations of the disease [5]. Moreover, decreased Th17 cell numbers are related to inflammatory response to Candida infections observed in patients with LOF mutations in STAT3 deficiency [22–24].

In these cases, the first attempt is to analyze regulatory and inflammatory cell ratios in the clinical immunology laboratory to clarify the cellular background of autoimmunity.

2.1 Analysis of Treg cells in PIDs

Treg cells are unique subset of T helper cells through its equilibrating functions on immune response to self and foreign antigens. Tregs suppress inflammatory T cell function and proliferation, therefore it plays critical roles to prevent autoimmune disorders. In PIDs with autoimmunity, impaired functions of Treg cells in parallel with decreased number of Treg cells are observed. IPEX is a well-known syndrome affecting Treg cell development due to mutations of FOXP3 which is a main transcription factor in the development of Treg cells. In patients with IPEX syndrome, loss of circulating and tissue associated Treg cells are thought to cause the multi-organ autoimmune manifestations [6, 20, 21]. Patients with CD25 (IL-2Rα) deficiency have IPEX-like phenotype as well as in patients with LRBA deficiency. Decreased Treg ratio is a significant laboratory characteristics in these PIDs [7, 25]. In patients with AIRE deficiency which is related to Autoimmune Poly Endocrinopathy, Candidiasis and Ectodermal Dystrophy (APECED) syndrome, decreased Treg cell ratio and function are associated with the occurrence of the disease [26].

Investigating Treg cell ratio by flow cytometry provides an important insight to understand autoimmunity from the benchside to bedside.

Below, it was described the Treg staining protocol and the gating strategy for human peripheral blood Treg cells (**Figure 1**).

2.1.1 Treg staining protocol

- Peripheral blood mononuclear cells (PBMCs) are separated by ficoll density gradient protocol from 4 ml of whole blood in tube with EDTA.

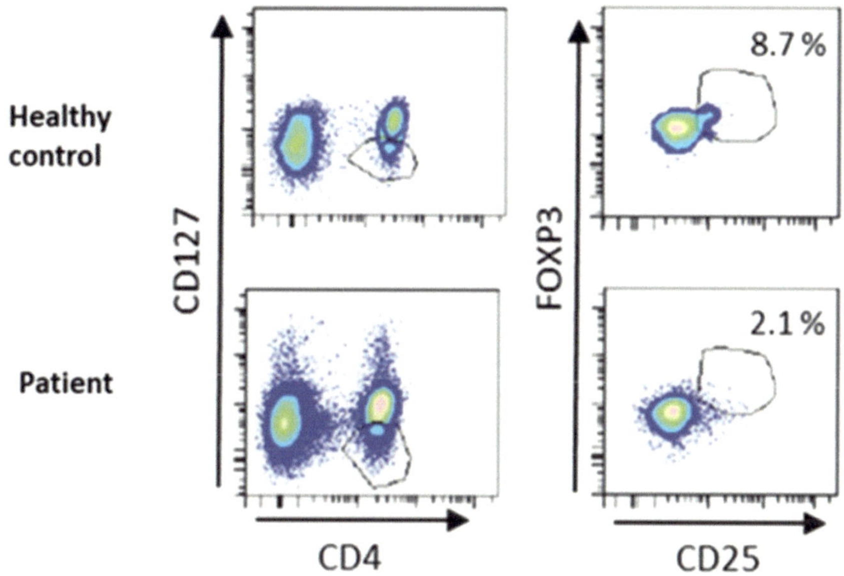

Figure 1.
Representative image of CD4+ CD127^{lo}CD25^{hi} FOXP3+ Treg cells in peripheral blood of healthy control and a patient.

- Wash PBMCs with Phosphate Buffer Saline (PBS) buffer, centrifuge at 300 g for 5 min and discard the supernatant
- Add appropriate volume of PBS and add 100 ul cell to flow cytometer tubes
- Add appropriate volume of CD4, CD127 and CD25 antibodies and incubate at room temperature and dark conditions for 20 min
- Following incubation wash with PBS, centrifuge at 500 g for 5 min and discard the supernatant
- Fix the cells with a fixation buffer for 10–20 min
- Wash with PBS, centrifuge at 500 g for 5 min and discard the supernatant
- Treat with the permeabilization buffer for 10–30 min
- Wash with PBS, centrifuge at 500 g for 5 min and discard the supernatant
- Add FOXP3 antibody for 30 min at room temperature and dark conditions
- Wash with PBS, centrifuge at 500 g for 5 min and discard the supernatant
- Add 300 ul PBS, vortex and analyze in flow cytometer

2.2 Analysis of circulating Tfh and TH17 cells in PIDs

Tfh cells are specialized Th cell subset which plays important role in B cell differentiation in lymph nodes, in producing high affinity antibodies and the development of memory cells. Therefore, Tfh provides help germinal center (GC) formation and selection of plasma cells [27–30]. Tfh cells have unique molecules that are expressed in cell surface and have special functions such as CXCR5. CXCR5 is a chemokine receptor and provides migration of Tfh cells to GC zone. Besides, Tfh expresses B Cell Lymphoma (BCL-6) and (Inducible T Cell Costimulator) ICOS or CD278 on their surfaces. Increased Tfh cell numbers in peripheral blood are investigated as an inflammatory marker of some PIDs such as LRBA deficiency [5].

Th17 cells are also a subset of helper T cells which are responsible for producing IL-17, a pro-inflammatory cytokine recruiting neutrophils to infection site to combat infection [22, 23, 31, 32]. IL-6 expression and STAT3 activation are required for the differentiation of Th17 cells from CD4+ T lymphocytes. Therefore in STAT3 deficiency caused by autosomal dominant loss of function mutations of STAT3 gene, decreased number of circulating Th17 cells are associated with susceptibility to Candida infections in STAT3 LOF deficiency which is a type of Autosomal Dominant- hyper IgE Syndrome (AD-HIES) [24].

Detection of Tfh and Th17 cell ratios in the peripheral blood of the patients with designated PIDs in clinical immunology laboratory by flow cytometry using various surface and intracellular markers which are unique to circulating Tfh and Th17 cells is important step to understand the inflammatory background of the autoimmune manifestations (**Figures 2** and **3**). See the Section 2.1.1. for the staining protocol.

Below, it was demonstrated Tfh and Th17 gating strategy.

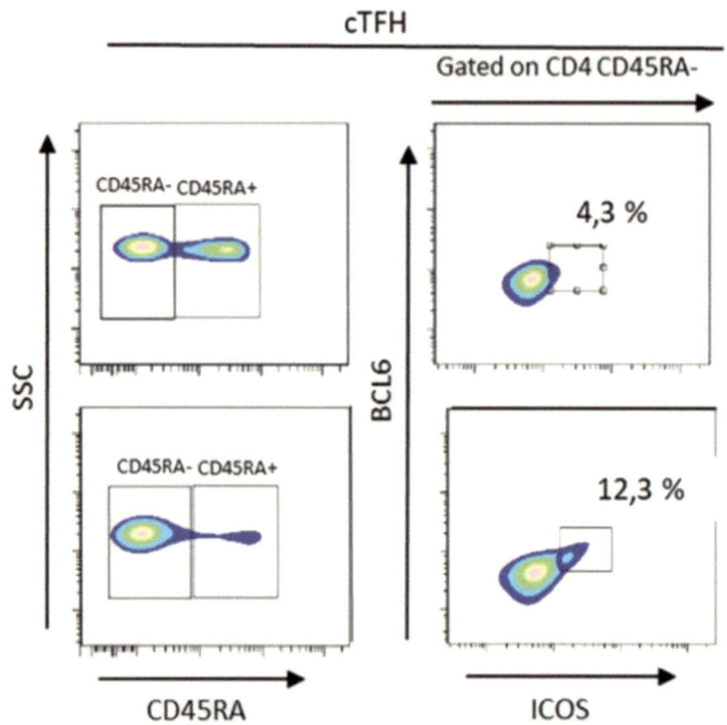

Figure 2.
Analysis of cTfh cells in a healthy control (top) and a patient with PID (below). In the patient, increased ratio of cTfh is observed.

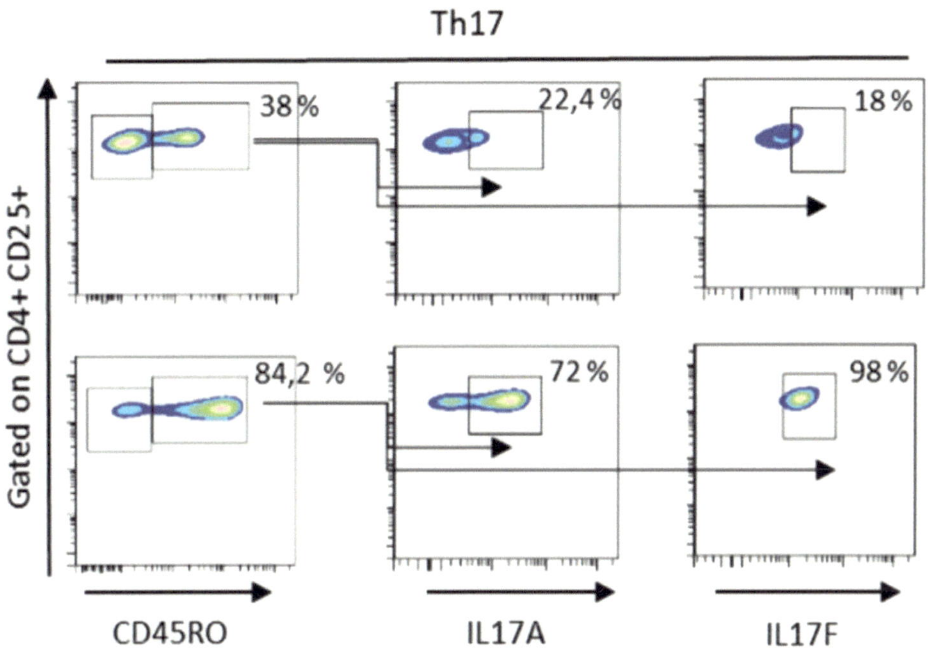

Figure 3.
Th17 gating strategy. Increased ratio of Th17 cells expressing IL17A and IL17F are observed in a patient (below) compared to healthy control (top).

3. Analysis of surface molecules in PIDs

3.1 Evaluation of molecules which are constitutively expressed on cell surface

In the diagnosis of suspicious patients for PID, flow cytometry is frequently applied to detect specific molecules which are expressed on specific subset of immune cells in clinical immunology research laboratory [2, 3]. It is used for immunophenotyping as well as in the detection of specific protein expression in cells. In the evaluation of constitutively expressed proteins on cell surface, activation with specific stimulus is not required. CD40 and CD55 deficiencies are the examples which are described in detail in Section 3.1.1. and 3.1.2 for the surface protein expression analysis in PIDs.

In the staining of surface proteins, fixation and permeabilization steps are not needed. Therefore staining protocol is easier and faster than intracellular staining of the proteins which is described in Section 4. Following staining protocol is used to detect surface protein expressions in PIDs:

- Add 100 ul of whole blood to flow cytometer tube.
- Add appropriate volume of specific antibodies to detect specific proteins and incubate at room temperature and dark conditions for 20–30 min.
- Lyse the erythrocytes using appropriate volume of lysis buffer and incubate for 10–15 min at room temperature and dark conditions.
- Centrifuge at 500 g for 5 min and discard the supernatant

- Wash with PBS, centrifuge at 500 g for 5 min and discard the supernatant
- Add 300 ul PBS, vortex and analyze at flow cytometer

3.1.1 CD40 deficiency in hyper IgM syndrome

CD40 is a costimulatory molecule which is expressed on antigen presenting cells such as B cells, macrophages and dendritic cells. CD40 interacts with CD40L on T cells in GC zones and is activated in the maturation of B cells and isotype switching [33, 34]. Similar to CD40L deficiency, CD40 deficiency is investigated for suspicious Hyper IgM syndromes. Decreased or unfunctional CD40 expression on B lymphocyte as well as CD40L expression defects on T cells in suspicious patients for Hyper IgM syndrome is related to disease occurrence [35, 36]. See the Section 3.1. for the staining protocol.

3.1.2 CD55 expression in CHAPLE syndrome

Decay-accelerating factor (DAF) or CD55 is an inhibitor molecule of complement system and it is related to various diseases and a recently described PID which is named as (CD55 deficiency with hyperactivation of complement, angiopathic thrombosis, and PLE) CHAPLE syndrome. Because CD55 acts as an inhibitor of complement system, low or loss of expressions due to mutations in its encoding gene, complement system is more active in patients than healthy individuals [37–39] (see the Section 3.1. for the staining protocol).

3.2 Analysis of the expression of induced surface proteins in PIDs

3.2.1 CD40L expression in T lymphocytes in hyper IgM syndrome

CD40L, also known as CD154, is expressed on T cells and responsible for the interaction with CD40 which is expressed on antigen presenting cells such as B cells. CD40L is a member of TNF-receptor superfamily and its interaction with CD40 on B cells is associated with Ig class switching, affinity maturation and GC formation. In most of the patients with CD40L deficiency, loss or decreased CD40L protein expression on T cells are associated with increased levels of soluble IgM levels and decreased IgG and IgA levels are investigated [35, 36]. Expression of CD40L protein on T cell surface is very low and increased by activation using Phorbol Myristate Acetate (PMA) and ionomycin inducing transcriptional activity of NFAT and AP-1 transcription factors in T cells following T cell receptor stimulation. Following 3 hours of activation of PBMCs, CD69 which is an early activation marker and CD40L expression are detected on T cell surface (**Figure 4**). Staining protocol of CD40L and CD69 on CD3+ CD8- T cells are as in below:

- Peripheral blood mononuclear cells (PBMCs) are separated by ficoll density gradient protocol from 1 to 2 ml of whole blood in tube with EDTA.
- Wash PBMCs with Phosphate Buffer Saline (PBS) buffer, centrifuge at 300 g for 5 min and discard the supernatant and resuspend the cells with serum free media.

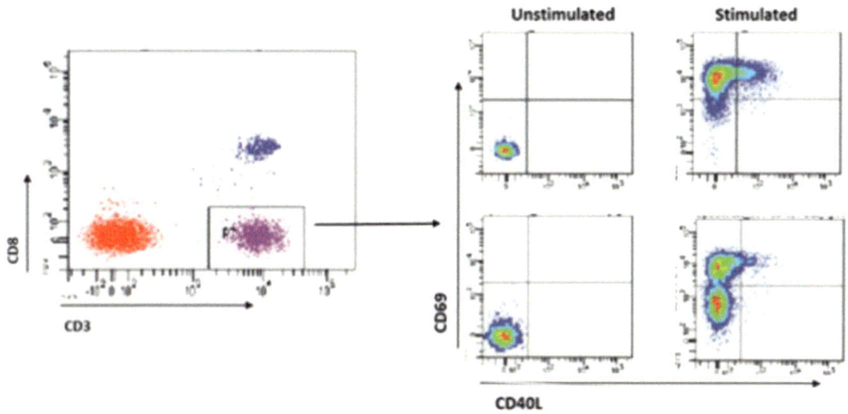

Figure 4.
Gating strategy for CD40L and CD69 expression on CD3+ CD8- T cells in unstimulated and stimulated samples from a healthy control (top) and a patient (below).

- Prepare two flasks for each sample to analyze unstimulated and stimulated samples
- Put the appropriate number of cells to culture flask. Add 1 ug/ml PMA and 500 ng/ml ionomycin to the stimulated culture flask
- Following 3 hours incubation in humidified incubator, wash the cells with PBS and centrifuge at 300 g for 5 min and discard the supernatant
- Resuspend the cells with 1 ml PBS and collect 100 ul of cell to a fresh flow cytometer tubes
- Add CD3, CD8, CD69 and CD40L antibodies at the appropriate concentrations
- Wash with PBS, centrifuge at 500 g for 5 min and discard the supernatant
- Add 300 ul PBS, vortex and analyze at flow cytometer

3.2.2 CD70 expression

CD27/CD70 signaling pathway is significant for the immune response to Epstein–Barr virus (EBV) infections. CD27 is expressed on T lymphocytes as well as B lymphocytes and whereas its ligand, CD70, is limited to induced T and B lymphocytes and dendritic cells. CD27-CD70 signaling is responsible for T cell survival, Treg activity, B cell differentiation and proliferation. Due to CD27-CD70 partnership in immune response against to EBV, similar clinical characteristics are monitored in patients with CD27 and CD70 deficiencies [40–42]. EBV-associated lymphoproliferative disorder, lymphoma, hypogammaglobulinemia and autoimmune manifestations are generalized clinical symptoms in both deficiencies [41, 42]. Therefore, analyzing of CD27 and CD70 proteins in PBMCs using flow cytometry due to its rapid and quantitative analysis guide to clinicians as a first step molecular diagnosis of patients with these clinical manifestations before sequencing. **Figure 5** shows the gating strategy for CD70 staining. Staining protocol for CD27 is as in Section 3.1.

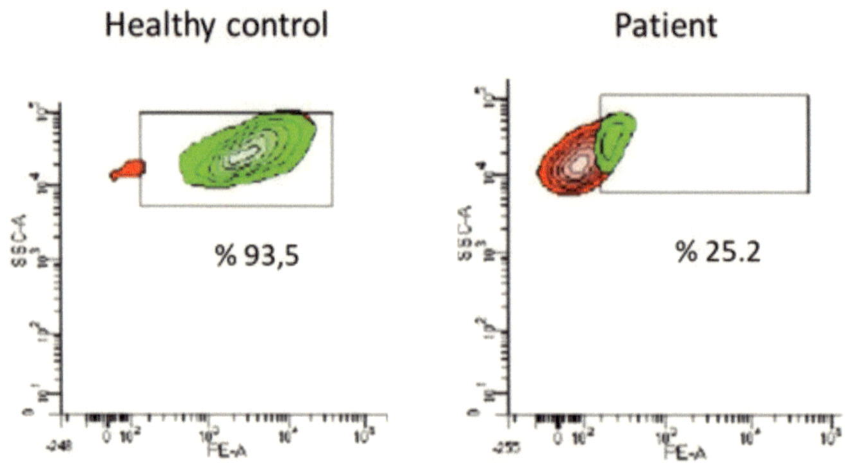

Figure 5.
A representative image of CD70 expression on CD19+ B lymphocyte in a healthy control and a patient.

CD70 activation and staining protocol is as below:

3.2.2.1 Activation of surface expression of CD70 and staining for flow cytometric analysis

- Peripheral blood mononuclear cells (PBMCs) are separated by ficoll density gradient protocol from 1 to 2 ml of whole blood in tube with EDTA
- Wash PBMCs with Phosphate Buffer Saline (PBS) buffer, centrifuge at 300 g for 5 min and discard the supernatant and resuspend the cells with serum free media
- Prepare two flasks for each sample to analyze unstimulated and stimulated samples
- Put the appropriate number of cells to culture flask and add 2,5 ug/ml phyto-hemagglutinin (PHA) in the completed culture media
- Incubate the cells in humidified incubator for 72 hours
- After 72 hours add appropriate volume of IL-2 to the cells
- At the day of 8, wash the cells with PBS
- Centrifuge at 500 g for 5 min and discard the supernatant
- Add appropriate volume of CD70 antibody and incubate for 30 min at room temperature
- Wash the cells with PBS and Centrifuge at 500 g for 5 min and discard the supernatant
- Resuspend the cells with 300 ul PBS and analyze at flow cytometer.

3.2.3 CTLA4 (CD152)

Cytotoxic T lymphocyte Antigen-4 (CTLA4) is an inhibitor ligand of T lymphocytes which bind to CD80/CD86 which is found on antigen presenting cells with higher affinity than a costimulator molecule CD28 [8–10]. CTLA4 ceases signaling axes in T lymphocytes due to its ITIM motifs in the intracytoplasmic domain. Therefore CTLA4 blocks T cell proliferation and act important function in homeostasis and peripheral tolerance. CTLA4 is constitutively expressed on T lymphocytes and it is expressed on cell surface only after stimulation via TCR and Ca+/Calcineurin pathway *in vitro*. In patients with autosomal dominant mutation of CTLA4, lymphadenopathy/splenomegaly, hypogammaglobulinemia, cytopenia and organ specific autoimmunity are observed. This disease is also called "haploinsufficiency with autoimmune infiltration (CHAI) disease" and characterized by unfunctional or loss of CTLA4 expression on T lymphocytes [8–10]. Using flow cytometric approach, suspicious patients with CHAI disease may be investigated for molecular diagnosis before sequencing. **Figure 6** demonstrates the gating strategy for CTLA4 expression in healthy control and a patient with PID. Flow cytometry protocol for CTLA4 activation and staining are below:

3.2.3.1 Staining protocol of CTLA4 in activated PBMCs

- Peripheral blood mononuclear cells (PBMCs) are separated by ficoll density gradient protocol from 1 to 2 ml of whole blood in tube with EDTA

- Wash PBMCs with Phosphate Buffer Saline (PBS) buffer, centrifuge at 300 g for 5 min and discard the supernatant and resuspend the cells with serum free media

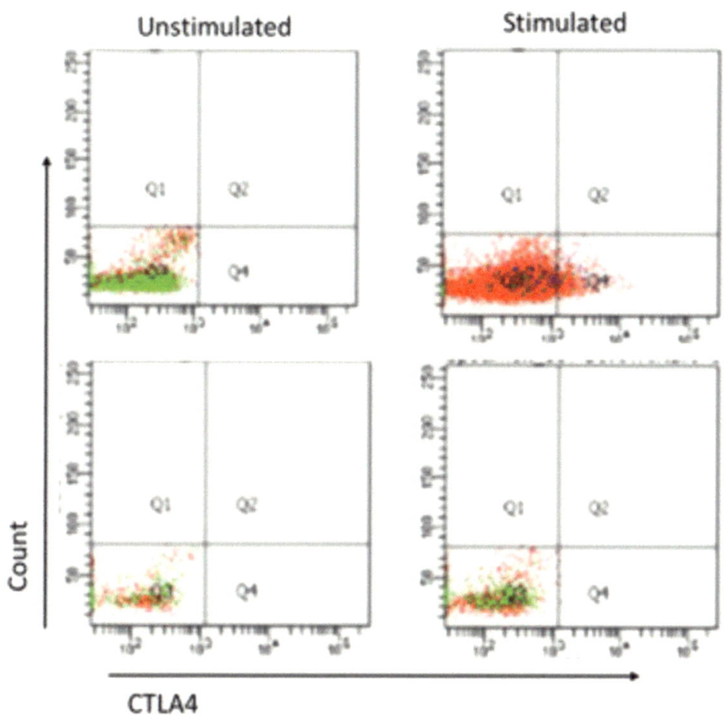

Figure 6.
A representative image of CTLA4 expression in unstimulated and stimulated PBMC samples obtained from in a healthy control (top) and a patient (below). Decreased CTLA4 expression was observed in the patient compared to the healthy control.

- Prepare two flasks for each sample to analyze unstimulated and stimulated samples
- Put the appropriate number of cells to culture flask and add 5 ug/ml (PHA) in the completed culture media
- Incubate the cells overnight in humidified incubator
- Wash the cells with PBS
- Centrifuge at 500 g for 5 min and discard the supernatant
- Add appropriate volume of CTLA4 antibody and incubate for 30 min at room temperature
- Wash the cells with PBS and centrifuge at 500 g for 5 min and discard the supernatant
- Resuspend the cells with 300 ul PBS and analyze at flow cytometer

4. Analysis of intracellular molecules in PIDs

4.1 Single protein evaluation in related cell population by flow cytometry

The following protocol is applied to the patients who have suggestive clinical history related to LRBA, STK4, DOCK8 and BTK deficiencies before and after sequencing to evaluate the alteration of designated protein expressions.

- Peripheral blood mononuclear cells (PBMCs) are separated by ficoll density gradient protocol from 1 to 2 ml of whole blood in tube with EDTA
- Wash PBMCs with Phosphate Buffer Saline (PBS) buffer, centrifuge at 300 g for 5 min and discard the supernatant and resuspend the cells with PBS
- Add appropriate volume of PBS and add 100 ul cell to flow cytometer tubes
- Add appropriate volume of antibodies related to cells which are interested for 30 min
- Following incubation wash with PBS, centrifuge at 500 g for 5 min and discard the supernatant
- Fix the cells with a fixation buffer for 10–20 min
- Wash with PBS, centrifuge at 500 g for 5 min and discard the supernatant
- Treat with the permeabilization buffer for 10–30 min
- Wash with PBS, centrifuge at 500 g for 5 min and discard the supernatant
- Incubate with related antibody for 30 min

- Wash the cells with PBS and centrifuge at 500 g for 5 min and discard the supernatant

- Resuspend the cells with 300 ul PBS and analyze at flow cytometer

4.1.1 LRBA deficiency

(Lipopolysaccharide responsive beige-like anchor protein) LRBA plays important roles in vesicle trafficking and receptor recycling. LRBA is responsible for CTLA4 trafficking from vesicular compartments to the cell membrane. In patients with LRBA mutations, an autosomal recessive form of combined immunodeficiency arises and this deficiency is associated with hypogammaglobulinemia, recurrent respiratory infections, multiple autoimmune manifestations and frequently susceptibility to inflammatory bowel disease and malignity in some cases [4, 6, 7, 43–45]. See the Section 4.1. for the staining protocol. **Figure 7** shows a representative image of LRBA expression in LRBA deficient patient and a healthy control.

4.1.2 STK4 (MST1) deficiency

STK4 (serine–threonine protein kinase 4), also known as MST1 (Macrophage Stimulating 1), was first found in *Drosophila* as a member of the Hippo pathway, which regulates proliferation and cell survival. Human STK4 is principally discovered as a constitutively expressed kinase, structurally homologous to the *Drosophila* Hippo, and plays roles in vital biologic processes such as morphogenesis, proliferation, apoptosis, and stress response [46–49]. STK4 deficiency was first defined in 2012 by 3 separate groups as causing a novel autosomal recessive CID, which is characterized by a profoundly decreased level of CD4+ T cells with the concomitant tendency to recurrent viral and bacterial infections and mucocutaneous candidiasis [46, 49]. Mutations in STK4 gene cause the lack of protein expression or severely reduced level of protein expression [50] (**Figure 8**). See the Section 4.1. for the staining protocol.

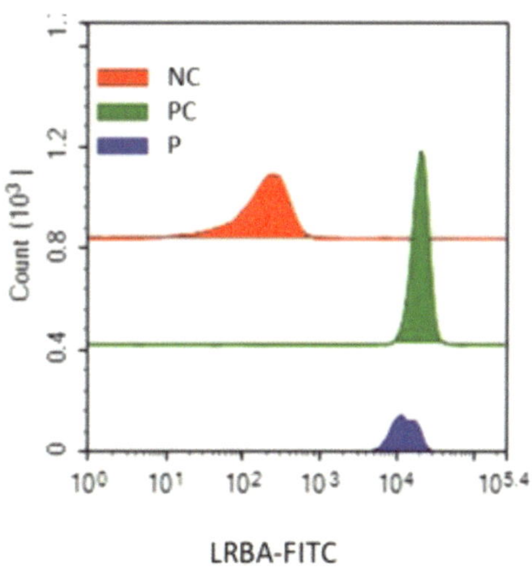

Figure 7.
A representative image of LRBA expression in a negative control (NC or isotype control), positive or healthy control (PC) and a patient (P). Decreased LRBA expression was observed in the patient compared the PC.

4.1.3 DOCK8 deficiency

DOCK8 is a member of DOCK-C family and is responsible for activation of GTPases such as CDC42 and RAC. Therefore it transmit the signals from the membrane to intracellular compartment of cells and involves the cytoskeletal rearrangement of the cells. Decreased expression or total loss of DOCK8 protein due to bi-allelic mutations of DOCK8 gene cause Autosomal-Recessive Hyper-IgE Syndrome (AR-HIES) which is associated with eosinophilia and elevated IgE levels in the effected patients [51–53] (**Figure 9**). See the Section 4.1. for the staining protocol.

4.1.4 BTK deficiency in XLA

BTK is a member of Tec family of non-receptor tyrosine kinases and plays a role in the transmission of the signals from the membrane into the cell. BTK localizes

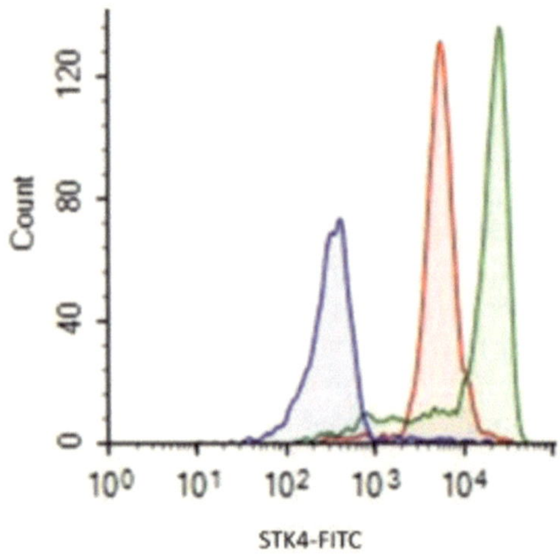

Figure 8.
A representative image of STK4 expression in isotype control (blue), healthy control (green) and the patient (red). Decreased STK4 expression was observed in the patient compared to the healthy control [50].

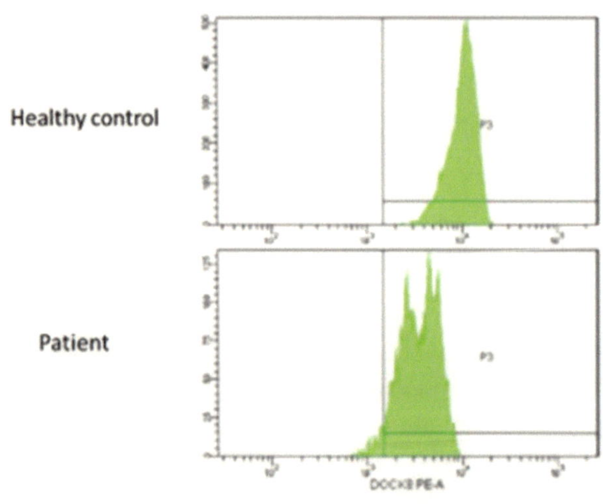

Figure 9.
A representative image of DOCK8 expression in healthy control (top) and the patient (below). Decreased DOCK8 expression was observed in the patient compared to the healthy control.

next to BCR in B cells, therefore it is important for B cell development. In mutations of BTK which is present on X-chromosome cause X-linked agammaglobulinemia in patients who suffered from recurrent bacterial infections due to low or nearly undetectable immunoglobulins and B lymphocytes [54]. Lymphocyte phenotyping is frequently used to diagnose the diseases in patients with suspicious clinical findings and BTK expression is analyzed for molecular diagnosis underlying the XLA. **Figure 10** demonstrates the BTK expression in a patients' and a healthy controls' samples. See the Section 4.1. for the staining protocol.

4.2 Pathway characterization in PIDs

4.2.1 PI3K pathway characterization

Activated phosphoinositide-3 kinase-δ syndrome (APDS) also known as p110δ-activating mutation causing senescent T cells, lymphadenopathy and immunodeficiency (PASLI) occurs in patients with combined immunodeficiency due to gain of function mutations of phosphoinositide 3-kinase (PI3K) genes PIK3CD and

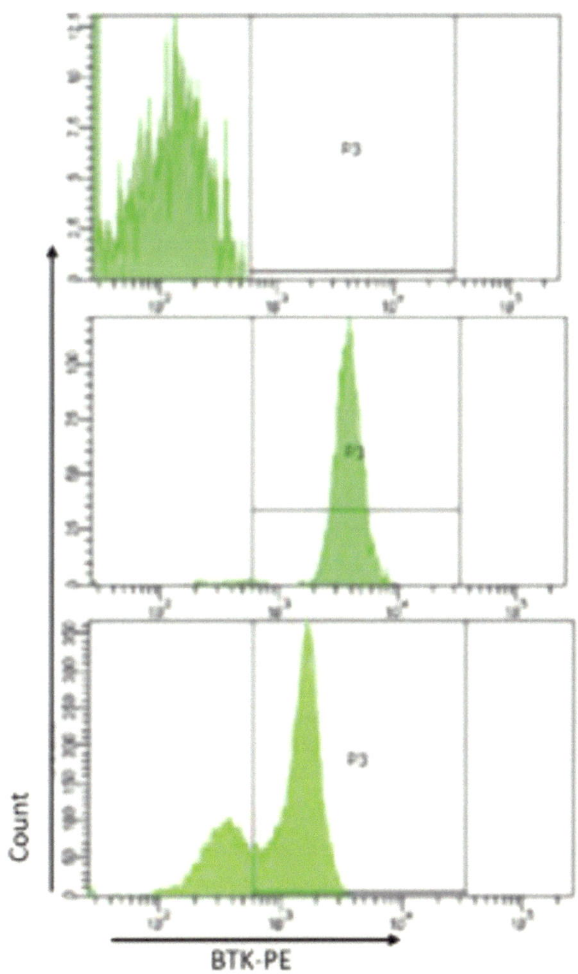

Figure 10.
BTK expression in isotype control (top) healthy control (middle) and the patient (below). BTK expression was lower in the patient than the healthy control.

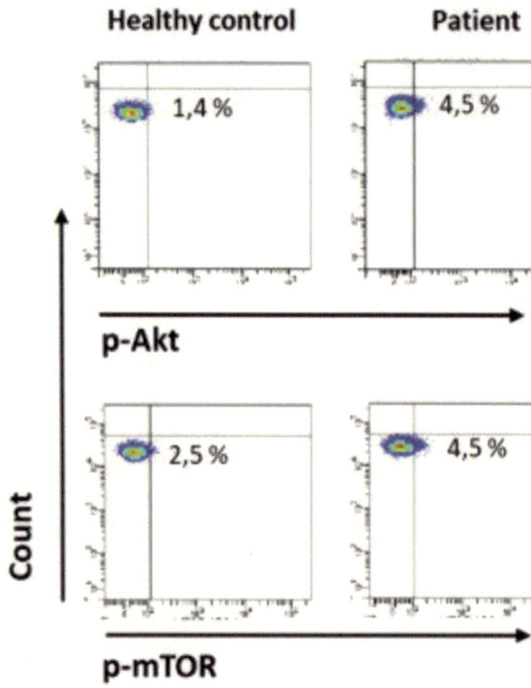

Figure 11.
Ratio of cells expressing p-Akt and p-mTOR in a patient with PIK3δ GOF deficiency and a healthy control following pathway stimulation as described in section 4.2.1.1.

PIK3R1 [14, 16–18]. Although clinical manifestations are heterogenous among the patients, recurrent and persistent infections with herpes family viruses, lymphoproliferation, immune cytopenia are observed in the majority of the patients. Investigating the pathway in patients with suggestive to APDS or PASLI, PI3K pathway analysis, downstream kinase phosphorylations with or without stimulation with specific receptors such as TCR or BCR are investigated by flow cytometry [16]. In the latter section, staining protocol of the PIK3δ, p-Akt and p-mTOR are summarized. **Figure 11** shows a representative image of p-Akt and p-mTOR expression in a patient with PIK3δ GOF deficiency and a healthy control sample.

4.2.1.1 PIK3δ and downstream pathway activation and staining protocol

- Peripheral blood mononuclear cells (PBMCs) are separated by ficoll density gradient protocol from 1 to 2 ml of whole blood in tube with EDTA
- Wash PBMCs with Phosphate Buffer Saline (PBS) buffer, centrifuge at 300 g for 5 min and discard the supernatant and resuspend the cells with serum free media
- Prepare two flasks for each sample to analyze unstimulated and stimulated samples
- Put the appropriate number of cells to culture flask and add an appropriate receptor activating agent to induce the pathway and incubate in humidified incubator in suggested time depend on the agent used in the activation
- Centrifuge at 500 g for 5 min and discard the supernatant

- Add appropriate volume of PIK3δ, p-Akt and p-mTOR antibodies and incubate for 30 min at room temperature
- Wash the cells with PBS and centrifuge at 500 g for 5 min and discard the supernatant
- Resuspend the cells with 300 ul PBS and analyze at flow cytometer

5. Analysis of cellular functions of immune cells

5.1 Cell proliferation

Severe combined immunodeficiencies (CIDs) including T-B + NK-, T-B-NK+, T-B-NK- and T-B + NK+ and/or isolated T cell deficiencies are severe forms of PIDs due to important roles of T lymphocytes to combat directly or indirectly protein and viral antigens [55]. T lymphocytes have specific subsets to achieve their superior roles on specific antigenic determinant. Their deficiencies due to specific molecular defects affect their activation, receptor editing, functions and proliferative capacity cause critically ill disease phenotype. They need to re-regulate their receptors and proliferate to expand agent-specific clones such an army to combat during various specific-infections. Therefore detecting cell proliferation is significant for the diagnosis and/or the course of the disease. Non-radioactive cell tracking dyes such as CFSE (carboxyfluorescein succinimidyl ester) has been started to use for the assessment of cell proliferation in flow cytometry. CFSE is a non-fluorescent dye and becomes permeable through its two acetate groups and passing through the cell membrane. After entering the cells, following the separation of acetate groups via esterases, it becomes fluorescent and its permeability is decreased. Succinimidyl group of CFSE reacts with amino groups of mostly from lysine residues of intracellular molecules such as cytoskeletal proteins and forms stable covalent bonds. In

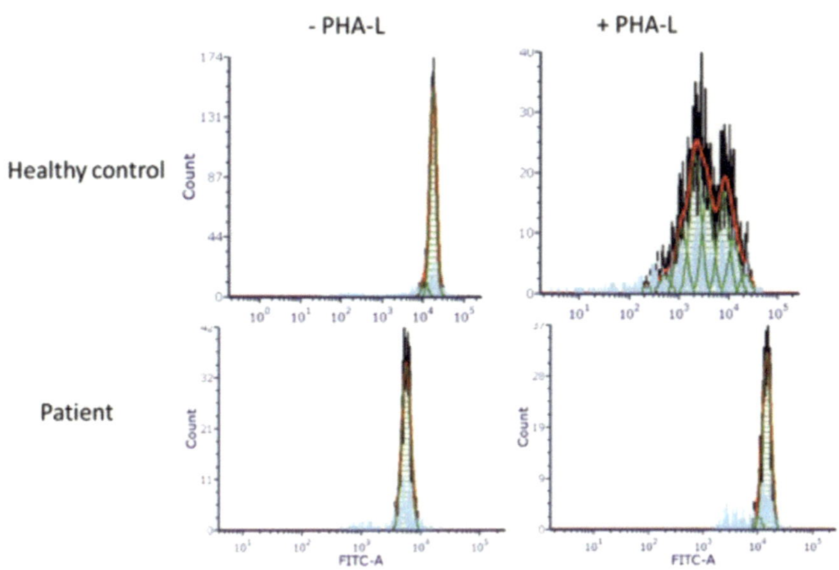

Figure 12.
Comparison of CD3+ T lymphocyte proliferation between a patient with SCID and a healthy control individual. Normal proliferation in the healthy control sample (top) and loss of CD3+ T lymphocyte proliferation in the patient with SCID (below).

each cell division its fluorescent density is decreased and this decrease in cells is evaluated in flow cytometry [56–58]. Severely affected lymphocyte proliferation in a patient with severe combined immunodeficiency is shown in **Figure 12**. See the CFSE cell staining protocol in Section 5.1.1.

5.1.1 CFSE staining protocol

- Peripheral blood mononuclear cells (PBMCs) are separated by ficoll density gradient protocol from 1 to 2 ml of whole blood in tube with EDTA
- Wash PBMCs with Phosphate Buffer Saline (PBS) buffer, centrifuge at 300 g for 5 min and discard the supernatant and resuspend the cells with serum free media
- Prepare two flasks to analyze the proliferation in unstimulated and stimulated cells
- Put the appropriate number of cells to culture flask and label them with the appropriate concentration of CFSE for 5–10 minutes in dark conditions
- Centrifuge at 500 g for 5 min and discard the supernatant for two times
- Add appropriate volume of T cell activator such as PHA (Phorbol Myristate Acetate) to stimulate the cells
- Incubate cells for 72–96 hours in humidified conditions
- Wash the cells with PBS and centrifuge at 500 g for 5 min and discard the supernatant
- Incubate with appropriate volume of anti-CD3 antibody
- Wash the cells with PBS and centrifuge at 500 g for 5 min and discard the supernatant
- Resuspend the cells with 300 ul PBS and analyze at flow cytometer.

Acknowledgements

I would like to express my sincere thanks to Prof. Ilhan Tezcan, MD, PhD and Prof. Deniz Cagdas Ayvaz, MD, PhD for their valuable supports. This study was supported by the grants with the number TSA-2018-17339 and 315S125 from Hacettepe University and TUBITAK, respectively. The authors would like to thank participants for being a part of this study.

Author details

Sevil Oskay Halacli
Department of Basic Sciences of Pediatrics, Division of Pediatric Immunology, Hacettepe University, Ankara, Turkey

*Address all correspondence to: seviloskay@gmail.com

References

[1] A. Bousfiha *et al.*, "Human Inborn Errors of Immunity: 2019 Update of the IUIS Phenotypical Classification," *J. Clin. Immunol.*, vol. 40, no. 1, 2020, doi: 10.1007/s10875-020-00758-x.

[2] H. Kanegane *et al.*, "Flow cytometry-based diagnosis of primary immunodeficiency diseases," Allergology International, vol. 67, no. 1. 2018, doi: 10.1016/j.alit.2017.06.003.

[3] T. Takashima *et al.*, "Multicolor Flow Cytometry for the Diagnosis of Primary Immunodeficiency Diseases," J. Clin. Immunol., vol. 37, no. 5, 2017, doi: 10.1007/s10875-017-0405-7.

[4] G. Lopez-Herrera *et al.*, "Deleterious mutations in LRBA are associated with a syndrome of immune deficiency and autoimmunity," Am. J. Hum. Genet., vol. 90, no. 6, 2012, doi: 10.1016/j.ajhg.2012.04.015.

[5] F. J. Alroqi *et al.*, "Exaggerated follicular helper T-cell responses in patients with LRBA deficiency caused by failure of CTLA4-mediated regulation," J. Allergy Clin. Immunol., vol. 141, no. 3, 2018, doi: 10.1016/j.jaci.2017.05.022.

[6] L. M. Charbonnier *et al.*, "Regulatory T-cell deficiency and immune dysregulation, polyendocrinopathy, enteropathy, X-linked-like disorder caused by loss-of-function mutations in LRBA," J. Allergy Clin. Immunol., vol. 135, no. 1, 2015, doi: 10.1016/j.jaci.2014.10.019.

[7] S. Eren Akarcan *et al.*, "Two male siblings with a novel LRBA mutation presenting with different findings of IPEX syndrome," *JMM Case Reports*, vol. 5, no. 10, 2018, doi: 10.1099/jmmcr.0.005167.

[8] A. M. Paterson *et al.*, "Deletion of CTLA-4 on regulatory T cells during adulthood leads to resistance to autoimmunity," *J. Exp. Med.*, vol. 212, no. 10, 2015, doi: 10.1084/jem.20141030.

[9] L. R. Watson et al., "Pitfalls of immunotherapy: lessons from a patient with CTLA-4 haploinsufficiency," Allergy, Asthma Clin. Immunol., vol. 14, no. 1, 2018, doi: 10.1186/s13223-018-0272-7.

[10] K. D., R. K., and U. G., "Multilineage cytopenias in CTLA4 deficiency due to autoimmune destruction: A retrospective review," *J. Clin. Immunol.*, vol. 38, no. 3, 2018.

[11] S. Jägle *et al.*, "Distinct molecular response patterns of activating STAT3 mutations associate with penetrance of lymphoproliferation and autoimmunity," *Clin. Immunol.*, vol. 210, 2020, doi: 10.1016/j.clim.2019.108316.

[12] F. Consonni, L. Dotta, F. Todaro, D. Vairo, and R. Badolato, "Signal transducer and activator of transcription gain-of-function primary immunodeficiency/immunodysregulation disorders," *Current Opinion in Pediatrics*, vol. 29, no. 6. 2017, doi: 10.1097/MOP.0000000000000551.

[13] W. Rae *et al.*, "Autoimmunity/inflammation in a monogenic primary immunodeficiency cohort," *Clin. Transl. Immunol.*, vol. 6, no. 9, 2017, doi: 10.1038/cti.2017.38.

[14] C. L. Lucas, A. Chandra, S. Nejentsev, A. M. Condliffe, and K. Okkenhaug, "PI3Kδ and primary immunodeficiencies," Nature Reviews Immunology, vol. 16, no. 11. 2016, doi: 10.1038/nri.2016.93.

[15] S. Preite, B. Huang, J. L. Cannons, D. B. McGavern, and P. L. Schwartzberg, "PI3K orchestrates T follicular helper cell differentiation in a context dependent manner: Implications for

autoimmunity," *Frontiers in Immunology*, vol. 10, no. JAN. 2019, doi: 10.3389/fimmu.2018.03079.

[16] C. L. Lucas *et al.*, "Heterozygous splice mutation in PIK3R1 causes human immunodeficiency with lymphoproliferation due to dominant activation of PI3K," *J. Exp. Med.*, vol. 211, no. 13, 2014, doi: 10.1084/jem.20141759.

[17] E. S. J. Edwards *et al.*, "Activating PIK3CD mutations impair human cytotoxic lymphocyte differentiation and function and EBV immunity," J. Allergy Clin. Immunol., vol. 143, no. 1, 2019, doi: 10.1016/j.jaci.2018.04.030.

[18] G. L. Dornan, B. D. Siempelkamp, M. L. Jenkins, O. Vadas, C. L. Lucas, and J. E. Burke, "Conformational disruption of PI3Kδ regulation by immunodeficiency mutations in PIK3CD and PIK3R1," *Proc. Natl. Acad. Sci. U. S. A.*, vol. 114, no. 8, 2017, doi: 10.1073/pnas.1617244114.

[19] F. Barzaghi, L. Passerini, and R. Bacchetta, "Immune dysregulation, polyendocrinopathy, enteropathy, X-linked syndrome: A paradigm of immunodeficiency with autoimmunity," *Front. Immunol.*, vol. 3, no. JUL, 2012, doi: 10.3389/fimmu.2012.00211.

[20] C. L. Bennett *et al.*, "The immune dysregulation, polyendocrinopathy, enteropathy, X-linked syndrome (IPEX) is caused by mutations of FOXP3," *Nat. Genet.*, vol. 27, no. 1, 2001, doi: 10.1038/83713.

[21] R. Bacchetta, F. Barzaghi, and M. G. Roncarolo, "From IPEX syndrome to FOXP3 mutation: A lesson on immune dysregulation," *Annals of the New York Academy of Sciences*, vol. 1417, no. 1. 2016, doi: 10.1111/nyas.13011.

[22] S. Al Khatib *et al.*, "Defects along the TH17 differentiation pathway underlie genetically distinct forms of the hyper IgE syndrome," J. Allergy Clin. Immunol., vol. 124, no. 2, 2009, doi: 10.1016/j.jaci.2009.05.004.

[23] H. D. Ochs, M. Oukka, and T. R. Torgerson, "TH17 cells and regulatory T cells in primary immunodeficiency diseases," Journal of Allergy and Clinical Immunology, vol. 123, no. 5. 2009, doi: 10.1016/j.jaci.2009.03.030.

[24] E. D. Renner *et al.*, "Novel signal transducer and activator of transcription 3 (STAT3) mutations, reduced TH17 cell numbers, and variably defective STAT3 phosphorylation in hyper-IgE syndrome," *J. Allergy Clin. Immunol.*, vol. 122, no. 1, 2008, doi: 10.1016/j.jaci.2008.04.037.

[25] A. A. Caudy, S. T. Reddy, T. Chatila, J. P. Atkinson, and J. W. Verbsky, "CD25 deficiency causes an immune dysregulation, polyendocrinopathy, enteropathy, X-linked-like syndrome, and defective IL-10 expression from CD4 lymphocytes," J. Allergy Clin. Immunol., vol. 119, no. 2, 2007, doi: 10.1016/j.jaci.2006.10.007.

[26] S. M. Laakso *et al.*, "Regulatory T cell defect in APECED patients is associated with loss of naive FOXP3+ precursors and impaired activated population," J. Autoimmun., vol. 35, no. 4, 2010, doi: 10.1016/j.jaut.2010.07.008.

[27] Y. Zhu, L. Zou, and Y. C. Liu, "T follicular helper cells, T follicular regulatory cells and autoimmunity," *International Immunology*, vol. 28, no. 4. 2016, doi: 10.1093/intimm/dxv079.

[28] C. S. Ma, "Human T Follicular Helper Cells in Primary Immunodeficiency: Quality Just as Important as Quantity," Journal of Clinical Immunology, vol. 36. 2016, doi: 10.1007/s10875-016-0257-6.

[29] D. Baumjohann, D. Baumjohann, and K. M. Ansel, "Identification of T

follicular helper (Tfh) cells by flow cytometry," *Protoc. Exch.*, 2013, doi: 10.1038/protex.2013.060.

[30] C. S. Ma and T. G. Phan, "Here, there and everywhere: T follicular helper cells on the move," *Immunology*, vol. 152, no. 3. 2017, doi: 10.1111/imm.12793.

[31] A. N. Kamali *et al.*, "A role for Th1-like Th17 cells in the pathogenesis of inflammatory and autoimmune disorders," Mol. Immunol., vol. 105, 2019, doi: 10.1016/j.molimm.2018.11.015.

[32] K. Yasuda, Y. Takeuchi, and K. Hirota, "The pathogenicity of Th17 cells in autoimmune diseases," Seminars in Immunopathology, vol. 41, no. 3. 2019, doi: 10.1007/s00281-019-00733-8.

[33] E. Castigli, R. Fuleihan, N. Ramesh, E. Tsitsikov, A. Tsytsykova, and R. S. Geha, "Cd40 ligand/cd40 deficiency," *Int. Arch. Allergy Immunol.*, vol. 107, no. 1-3, 1995, doi: 10.1159/000236923.

[34] L. Murguia-Favela *et al.*, "CD40 DEFICIENCY: A UNIQUE ADULT PATIENT WITH HYPER IMMUNOGLOBULIN M SYNDROME AND NORMAL EXPRESSION OF CD40," *LymphoSign J.*, 2017, doi: 10.14785/lymphosign-2017-0004.

[35] O. Cabral-Marques *et al.*, "Expanding the clinical and genetic spectrum of human CD40L deficiency: The occurrence of paracoccidioidomycosis and other unusual infections in brazilian patients," J. Clin. Immunol., vol. 32, no. 2, 2012, doi: 10.1007/s10875-011-9623-6.

[36] X. Du *et al.*, "Clinical, genetic and immunological characteristics of 40 Chinese patients with CD40 ligand deficiency," *Scand. J. Immunol.*, vol. 90, no. 4, 2019, doi: 10.1111/sji.12798.

[37] A. Ozen, "CHAPLE syndrome uncovers the primary role of complement in a familial form of Waldmann's disease," *Immunological Reviews*, vol. 287, no. 1. 2019, doi: 10.1111/imr.12715.

[38] A. Ozen *et al.*, "CD55 Deficiency, Early-Onset Protein-Losing Enteropathy, and Thrombosis," *N. Engl. J. Med.*, vol. 377, no. 1, 2017, doi: 10.1056/nejmoa1615887.

[39] O. A. *et al.*, "Inherited CD55 deficiency in patients with early onset protein-losing enteropathy and thrombosis," *J. Pediatr. Gastroenterol. Nutr.*, vol. 64, 2017.

[40] H. Abolhassani *et al.*, "Combined immunodeficiency and Epstein-Barr virus- induced B cell malignancy in humans with inherited CD70 deficiency," *J. Exp. Med.*, vol. 214, no. 1, 2017, doi: 10.1084/jem.20160849.

[41] K. Izawa *et al.*, "Inherited CD70 deficiency in humans reveals a critical role for the CD70-CD27 pathway in immunity to Epstein-Barr virus infection," *J. Exp. Med.*, vol. 214, no. 1, 2017, doi: 10.1084/jem.20160784.

[42] S. Ghosh *et al.*, "Extended clinical and immunological phenotype and transplant outcome in CD27 and CD70 deficiency," *Blood*, 2020, doi: 10.1182/blood.2020006738.

[43] E. Lévy *et al.*, "LRBA deficiency with autoimmunity and early onset chronic erosive polyarthritis," Clin. Immunol., vol. 168, 2016, doi: 10.1016/j.clim.2016.03.006.

[44] F. Salami *et al.*, "Leishmaniasis and Autoimmunity in Patient with LPS-Responsive Beige-Like Anchor Protein (LRBA) Deficiency," *Endocrine, Metab. Immune Disord. - Drug Targets*, vol. 20, no. 3, 2019, doi: 10.2174/1871530319666190807161546.

[45] D. Cagdas *et al.*, "A Spectrum of Clinical Findings from ALPS to

CVID: Several Novel LRBA Defects," J. Clin. Immunol., 2019, doi: 10.1007/s10875-019-00677-6.

[46] S. O. Halacli *et al.*, "STK4 (MST1) deficiency in two siblings with autoimmune cytopenias: A novel mutation," Clin. Immunol., vol. 161, no. 2, 2015, doi: 10.1016/j.clim.2015.06.010.

[47] H. Abdollahpour *et al.*, "The phenotype of human STK4 deficiency," Blood, vol. 119, no. 15, 2012, doi: 10.1182/blood-2011-09-378158.

[48] B. Al-Saud *et al.*, "STK4 Deficiency in a Patient with Immune Complex Glomerulonephritis, Salt-Losing Tubulopathy, and Castleman's-Like Disease," Journal of Clinical Immunology, vol. 39, no. 8. 2019, doi: 10.1007/s10875-019-00682-9.

[49] C. Schipp *et al.*, "EBV negative lymphoma and autoimmune lymphoproliferative syndrome like phenotype extend the clinical spectrum of primary immunodeficiency caused by STK4 deficiency," *Front. Immunol.*, vol. 9, no. OCT, 2018, doi: 10.3389/fimmu.2018.02400.

[50] S. OSKAY HALACLI, D. CAGDAS, and I. TEZCAN, "Flow Cytometry is a Reliable Tool in the Diagnosis of STK4 Deficiency," *Asthma Allergy Immunol.*, 2020, doi: 10.21911/aai.518.

[51] S. Haskologlu *et al.*, "Clinical, immunological features and follow up of 20 patients with dedicator of cytokinesis 8 (DOCK8) deficiency," *Pediatr. Allergy Immunol.*, vol. 31, no. 5, 2020, doi: 10.1111/pai.13236.

[52] S. E. Aydin *et al.*, "DOCK8 Deficiency: Clinical and Immunological Phenotype and Treatment Options - a Review of 136 Patients," J. Clin. Immunol., vol. 35, no. 2, 2015, doi: 10.1007/s10875-014-0126-0.

[53] C. M. Biggs, S. Keles, and T. A. Chatila, "DOCK8 deficiency: Insights into pathophysiology, clinical features and management," Clinical Immunology, vol. 181. 2017, doi: 10.1016/j.clim.2017.06.003.

[54] S. Hashimoto *et al.*, "Identification of Bruton's tyrosine kinase (Btk) gene mutations and characterization of the derived proteins in 35 X-linked agammaglobulinemia families: A nationwide study of Btk deficiency in Japan," *Blood*, vol. 88, no. 2, 1996, doi: 10.1182/blood.v88.2.561.bloodjournal882561.

[55] R. Kumrah *et al.*, "Genetics of severe combined immunodeficiency," Genes and Diseases, vol. 7, no. 1. 2020, doi: 10.1016/j.gendis.2019.07.004.

[56] M. Koyanagi, S. Kawakabe, and Y. Arimura, "A comparative study of colorimetric cell proliferation assays in immune cells," Cytotechnology, vol. 68, no. 4, 2016, doi: 10.1007/s10616-015-9909-2.

[57] I. Terrén, A. Orrantia, J. Vitallé, O. Zenarruzabeitia, and F. Borrego, "CFSE dilution to study human T and NK cell proliferation in vitro," in *Methods in Enzymology*, vol. 631, 2020.

[58] E. Azarsiz, N. Karaca, B. Ergun, M. Durmuscan, N. Kutukculer, and G. Aksu, "In vitro T lymphocyte proliferation by carboxyfluorescein diacetate succinimidyl ester method is helpful in diagnosing and managing primary immunodeficiencies," *J. Clin. Lab. Anal.*, vol. 32, no. 1, 2018, doi: 10.1002/jcla.22216.

Chapter 2

Organoids Models for the Study of Cell-Cell Interactions

Margarita Jimenez-Palomares, Alba Cristobal and Mª Carmen Duran Ruiz

Abstract

Organoids have arisen as promising model systems in biomedical research and regenerative medicine due to their potential to reproduce the original tissue architecture and function. In the research field of cell–cell interactions, organoids mimic interactions taking place during organogenesis, including the processes that conduct to multi-lineage differentiation and morphogenetic processes, during immunology response and disease development and expansion. This chapter will address the basis of organoids origin, their importance on immune system cell–cell interactions and the benefits of using them in biomedicine, specifically their potential applications in regenerative medicine and personalized therapy. Organoids might represent a personalized tool for patients to receive earlier diagnoses, risk assessments, and more efficient treatments.

Keywords: organoids, cell-interactions, disease development, regenerative medicine, personalized therapy

1. Introduction

Most multicellular living organisms, especially vertebrates, develop from a single totipotent cell to a multicellular complex adult organism, reflecting an outstanding coordination and organization capacity. Furthermore, in some cases, after organ dissociation, cells can recombine and reconstruct the original structure. Researchers have used that feature to create organ-like structures from stem cells or tissues samples, leading to the formation of structures currently known as organoids [1].

Thus, organoids are self-organizing 3D structures derived from stem cells highly similar in structure and function to actual human organs. The different cell types and interactions guide and make possible this organization process. These structures resemble crucial aspects of the tissues from which derived and thereby organoids allow for biological relevant cell–cell and cell-matrix interactions. Those attributes make organoids technology a valuable tool in multiple applications such as developmental biology, molecular biology, and health studies like pharmacology, disease development and therapy, among others [2, 3].

The organoids field has exponentially accelerated in the last years, mainly after the application of appropriate culturing conditions that allow stem cells to differentiate and participate in cell–cell interactions responsible of the community effect required for optimal resembling of self-organized tissue-like structures.

For instance, the use of Matrigel, a gel protein mixture that mimics the complex extracellular environment found in many tissues [4], has allowed the establishment of the right culture conditions required to achieve 3D cell cultures *in vitro*.

Organoids technology constitutes a step-forward approach for conventional cell-based research, full-filling the gap between 2D cultures and *in vivo* mouse/human models. Organoids are physiologically more relevant than monolayer culture models, and allow easier manipulation of niche components, signaling pathways and genome editing than *in vivo* models [5].

Therefore, organoids represent a needed and also an advantageous approach in many senses. The organoids technology brings the opportunity to work with 3D-tissue models at a "bench-side" level, opening a wide range of opportunities in basic and clinical research. Moreover, organoids also overcome the problems derived from using animal models to study human physiology and related-diseases. Although many results obtained in animal models can be easily extrapolated, some biological processes are specific to humans [6].

2. Organoids origin, structures and culture

Organoids can be derived from either [1] pluripotent stem cells (PSCs), including embryonic stem cells (ESCs) and induced pluripotent stem cells (iPSCs) or [2] multipotent organ-specific adult stem cells (AdSCs). Both approaches take advantage of the endless expansion potential of stem cells in culture. Also, when PSCs and AdSCs are allowed to differentiate in culture, they display a remarkable capacity to self-organize into structures that reflect similar characteristics of the organ they attempt to mimic [7].

PSCs can be differentiated into different cell types and grown *ex vivo* as organoid models by the treatment with defined developmental stimuli. PSCs isolated both from mouse and human tissues have given rise to brain, retina, inner ear, stomach, intestine, thyroid, lung, liver, and kidney organoids. ESCs or iPSCs can be derived *in vitro* into endoderm, mesoderm and ectoderm, with specific procedures involving multiple differentiation steps. Thus, human iPSCs are sequentially exposed to a progression of differentiation signals in order to simulate the stages of a human developmental process. Once the initial germ layer has arranged, cells are transferred into 3D systems [8], where differentiated iPSCs aggregate to form an organ bud and, later on, organoids. These organoids contain multiple cell types and faithfully mimic the mature organ structure, and the interactions between them.

As an example, embryoid bodies (EBs), 3D aggregates of PSCs, originate cerebral organoids and develop into a forebrain region in the presence of growth factors (i.e., hFGF basic, ROCK inhibitor, N2, Heparin, MEM-NEAA, etc.). For other organs, the addition of Activin A to PSCs specifies them towards an endodermal fate. These cells are further cultured as 3D organoids in Matrigel with medium containing tissue-specific growth factors [9].

On the other hand, AdSCs-organoids can be originated from isolated adult stem/progenitor cells or from isolated tissue fragments of the corresponding organ (e.g. intestinal crypts, liver or pancreas ducts) [8]. These structures can be generated from biopsies isolated directly from the organ of interest or from diseased patient tissue without the complicated process of reprogramming and differentiation required in iPSC organoids. In general, human AdSCs-derived organoids are composed mainly of cell types found in the epithelium.

AdSCs were long believed to be unable to proliferate outside the body, but the culture with specific growth factor cocktails mimicking stem cell niches, has helped to sort out such obstacle. These niche factors are essential to support stem cell activity and vary depending on the tissue of origin. Also, 3D Matrigel-based cultures have provided the appropriate culture conditions to generate AdSCs-derived organoids from various mouse and human tissues including the colon, stomach, liver, lung, prostate, pancreas, ovaries, taste buds, and lingual epithelium.

Thus, to generate AdSCs-organoids a tissue biopsy is cut into fine particles and then incubated with enzymes (i.e., collagenase, elastase, or dispase) to obtain a single cell suspension. Next, cells are grown in Matrigel and culture medium supplemented with specific tissue growth factors [9]. For example, intestinal organoids need Noggin, R-spondin, Epidermal growth factor (EGF), and WNT [10–12]; retina organoids need IWR1e and Smoothened agonist CHIR99021 [13, 14]; prostate organoids require Noggin, R-spondin and EGF [15], while pancreas organoids require Noggin, R-spondin, EGF, fibroblast growth factor (FGF) and Nicotinamine [16].

AdSCs organoids do not require genetic transduction with transcription factors, as it happens with those with PSCs. This situation makes organoids physiologically well-suited with the host tissue, leading to an improved stem cell transplantation. Moreover, molecular techniques such as clustered regularly interspaced short palindromic repeats (CRISPR)/CRISPR-associated system (CRISPR-Cas9) genome technology and single-cell RNA sequencing, can be applied to organoids [7, 9]. On the other hand, the establishment of human AdSC-derived organoids is limited by the accessibility to the tissue and prior knowledge of the culture conditions for that tissue. However, an iPSC line, once established from a patient, can generate different tissue models without any time limit, beyond the patient's lifespan [17, 18].

3. Organoids in the immunology field

The knowledge concerning the interactions of the immune system with other tissues has been gained mainly from animal models and/or cell lines co-cultures. Nevertheless, some interactions between human cells cannot be addressed with murine models or cell lines which are usually transformed or genetically modified [19]. For instance, a specific immune cell morphology is required to maintain the tissue properties and, moreover, the immune system needs of multiple cell types interactions for appropriate functioning. Similarly, there are some aspects that cannot be extrapolated in mice due to, for example, different protein pattern expressions in human and mice. Thus, immunology researchers are starting to get the benefits of using organoids, for a better comprehension of the immune cell interactions with other tissues, its development, homeostasis and in the bout of disease. The organoids approach maintains those cells in a near-native state, mimicking more accurately its original state and environment, providing researchers with a new effective tool.

The main challenge in the use of organoids in immunology resides in the fact that the organoids technology cultures only epithelial cells. However, a more complete resource for immunological research can be developed by co-culturing these organoids with other elements.

The number of publications showing multiple co-cultures has spiked up in the past decade, particularly in the last five years [20–22]. In order to develop effective interventions to preserve health and defeat diseases it is necessary to know how immune cells coordinate their activities to initiate, modulate, and terminate inflammation. Immune cells and molecules released by immune cells promote inflammation processes that are mediating the interactions between these cells [23].

These studies have revealed not only the importance of the presence (or absence) of immune cell derived factors in the epitheliums in culture, but also the need of the reciprocal communication with the immune system. A work concerning the role of macrophages and fibroblasts on myoblast proliferation and migration highlights the importance of multicellular communication [24]. Thus, co-culture of either macrophages or fibroblasts with myoblasts prompted a significant increase in myoblast proliferation. Conversely, in the triple co-culture, although macrophages continued promoting myoblast proliferation, they had a negative effect over the ability of fibroblasts to enhance myoblast migration [25]. Another study, using single-cell transcriptomics, highlighted that intestinal stem cells can function as non-classical antigen-presenting cells for CD4+ T cells. Moreover, these interactions, directly or through activated T cell-derived cytokines, seem to play a role in the intestinal epithelium differentiation [26].

The intestinal mucosal barrier function and the immune responses against invading pathogens seem to be regulated by the interaction between intestinal epithelial cells (IECs) and intraepithelial lymphocytes (IELs) [27]. IELs represent a heterogeneous population of activated and antigen-experienced T cells. A novel culture system of intestinal 'enteroids' has allowed the study of the complex interactions between IECs and immune peripheral T cells in long-term co-cultures. The development of these long-term co-cultures allowed the study of cell survival, proliferation, differentiation and IECs behavior. Moreover, IECs and T cells co-cultures revealed that peripheral T cells activated in the presence of enteroids acquire several features of IELs, including morphology, membrane markers and movement in the epithelial layer [27]. Similarly, mouse-derived enteroids co-cultured with intestinal myofibroblasts and macrophages boosted their growth and differentiation [28].

In the same line, another study with intestinal organoids underlined the importance of the interactions between immune cells and other tissues for optimal maturation. In this work, the inclusion of the immune component (co-cultured with human T lymphocytes) into the differentiation protocol to form human pluripotent stem cell-derived intestinal organoids (hIOs) from hPSCs, enabled hIOs maturation. hIOs co-cultured with human T lymphocytes displayed expression levels of mature intestinal markers equivalent to adult intestinal epithelium, as well as increased intestine-specific functional activities, retaining their maturation status even after their *in vivo* engraftment. This study has proven the needless for animal models and *in vivo* maturation when working with organoids [29].

Holokai, L. *et al.* were among the first researchers to successfully obtain a multiple organoid-co-culture involving cytotoxic T lymphocytes (CTLs) and Helicobacter pylori-infected gastric organoids. CTLs express programmed death 1 (PD1) on the surface. When PD1 interacts with its ligand, CTLs cannot induce apoptosis. Thanks to this approach they discovered that PD-L1 signaling induces cellular proliferation and survival, leading to an increased expression of PD-1, IL-2 and IFNγ in lymphocytes [30].

Overall, epithelial organoid cultures, whether derived from iPSCs or AdSCs, constitute a promising platform for immunological research for several applications, allowing, among others, to study immune cell–epithelial cell interactions in the context of pathogenic infections or sterile tissue damage [19]. In this sense, the vast majority of organoid studies about the immune system and its effects on epithelial differentiation and function have been performed on intestine-like structures. However, it would be useful to have similar works with different organoid systems such as skin or lung, which also interact with both immune cells and commensal microorganisms [31].

Despite the amount of work already accomplished regarding the immune system, there is still a long way to go in inflammation research, due to the current lack of optimal immune cells organoids cultures.

4. Organoids in the study and treatment of disease

Recent advances in the development of human patient-derived organoids have allowed a more accurate study of diseases. This technology has opened a new horizon in biomedical research, and provides unprecedented opportunities in translational medicine, and personalized therapy [32].

4.1 Disease modeling and drug screening approaches

Recent discoveries involving organoids as a disease model reflect that researchers have started to unravel the potential of this tool. To date, organoids have been mostly applied in cancer, cystic fibrosis and studies on host–microbe interactions. However, a growing interest in this field has promoted an exponential increase of publications using organoids technology to study many other diseases (**Table 1**). The fact that organoids are 3D structures originated from stem cells with similar architecture, multi-lineage differentiation and many of the original tissue functions, make them the perfect candidate for disease pathogenesis studies [33, 34].

Organoids can be designed to reproduce patient conditions of disease-relevant genetic and epigenetics. Thanks to the development of new techniques like the CRISPR/CRISPR -Cas9 genome engineering tool, is currently feasible to efficiently manipulate genomic sequences in hESCs and hiPSCs [35, 36]. In the case of host–microbe interactions, organoids can also reproduce the infection process allowing its study in more life-like manner.

Organoids can also be applied to study cellular dysfunction in diseased tissues, as well as to identify strategies for its restoration. For example, Dekkers *et al*. used organoids to study cystic fibrosis (CF), a disease caused by mutations in the CF transmembrane conductance regulator (CFTR) gene, severely reducing the CFTR protein function [37]. Thus, rectal organoids from CF patients were used to evaluate CFTR function as well as the response to CFTR-modulating drugs. Their results demonstrated the pharmacological restoration of CFTR function in the rectal organoids of individual donors, suggesting that *in vitro* functional measurements of CFTR may be used to preclinically identify CF patients who would benefit from CFTR-modulating treatments, independent of their CFTR mutation [38].

A major challenge in clinical practice is the absence of appropriate models for drug screening and pre-evaluation of the pharmacological effects prior administration to patients. For cancer research, the development of tumor organoids, also known as tumoroids, represents an overwhelming step to be able to reproduce *in vitro* such a heterogeneous microenvironment. Tumor organoids can be generated *in vitro* for the analysis of cancer phenotypes [39], anticancer drug discovery, and to evaluate the response of patient cancer cells to a specific treatment [39, 40]. Lazzari *et al*. reported a triple co-culture of pancreatic cancer cells fibroblasts and endothelial cells. As a result, cells assembled in a hetero-type multicellular tumor-spheroid (MCTS) that reliably reproduced the impact of the surrounding environment, on the sensitivity of cancer cells to chemotherapy. This approach can be successfully applied as a predictive tool of various therapeutic strategies [41]. In this sense, the establishment of patient-derived tumor organoids (PDTO) biobanks provides exciting new insights into developmental biology. Different researchers have started to develop methods for generating and bio-banking PDTO. Among them, a non-profit organization called HUB (Hubrecht Organoid Technology) has initiated and established "Living Biobank", a collection of organoids representing different organs and disease models (huborganoids.nl). Overall, these biobanks maintain the key features to resemble the parental tumors and can be therefore used to evaluate patient-specific treatment approaches [42].

Tissue of origin	AdSC-derived		PSC-derived		Cancer biobank	Disease	References
	Human	Mouse	Human	Mouse			
Brain			✓	✓	✓	Autism,	[6]
						Microcephaly, Macrocephaly	[43]
						Neurodegenerative disorders	[44]
						Infectious diseases	[45]
						Psychiatric disorders	
						Cancer	
Retina/Optic cup			✓	✓		Leber congenital amaurosis (LCA)	[13]
						Retinitis pigmentosa	[14]
						Age-related macular degeneration	
						Retinal degeneration	
Salivary gland	✓		✓	✓		Hyposalivation	[46]
Thyroid			✓	✓	✓	Cancer	[47]
Lung	✓	✓	✓	✓	✓	Cystic fibrosis	[24]
						Cancer	[48]
Breast	✓	✓	✓	✓	✓	Cancer	[49]
Esophagus	✓	✓		✓		Barrett's esophagus	[50]
						Cancer	
Stomach	✓	✓	✓	✓	✓	Infectious diseases	[10]
						Cancer	[35, 51]

DOI: http://dx.doi.org/10.5772/intechopen.94562

Tissue of origin	AdSC-derived		PSC-derived		Cancer biobank	Disease	References
	Human	Mouse	Human	Mouse			
Intestine	✓	✓	✓	✓	✓	Infectious diseases	[10–12]
						Cystic fibrosis	[26, 28, 29]
						Cancer	[52, 53]
Colon	✓	✓	✓	✓	✓	Infectious diseases	[52–54]
						Ulcerative colitis	
						Crohn's disease	
						Cancer	
Pancreas	✓	✓	✓	✓	✓	Cystic fibrosis	[16, 41, 55]
						Pancreatic ductal adenocarcinoma	
						Diabetes mellitus	
						Cancer	
Liver	✓	✓	✓	✓	✓	Alagille syndrome	[56–58]
						Nonalcoholic fatty liver disease	
						Cystic fibrosis	
						Lethal liver failure	
						Cancer	
Kidney	✓	✓	✓	✓	✓	Polycystic kidney disease	[59, 60]
						Cancer	
Female reproductive tract	✓	✓	✓	✓	✓	Cancer	[36, 61]
Prostate	✓	✓	✓	✓	✓	Cancer	[15, 62]

Table 1.
Use of organoids as disease models.

Jacob *et al.* reported the generation of patient-derived glioblastoma organoids (GBOs) biobanks [42]. The authors successfully transplanted the GBOs into adult rodent brains and performed personalized tests. Calandrini *et al.* have recently established the first pediatric cancer organoid biobank [59]. This biobank contains a collection of over 50 tumors matching normal kidney organoids and also covers a diversity of tumor subtypes. Similarly, a primary gastric cancer organoid (GCO) biobank was established by Leung and coworkers [63], including a total of 34 patients with different gastric cancer subtypes. In this study, whole-exome sequencing and transcriptome analysis were performed, as well as large-scale drug screening studies. Overall, the establishment of organoids biobanks provides a rich resource for cancer cell biology and drug-screening studies to test personalized therapies. Patient-specific drug sensitivities could be achieved as the organoids closely resemble the *in vivo* tumors. Furthermore, these biobanks could play a prominent role in biomarker discovery and represent a powerful tool to predict disease development, recurrence and progression [42, 51, 64].

4.2 Applications in regenerative medicine

Several of the most life-threatening diseases require organ transplantation in order to save patients life. Nevertheless, transplantations are not always an option due to the high cost, organ availability or potential organ rejection. Therefore, other alternatives needed to be explored in order to overcome this challenge. The development of organoids brought hope to the scientific community and patients themselves. This technology could potentially serve as an unlimited source for replacing damaged tissues. Furthermore, the transplantation of organoids derived from healthy tissue of the same patient would prevent immune responses related to non-autologous transplants. In this sense, diseases involving dysfunctional organs such as kidneys or the liver, can significantly benefit from the opportunity that liver-derived organoids bring. Researchers have already developed strategies to allow long-term *in-vitro* expansion of liver progenitors into "liver organoids" [56]. The huge expansion and differentiation potential of liver organoids cultures has facilitated the engraftment [56] and survival of livers in murine models, as it happened in a study with liver organoids transplanted to a tyrosinemia type I liver disease model, partially restoring the hepatic function [57]. Similarly, transplantation of human adult stem cell-derived liver organoids into chemically damaged-liver immune-deficient mice produced functional hepatocytes containing grafts [58]. Cultured organoids have also shown the potential to expand, engraft, reconstitute and recover the colon and intestinal epithelia as well as their function in several murine models [52–54].

Despite all the advances in the field, there is still a long way before organoids transplantation becomes a reality. Current resources and techniques do not provide a suitable organ niche, limiting the formation of optimal organ sizes and tissue structures *in vivo*, as well as the appropriate intercellular communication required for functional restoration. Thus, alternative approaches are required, such as the combination of organoids with gene therapy, to implement organ transplantation [65]. Experts on the field will still have to poise excitement with reality before organoid research can be successfully translated to clinical practice and become a real therapy option [66].

4.3 Personalized medicine

Over the past decades, medicine has discovered novel ways of changing the course of many human diseases [67–70]. Nowadays, researches all over the globe are discovering new therapies which bring new options for previously untreatable diseases [71, 72]. However, the key challenge is that the efficacy of most of these new treatments will depend on the complex and unique nature of each individual human being. Lastly, the efficacy of a treatment is significantly determined by behavioral factor, environmental influences as well as genetic particularities.

Moreover, currently available therapies might cause a high impact on patient's quality of life due to the unpleasant side effects directly related to the treatment. Thus, research groups and pharmaceutical companies are developing strategies to personalize their treatments in order to predict the outcome of the proposed therapy and avoid unnecessary aggressive treatments. These aspects are key to achieve the ultimate goal of any therapy: to ensure patients´ health and integrity.

The concept of Personalized Medicine arose with the aim of tailoring the best response and highest safety standards to preserve patient's well-being. This optimized health care strategy would also lead to reduced treatment costs and shorter diagnosis times required for each patient [73–75].

Organoids have revolutionized personalized medicine due to their unique ability of simulate, even mimic, specific cellular microenvironments with remarkable similarity to *in vivo* organs/tissues under normal or pathological conditions [76]. Such models have started to be used in the clinic, mainly in cancer research, to evaluate the response to experimental therapies prior administration of certain drugs or other treatments to patients [77]. The possibility of using accessible models of organ diseases allows to understand the effect of experimental therapy in a deeper manner than in a traditional culture assay or "sphere" culture assays, applied over the last decades [78].

Personalized medicine could also be linked to regenerative medicine which is based on the capacity of the stem cells to derive into many different cell subtypes. Currently, this basic characteristic is key for the understanding of normal and abnormal cell behavior and organization, and is leading to improved tissue engineering approaches [60].

In this scenario, organoids constitute a solid foundation on which personalized and regenerative medicine is taking long steps forward.

One of the best examples of this input on current society is the novel application of organoids cultures to optimize treatment to cancer patients [55, 79]. Oncologic centers are developing translational procedures to understand as much as possible the specific characteristics of each tumor in order to optimize the therapy approach.

Once the tumor is detected, a biopsy of the mass is obtained to culture organoids derived from patient's tumor cells. A complete biological profile of the tumor could be developed combining this information together with histopathological analysis of primary tumor sample, histopathological analysis of the organoids, gene sequencing and cytotoxicity analysis from *in vitro* drug assays or studies using avatar mice.

This complete analysis only takes 2–4 weeks and it could provide physicians relevant information regarding the best treatment for the patient according to the characteristics of the tumor.

Furthermore, in cases of progressive disease or metastasis, new tumor biopsies could be collected, new organoids lines could be established, and new therapeutic strategies could be carried out giving a new opportunity and new hopes to the patient [80–82].

According to the website ClinicalTrials-gov, by 2019 there were 30 projects related to cancer organoids. Most of them (73%) focused on studying anti-cancer therapies, including among others T-cell immunotherapy, or evaluation of radiotherapy sensitivity. The rest aimed to generate patient-derived cancer organoid models (13%) or to evaluate the mechanisms related to cancer progression [83].

A large number of cancer patients are insensitive to immunotherapy due to the heterogeneity of the T cell repertoire [83]. Thus, the use of cancer organoids allows studying the effectiveness of combining immune therapy with specific anti-cancer drugs. To date, two clinical trials involving cancer organoids for immunotherapy have been registered (NCT03778814, NCT02718235). Overall, the inclusion of PDTO into clinic represents an enormous potential to understand the onset of diseases such as cancer and, moreover, to evaluate the individual response to specific therapies for personalized approaches.

5. Limitations and future perspectives

Regardless of the advances made in this emerging technology, a series of limitations still need to be addressed in order to fully exploit its potential. For instance, despite the development of specific culture conditions and growing techniques, there are still tissues that withstand to organoid derivation [84, 85]. Organoid technology requires further advances to achieve less laborious protocols as well as the establishment of standardized conditions for proper differentiation and maturation. A reduction of the heterogeneity seen in organoids size and shape should also be achieved [85]. In addition, it requires the co-induction of the essential cell types, the associated extracellular matrices and native microenvironment that will allow the recapitulation of the *in vivo* tissue sizes, structures, organization, inter-cellular communication and functionality. Also, shorter processes and more affordable culture conditions are required to ensure that the organoids system becomes accessible to a large number of academic and clinical researchers, thereby helping to maximize its potential [5]. Moreover, the protocols used for generating one specific type of organoid are usually not transferable to another organ system. Due to this drawback, scalable and cross-system parameters are challenging to generate via bioengineering tools. Computational prediction models are also difficult to design limiting the capacity to predict phenotypic, toxicological and drug screening results. Lastly, organoids technology requests the development of a complex vasculature network to provide not only oxygen, nutrient and waste exchange, but also an inductive biochemical exchange and a structural template for growth. The advances in microvascular patterning and organ-on-a-chip microfluidic technology would bridge this limitation supporting the use of perfusable organoids generation [86, 87].

In this context, different strategies are currently under research and new ideas have arisen to implement the potential use of organoids. As stated before, the development of organoid biobanks constitutes an important step in this direction. Currently, there are organoid biobanks with healthy organoids as well as patient-derived intestine, liver, pancreatic, lung and mammary gland organoids related to cancer, cystic fibrosis or inflammatory bowel disease [88]. Thus, organoid biobanks are becoming a demanding business and several companies worldwide have already started to commercialize organoids after the establishment of optimized organoid biobanks [88]. Advantages of organoid biobanks include immediate accessibility or cost-effectiveness, as well as the possibility to access a large repository of data related to patient's diseases [83, 88]. This, however, involves some ethical and regulatory challenges that need to be addressed such as donor confidential information or the organoid source itself [89].

The development of microfluidic organoid-on-a-chip platforms [90] and 3D bioprinting [91, 92] constitute two major advances in the last years that are contributing to speed up organoid manufacturing and commercialization [88]. Organ-on-a-chips are devices containing living cells, extra-cellular matrix (ECM) and microstructures emulating key features of organs or issues, and their functions [83, 93]. These devices aim to provide continuous flow perfusion culture to simulate organ microenvironments. Nevertheless, most of these systems are made of primary cell lines or stem-cell-derived cells to mimic organs, but they are unable yet to imitate the cellular interactions taking place in the native sources [94].

Similarly, advances in 3D printing technology and biomaterials research have led to the creation of 3D bioprinting, with the aim to resemble *in vitro* the interactions between tumor cells, ECM and the 3D tumor microenvironment [83, 95]. With this technology, different cell types can be printed in hydrogels and mixed with other cells and/or specific factors to simulate a healthy or pathological microenvironment. Increasing evidence has pointed to the tumor microenvironment as a major modulator of the tumorgenic process [96]. Thus, in order to understand the mechanisms by which tumor cells become metastatic, different studies are benefiting with the use of 3D bioprinting strategy. For example, Grolman JM *et al.* designed a 3D environment with breast adenocarcinoma and macrophage cell lines printed in hydrogel to evaluate the effect of paracrine signals in the regulation of breast cancer metastasis [97]. In the same way, Pang Y *et al*. developed an *in vitro* cervical tumor model to demonstrate the epithelial-to mesenchymal transition (EMT), by mixing *HeLa* cells with hydrogel. These authors were able to evaluate the effect of different activators and inhibitors over the EMT in the 3D system designed [98].

Despite the benefits of using these techniques, there are still several factors that need to be optimized. For instance, biomaterials represent a limiting feature for 3D bioprinting, and the development of improved materials is required. A consensus in the best printing strategy (i.e. polymerization steps, light-based 3D bioprinting *vs* inkjet printing) should be also reached.

6. Conclusions

This chapter focused on the advantages of using organoids to expand our knowledge in the field of cellular interactions. We have focused specifically in immunology and disease-related research, going through some of the latest or more relevant publications involving organoids. Overall, organoids constitute an efficient tool to study immune cells′ interactions *in vitro* in 3D-tissue models that provide a closer view of the interactions taking place *in vivo*. Moreover, organoids represent a promising approach in the development of autologous tissue-based cellular therapies, especially in life-threatening diseases. Nevertheless, despite the organoids relevance, the growing interest in these structures and their potential applications, there is still a long way to go to achieve the translation of organoids into clinical practice. The development of bioengineering tools such as microfluidic organ-on-a-chip platforms or 3D bioprinting systems represents a huge step in this direction. These strategies could provide consistent nutrients and factors required to emulate 3D tissue physiology *in vivo*. The optimal conditions are not yet established and further research is required before results can be undoubtedly extrapolated and clinical applications implemented. Nevertheless, the growing interest in organoids commercialization will probably help to speed up the translation of organoids to the clinic.

Acknowledgements

This study was supported by the Institute of Health Carlos III, ISCIII (PI16-00784), and the Programa Operativo de Andalucia-FEDER, Iniciativa Territorial

Integrada ITI 2014-2020, Consejería de Salud, Junta de Andalucía (PI0026-2017), co-funded by European Regional Development Fund "A way to make Europe".

Conflict of interest

The authors declare no conflict of interest.

Author details

Margarita Jimenez-Palomares[1], Alba Cristobal[2] and Mª Carmen Duran Ruiz[1*]

1 University of Cádiz, Puerto Real, Spain

2 University Medical Center Utrecht, Utrecht, The Netherlands

*Address all correspondence to: maricarmen.duran@gm.uca.es

References

[1] Lancaster MA, Knoblich JA. Organogenesis in a dish: modeling development and disease using organoid technologies. Science. 2014;345(6194): 1247125.

[2] Kretzschmar K, Clevers H. Organoids: Modeling Development and the Stem Cell Niche in a Dish. Dev Cell. 2016;38(6):590-600.

[3] Lehmann R, Lee CM, Shugart EC, et al. Human organoids: a new dimension in cell biology. Mol Biol Cell. 2019;30(10):1129-37.

[4] Benton G, George J, Kleinman HK, Arnaoutova IP. Advancing science and technology via 3D culture on basement membrane matrix. J Cell Physiol. 2009;221(1):18-25.

[5] Fatehullah A, Tan SH, Barker N. Organoids as an in vitro model of human development and disease. Nat Cell Biol. 2016;18(3):246-54.

[6] Sidhaye J, Knoblich JA. Brain organoids: an ensemble of bioassays to investigate human neurodevelopment and disease. Cell Death Differ. 2020.

[7] Clevers H. Modeling Development and Disease with Organoids. Cell. 2016;165(7):1586-97.

[8] Huch M, Koo BK. Modeling mouse and human development using organoid cultures. Development. 2015;142(18):3113-25.

[9] Dutta D, Heo I, Clevers H. Disease Modeling in Stem Cell-Derived 3D Organoid Systems. Trends Mol Med. 2017;23(5):393-410.

[10] Dedhia PH, Bertaux-Skeirik N, Zavros Y, Spence JR. Organoid Models of Human Gastrointestinal Development and Disease. Gastroenterology. 2016;150(5):1098-112.

[11] Pleguezuelos-Manzano C, Puschhof J, van den Brink S, et al. Establishment and Culture of Human Intestinal Organoids Derived from Adult Stem Cells. Curr Protoc Immunol. 2020;130(1):e106.

[12] Spence JR, Mayhew CN, Rankin SA, et al. Directed differentiation of human pluripotent stem cells into intestinal tissue in vitro. Nature. 2011;470(7332):105-9.

[13] Eiraku M, Takata N, Ishibashi H, et al. Self-organizing optic-cup morphogenesis in three-dimensional culture. Nature. 2011;472:51-6.

[14] Nakano T, Ando S, Takata N, et al. Self-Formation of Optic Cups and Storable Stratified Neural Retina from Human ESCs. Cell Stem Cell. 2012;10(6):771-85.

[15] Gao D, Vela I, Sboner A, et al. Organoid cultures derived from patients with advanced prostate cancer. Cell. 2014;159(1):176-87.

[16] Boj SF, Hwang C-I, Baker LA. Organoid Models of Human and Mouse Ductal Pancreatic Cancer. Cell. 2015;160(1-2):324-38.

[17] Kim J, Bon-Kyoung K, Knoblic JA. Human organoids: model systems for human biology and medicine. Nature Reviews Molecular Cell Biology. 2020;21:571-84.

[18] Lancaster MA, Huch M. Disease modelling in human organoids. Dis Model Mech. 2019;12(7).

[19] Bar-Ephraim YE, Kretzschmar K, Clevers H. Organoids in immunological research. Nat Rev Immunol. 2020;20(5):279-93.

[20] Cattaneo CM, Dijkstra KK, Fanchi LF, et al. Tumor organoid-T-cell

coculture systems. Nat Protoc. 2020;15(1):15-39.

[21] Dijkstra KK, Cattaneo CM, Weeber F, et al. Generation of Tumor-Reactive T Cells by Co-culture of Peripheral Blood Lymphocytes and Tumor Organoids. Cell. 2018;174(6):1586-98 e12.

[22] Ye W, Luo C, Li C, Huang J, Liu F. Organoids to study immune functions, immunological diseases and immunotherapy. Cancer Lett. 2020;477:31-40.

[23] Daniel Irimia XW. Inflammation-on-a-chip: probing the immune system ex vivo. Trends in biotechnology. 2018 36(9):923-37.

[24] Longmire TA, Ikonomou L, Hawkins F, et al. Efficient derivation of purified lung and thyroid progenitors from embryonic stem cells. Cell Stem Cell. 2012;10(4):398-411.

[25] Venter C, Niesler C. A triple co-culture method to investigate the effect of macrophages and fibroblasts on myoblast proliferation and migration. BIOTECHNIQUES. 2018;64(2).

[26] Biton Mea. T helper cell cytokines modulate intestinal stem cell renewal and differentiation. Cell. 2018;175(e22):1307-20.

[27] Rogoz A, Reis BS, Karssemeijer RA, Mucida D. A 3-D enteroid-based model to study T-cell and epithelial cell interaction. J Immunol Methods. 2015; 421:89-95.

[28] Shaffiey SA, Jia H, Keane T, et al. Intestinal stem cell growth and differentiation on a tubular scaffold with evaluation in small and large animals. Regenerative Medicine. 2016;1(11):45-61.

[29] Jung KB, Lee H, Son YS, et al. Interleukin-2 induces the in vitro maturation of human pluripotent stem cell-derived intestinal organoids. Nature communications. 2018;9(3039).

[30] Holokai L, Chakrabarti J, Broda T, et al. Increased programmed death-ligand 1 is an early epithelial cell response to Helicobacter pylori infection. PLOS Pathogens. 2019;15(1):e1007468.

[31] Bar-Ephraim YE, Kretzschmar K, Clevers H. Organoids in immunological research. Nature reviews. 2020;20:279-93.

[32] Schutgens F, Clevers H. Human Organoids: Tools for Understanding Biology and Treating Diseases. Annu Rev Pathol. 2020;15:211-34.

[33] Rena W, Lewandowski BC, Watson J, et al. Single Lgr5- or Lgr6-expressing taste stem/progenitor cells generate taste bud cells ex vivo. PNAS. 2014;111(46):16401-6.

[34] Bartfeld S, Bayram T, van de Wetering M, et al. In Vitro Expansion of Human Gastric Epithelial Stem Cells andTheir Responses to Bacterial Infection. Gastroenterology. 2015;148:126-36.

[35] Sun Y, Ding Q. Genome engineering of stem cell organoids for disease modeling. Protein Cell. 2017;8(5):315-27.

[36] Lohmussaar K, Kopper O, Korving J, et al. Assessing the origin of high-grade serous ovarian cancer using CRISPR-modification of mouse organoids. Nat Commun. 2020;11(1):2660.

[37] Dekkers JF, Berkers G, Kruisselbrink E, et al. Characterizing responses to CFTR-modulating drugs using rectal organoids derived from subjects with cystic fibrosis. Sci Transl Med. 2016;8(344):344ra84.

[38] Dekkers JF, Berkers G, Kruisselbrink E, et al. Characterizing responses to CFTR-modulating

drugs using rectal organoids derived from subjects with cystic fibrosis. Science Translational Medicine. 2016;8(344):344ra84.

[39] Sachs N, Clevers H. Organoid cultures for the analysis of cancer phenotypes. Current Opinions in Genetic Development. 2014;24:68-73.

[40] Driehuis E, Kretzschmar K, Clevers H. Establishment of patient-derived cancer organoids for drug-screening applications. Nat Protoc. 2020;15(10):3380-409.

[41] Lazzari G, Nicolas V, Matsusak M, et al. Multicellular spheroid based on a triple co-culture: A novel 3D model to mimic pancreatic tumor complexity. Acta Biomaterialia. 2018;78:296-307.

[42] Jacob F, Salinas RD, Zhang DY, et al. A Patient-Derived Glioblastoma Organoid Model and Biobank Recapitulates Inter- and Intra-tumoral Heterogeneity. Cell. 2020;180(1):188-204 e22.

[43] Lancaster MA, Knoblich JA. Generation of cerebral organoids from human pluripotent stem cells. Nat Protoc. 2014;9(10):2329-40.

[44] Kelava I, Lancaster MA. Dishing out mini-brains: Current progress and future prospects in brain organoid research. Dev Biol. 2016;420(2):199-209.

[45] Raja WK, Mungenast AE, Lin YT, et al. Self-Organizing 3D Human Neural Tissue Derived from Induced Pluripotent Stem Cells Recapitulate Alzheimer's Disease Phenotypes. PLoS One. 2016;11(9):e0161969.

[46] Serrano Martinez P, Cinat D, van Luijk P, Baanstra M, de Haan G, Pringle S, et al. Mouse parotid salivary gland organoids for the in vitro study of stem cell radiation response. Oral Dis. 2020.

[47] He A, Powell S, Kyle M, et al. Cryopreservation of Viable Human Tissues: Renewable Resource for Viable Tissue, Cell Lines, and Organoid Development. Biopreserv Biobank. 2020;18(3):222-7.

[48] Mejias JC, Nelson MR, Liseth O, Roy K. A 96-well format microvascularized human lung-on-a-chip platform for microphysiological modeling of fibrotic diseases. Lab Chip. 2020;20(19):3601-11.

[49] Srivastava V, Huycke TR, Phong KT, Gartner ZJ. Organoid models for mammary gland dynamics and breast cancer. Curr Opin Cell Biol. 2020;66:51-8.

[50] Derouet MF, Allen J, Wilson GW, et al. Towards personalized induction therapy for esophageal adenocarcinoma: organoids derived from endoscopic biopsy recapitulate the pre-treatment tumor. Sci Rep. 2020;10(1):14514.

[51] Tentler JJ, Tan AC, Weekes CD, et al. Patient-derived tumour xenografts as models for oncology drug development. Nature Reviews Clinical Oncology. 2012;9:338-50.

[52] Fukuda M, Mizutani T, Mochizuki W, et al. Small intestinal stem cell identity is maintained with functional Paneth cells in heterotopically grafted epithelium onto the colon. Genes & Development. 2014;28:1752-7.

[53] Fordham RP, Yui S, Hannan NRF, et al. Transplantation of expanded fetal intestinal progenitors contributes to colon regeneration after injury. Cell Stem Cell. 2013;13:734-44.

[54] Yui S, Nakamura T, Sato T, et al. Functional engraftment of colon epithelium expanded in vitro from a single adult Lgr5+ stem cell. Nature Medicine. 2012;18:618-23.

[55] Driehuis E, van Hoeck A, Moore K, et al. Pancreatic cancer organoids recapitulate disease and allow personalized drug screening. PNAS. 2019.

[56] Huch M, Dorrell C, Boj SF, et al. In vitro expansion of single Lgr5+ liver stem cells induced by Wnt-driven regeneration. Nature. 2013; 494(7436):247-50.

[57] Huch M, Boj SF, Clevers H. Lgr5(+) liver stem cells, hepatic organoids and regenerative medicine. Regenerative Medicine. 2013;8:385-7.

[58] Huch M, Gehart H, van Boxtel R, et al. Long-Term Culture of Genome-Stable Bipotent Stem Cells from Adult Human Liver. Cell. 2015;160(1-2):299-312.

[59] Calandrini C, Schutgens F, Oka R, et al. An organoid biobank for childhood kidney cancers that captures disease and tissue heterogeneity. Nat Commun. 2020;11(1):1310.

[60] Grassi L, Alfonsi R, Francescangeli F, et al. Organoids as a new model for improving regenerative medicine and cancer personalized therapy in renal diseases. Cell Death Dis. 2019;10(3):201.

[61] Gu ZY, Jia SZ, Liu S, Leng JH. Endometrial organoids: A new model for the research of endometrial related diseasesdagger. Biol Reprod. 2020.

[62] Beshiri ML, Tice CM, Tran C, et al. A PDX/Organoid Biobank of Advanced Prostate Cancers Captures Genomic and Phenotypic Heterogeneity for Disease Modeling and Therapeutic Screening. Clin Cancer Res. 2018;24(17):4332-45.

[63] Yan HHN, Siu HC, Law S, et al. A Comprehensive Human Gastric Cancer Organoid Biobank Captures Tumor Subtype Heterogeneity and Enables Therapeutic Screening. Cell Stem Cell. 2018;23(6):882-97 e11.

[64] Tentler JJ, Nallapareddy S, Tan AC, et al. Identification of predictive markers of response to the MEK1/2 inhibitor selumetinib (AZD6244) in K-ras-mutated colorectal cancer. Molecular Cancer Theraphy. 2010;12(9):3351-62.

[65] Drost J, Clevers H. Translational applications of adult stem cell-derived organoids. Development. 2017;144:968-75.

[66] Xinaris C. Organoids for replacement therapy: expectations, limitations and reality. Current Opinion in Organ Transplantation. 2019;24(5):555-61.

[67] Zakrzewski W, Dobrzynski M, Szymonowicz M, Rybak Z. Stem cells: past, present, and future. Stem Cell Res Ther. 2019;10(1):68.

[68] Chang EH, Harford JB, Eaton MA, et al. Nanomedicine: Past, present and future - A global perspective. Biochem Biophys Res Commun. 2015;468(3):511-7.

[69] Cao Y, DePinho RA, Ernst M, Vousden K. Cancer research: past, present and future. Nat Rev Cancer. 2011;11(10):749-54.

[70] Cohen H, Salmon A, Tietel Z, Hacham Y, Amir R. The relative contribution of genes operating in the S-methylmethionine cycle to methionine metabolism in Arabidopsis seeds. Plant Cell Rep. 2017;36(5):731-43.

[71] Broadley SA, Barnett MH, Boggild M, et al. A new era in the treatment of multiple sclerosis. Med J Aust. 2015;203(3):139-41, 41e 1.

[72] Spencer B, Masliah E. Immunotherapy for Alzheimer's disease:

past, present and future. Front Aging Neurosci. 2014;6:114.

[73] Vogenberg FR, Baras CI, Pursel M. Personalized Medicine. Part 1: Evolution and Development into Theranostics. Pharmacy & Therapeutics. 2010;35(10):560-2, 5-7, 76.

[74] Vogenberg FR, Barash CI, Pursel M. Personalized Medicine. Part 2: Ethical, Legal, and Regulatory Issues. Pharmacy and therapeutics. 2010;35(11):624-6, 8-31, 42.

[75] Vogenberg FR, Barash CI, Pursel M. Personalized medicine. Part 3: challenges facing health care plans in implementing coverage policies for pharmacogenomic and genetic testing. Pharmacy and therapeutics. 2010;35(12):670-5.

[76] Bartfeld S, Clevers H. Stem cell-derived organoids and their application for medical research and patient treatment. J Mol Med (Berl). 2017; 95(7):729-38.

[77] Clevers HC. Organoids: Avatars for Personalized Medicine. Keio J Med. 2019;68(4):95.

[78] Perkhofer L, Frappart P-O, Müller M, et al. Importance of organoids for personalized medicine. Personalized Medicine. 2018;15(6):461-5.

[79] Yao Y, Xu X, Yang L, et al. Patient-Derived Organoids Predict Chemoradiation Responses of Locally Advanced Rectal Cancer. Cell Stem Cell. 2020;26(1):17-26 e6.

[80] Clever H. Modeling Development and Disease with Organoids. Cell. 2016;165(7):1586-97.

[81] Grassi L, Alfonsi R, Francescangeli F, et al. Organoids as a new model for improving regenerative medicine and cancer personalized therapy in renal diseases. Cell Death & Disease. 2019;10(201).

[82] Clevers H. Organoids: Avatars for Personalized Medicine. The Keio journal of medicine. 2019;68(4):95.

[83] Fan H, Demirci U, Chen P. Emerging organoid models: leaping forward in cancer research. J Hematol Oncol. 2019;12(1):142.

[84] Fatehullah A, Tan SH, Barker N. Organoids as an in vitro model of human development and disease. Nature Cell Biology. 2016;18:246-54.

[85] Lou YR, Leung AW. Next generation organoids for biomedical research and applications. Biotechnol Adv. 2018;36(1):132-49.

[86] Grebenyuk S, Ranga A. Engineering Organoid Vascularization. Frontiers in Bioengineering and Biotechnology. 2019;7(39).

[87] Ashok A, Choudhury D, Fang Y, Hunziker W. Towards manufacturing of human organoids. Biotechnol Adv. 2020;39:107460.

[88] Choudhury D, Ashok A, Naing MW. Commercialization of Organoids. Trends Mol Med. 2020;26(3):245-9.

[89] Bredenoord AL, Clevers H, Knoblich JA. Human tissues in a dish: The research and ethical implications of organoid technology. Science. 2017;355(6322).

[90] Yu F, Hunziker W, Choudhury D. Engineering Microfluidic Organoid-on-a-Chip Platforms. Micromachines (Basel). 2019;10(3).

[91] Peng W, Unutmaz D, Ozbolat IT. Bioprinting towards Physiologically Relevant Tissue Models for Pharmaceutics. Trends Biotechnol. 2016; 34(9):722-32.

[92] Lee H, Cho DW. One-step fabrication of an organ-on-a-chip with spatial heterogeneity using a 3D bioprinting technology. Lab Chip. 2016; 16(14):2618-25.

[93] Park SE, Georgescu A, Huh D. Organoids-on-a-chip. Science. 2019;364(6444):960-5.

[94] Takebe T, Zhang B, Radisic M. Synergistic Engineering: Organoids Meet Organs-on-a-Chip. Cell Stem Cell. 2017;21(3):297-300.

[95] Albritton JL, Miller JS. 3D bioprinting: improving in vitro models of metastasis with heterogeneous tumor microenvironments. Dis Model Mech. 2017;10(1):3-14.

[96] Bissell MJ, Hines WC. Why don't we get more cancer? A proposed role of the microenvironment in restraining cancer progression. Nat Med. 2011;17(3):320-9.

[97] Grolman JM, Zhang D, Smith AM, Moore JS, Kilian KA. Rapid 3D Extrusion of Synthetic Tumor Microenvironments. Adv Mater. 2015;27(37):5512-7.

[98] Pang Y, Mao SS, Yao R, et al. TGF-beta induced epithelial-mesenchymal transition in an advanced cervical tumor model by 3D printing. Biofabrication. 2018;10(4):044102.

Chapter 3

Innovative *In Vitro* Models for the Study of Lung Diseases

Vittorio Picchio, Vittoria Cammisotto, Francesca Pagano, Roberto Carnevale and Isotta Chimenti

Abstract

Basic and translational research on lung biology and pathology can greatly benefit from the development of 3D *in vitro* models with physiological relevance. Lung organoids and lungs-on-chip allow the creation of different kinds of *in vitro* microenvironments, that can be useful for the elucidation of novel pathogenetic pathways, for example concerning tissue fibrosis in chronic diseases. Moreover, they represent important translational models for the identification of novel therapeutic targets, and for preliminary testing of new drugs. In this chapter, we provide a selected overview of recent studies on innovative 3D *in vitro* models that have enhanced our knowledge on chronic obstructive pulmonary disease (COPD) and idiopathic pulmonary fibrosis (IPF), particularly concerning oxidative stress and pro-fibrotic pathogenetic mechanisms. Despite several limitations, these complex models must be considered as complementary in all respects to in vivo studies on animal models and clinical research.

Keywords: organoids, organ-on-chip, oxidative stress, lung fibrosis, cell spheroids

1. Lung chronic diseases: a brief overview

The primary function of the lungs is the exchange of gas occurring at the level of alveoli, which are arranged as acini in the lung parenchyma. There is strong need to understand the mechanisms of alveolar maintenance and repair because damage to this region underlies many chronic adult lung diseases, such as chronic obstructive pulmonary disease, idiopathic pulmonary fibrosis, acute respiratory failure in pneumonia, and acute respiratory distress syndrome. Additionally, insufficient development of alveoli results in various neonatal and childhood diseases including bronchopulmonary dysplasia [1]. Despite the pivotal role of the alveoli in the development of lung diseases, the pathogenesis of these various conditions is still largely unknown, and treatment options for patients remain limited.

There is clear evidence that environmental exposures and genetic predisposition contribute to the pathogenesis of idiopathic pulmonary fibrosis (IPF). IPF is defined as a specific form of progressive chronic fibrotic interstitial pneumonia, that is occurring mainly in older adults, and is limited to the lungs [2]. IPF remains relatively rare, with an estimated incidence of roughly 10 cases per 100,000 person-years. Nonetheless, IPF is a lethal lung disorder with a predicted survival of 3–6 years from the onset of symptoms. Most of the deaths among patients with IPF are due to respiratory failure or complicating comorbidities [3]. The pathogenesis of IPF is characterized by continuous insults or micro-lesions to the alveolar epithelium, which result in abnormal activation of both epithelial cells and fibroblasts. Finally, there is an alteration in the deposition of collagen, which contributes to the irreversible fibrosis typical of the disease [4]. Various risk factors have been identified in the development of IPF, that can be divided between intrinsic and extrinsic [5]. Intrinsic risk factors include genetics, aging, sex, lung microbiome [6–9], while extrinsic risk factors comprise cigarette smoking, environmental exposures, and air pollution [10, 11]. Moreover, studies of familial clustering of pulmonary fibrosis provided evidence that IPF is associated with genetic susceptibility. Multiple genes can affect alveolar stability, for example, genes encoding surfactant proteins A and C, genes associated with enhanced cell senescence by disruption of telomerase function, with the integrity of the epithelial barrier, and with mutant desmosome proteins [12–15].

Chronic obstructive pulmonary disease (COPD) is another chronic lung pathology, representing a serious and growing global health problem, as it is currently a leading cause of death worldwide [16]. COPD is a disease characterized by irreversible airflow reduction, associated with a decline in lung function and increased inflammatory response [17]. It represents a massive health problem, and it is estimated to affect around 200 million people worldwide, with a projected estimate towards further increase in the near future [18]. COPD is the result of the interaction between genetic susceptibility and environmental factors [19]. A well acknowledged genetic cause is α1-antitrypsin deficiency [20], while among environmental factors cigarette smoking represents the main cause; nonetheless, environmental pollution, occupational exposure to dust and fumes, and exposure to passive smoke can induce an increased risk in non-smokers, as well [21, 22]. Exposure to cigarette smoke, which contains a large number of pro-oxidant molecules [23], causes direct damage to the epithelial cells of the airways, leading to increased inflammation and activation of neutrophils, macrophages and lymphocytes in the airways [24]. There is currently no cure for COPD, but fortunately most symptoms can be treated and controlled mostly pharmacologically, at least delaying its progression and worsening. It represents indeed the most common indication for lung transplantation, that is the only conclusive therapeutic option for severe COPD, particularly in younger patients.

2. Smoke, oxidative stress and fibrosis in lung pathogenesis

Cigarette smoking likely represents the single most significant risk factor for several lung conditions, and it is strongly associated with COPD and IPF, both familial and sporadic. Although observations about environmental risk factors have many biases and limitations [25], increasing knowledge on the underlying causes of lung diseases is evidencing how oxidative stress (OXS) and reactive oxygen species (ROS) play a crucial pathogenetic role (**Figure 1**).

The lungs are indeed highly susceptible to ROS-induced injuries. ROS are commonly thought to be a harmful by-product generated in cellular systems. However, recent studies have suggested that ROS physiological levels regulate important biological functions in cellular processes [2, 26]. Normally, ROS are tightly controlled by enzymes and antioxidant molecules. Nonetheless, excessive ROS accumulation may occur under certain conditions, thus making detoxification by the antioxidant system difficult. The result is indeed a condition called OXS that can affect cell proliferation, differentiation, aging, and death [27]. Cigarette smoke is responsible for significant oxidant burden and decreased antioxidant capacity even in plasma [28, 29]. ROS produced from cigarette smoke, combustion of organic matter and gases, like ozone and nitrogen dioxide, are featured on the lung epithelium [30], and could decrease antioxidant defenses, increasing OXS in the lungs [31].

DOI: http://dx.doi.org/10.5772/intechopen.95300

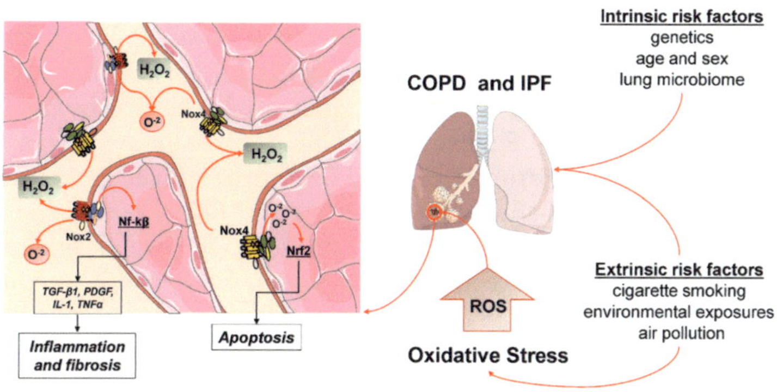

Figure 1.
Oxidative stress in lung pathogenesis. Intrinsic and extrinsic risk factors contribute to the progression of lung damage in the development of idiopathic pulmonary fibrosis (IPF) and obstructive pulmonary disease (COPD). In particular, cigarette smoking, environmental exposures, and air pollution induce an increase of reactive oxygen species (ROS) and oxidative stress condition, which play a crucial pathogenetic role in lung diseases. ROS, such as H_2O_2 and O^{2-}, generated from NOX2 and NOX4, have a central role in the pathogenesis of pulmonary diseases. Indeed, ROS produced by these enzymes are involved in alveolar epithelial cell apoptosis, activation of inflammation, and induction of tissue fibrosis, that are all mechanisms underlying the progression of IPF and COPD. Figure was prepared using images from Servier Medical Art by Servier (https://smart.servier.com), which are licensed under a Creative Commons Attribution 3.0 Unported License.

A disrupted function of the redox system can consequentially impact on key cell signaling pathways involved in disease progression. Conversely, several signals can alter the oxidative state of lung cells. For example, the lung is constantly exposed to biomechanical forces, such as fluid shear stress, cyclic stretch, and pressure, due to the blood flowing through the pulmonary vessels, and the distension of the lungs during the breathing cycle. It is indeed known that cells within the lung respond to these changes by activating signal transduction pathways that can also alter their redox state with pathophysiological consequences [32]. Particularly in the vasculature, the two types of biomechanical stimuli, such as frictional force known as shear stress (SS), or wall shear stress (WSS) that acts tangentially to the vessel, could determinate dysregulation of the cellular redox status, that in turn could have effects on intracellular signaling pathways involved in disease progression [33]. For example, exposure of endothelial cells to laminar SS can induce a suppression of ROS levels [34, 35]. Conversely, exposure of endothelial cells to WSS using an irregular flow induces an increase of ROS levels and a reduced bioavailability of the vasodilator molecule NO [36], which is involved in preventing the activation and adhesion of platelets and leukocytes to the wall of the injured vessel [37].

A significant role in the pathogenesis of COPD is precisely the imbalance of ROS production and antioxidant capacity [38]. Changes in the redox balance in the lungs and circulatory system, genetic polymorphisms, and activation of transcription factors, such as the nuclear factor kappa B (NF-κB), lead to the molecular pathogenesis of COPD [39, 40]. Oxidized proteins and lipid products, such as isoprostanes and carbonylated proteins, can be identified in exhaled air, bronchoalveolar lavage fluid, and lung tissue from patients with fibrotic lung diseases and COPD [41, 42]. Furthermore, clinical worsening of COPD is often associated with down-regulation of the antioxidant system, thus a possible therapeutic method for COPD could be the administration of redox-protective antioxidants [38]. Finally, it is possible that maintaining a balance between oxidant and antioxidant species in COPD affected smokers may slow down disease progression [43].

As discussed, smoking, occupational exposures like asbestos or silica, and radiation are the principal sources of OXS with overproduction of ROS, that could lead and contribute to pulmonary fibrosis [44]. Indeed, OXS is an important molecular mechanism underlying fibrosis in a variety of organs, including lungs. Bleomycin-induced pulmonary fibrosis, the most commonly used experimental animal model, has been shown to be associated with marked increase in the level of ROS, oxidized proteins, DNA and lipids [45]. Following lung injury three main mechanisms (i.e. inflammation, coagulation disturbances, and OXS) are involved and alter the lung interstitial cell compartment and extracellular matrix (ECM) homeostasis, resulting overall in pulmonary fibrosis. ROS can be produced by several cellular types involved in fibrosis including alveolar macrophages [46–48] and lung epithelial cells [49]. In particular, ROS generated from the mitochondria of stressed or damaged epithelial cells are very important; their mitochondrial dysfunction results in the generation and release of ROS, such as H_2O_2 and O^{2-}, further enhancing OXS and cell damage [50]. As previously discussed, NAD(P)H oxidase is the main source of ROS, and isoforms NOX1, NOX2, and NOX4 have a central role in the pathogenesis of pulmonary fibrosis [51] (**Figure 1**). For example, NOX4 is strongly expressed in the hyperplastic alveolar epithelium of IPF patients [52], and ROS produced by NOX4 are involved in alveolar epithelial cell apoptosis. Continuous epithelial apoptosis further supports activation of inflammatory processes and cytokine release, including myofibroblast activating molecules, such as TGF-β1, PDGF, IL-1, and TNFα (**Figure 1**). There is also evidence of direct pathogenetic involvement of these enzymes in IPF, for example for NOX2: in fact, supporting data have shown that mice genetically deficient in NOX2 do not develop IPF after bleomycin or carbon nanotubes exposure [51, 53]. Finally, the interplay between oxidative stress and TGF-β1 signaling is of great importance in promoting fibrosis. In fact, TGF-β1 is the most profibrogenic protein and can directly stimulate NOX-mediated ROS production, while OXS in turn can activate latent TGF-β1, setting up a vicious profibrogenic positive feedback loop [54].

3. Modelling lung diseases in 3D with organoids

As mentioned previously, fibrosis and oxidative stress are linked to a dysregulation of cellular homeostasis and impaired alveolar structure in chronic lung diseases [55]. Despite the paramount importance of animal models in biomedical and clinical research, they often do not fully recapitulate the pathogenesis of human IPF [56]. Moreover, there is increasing social and political pressure on reducing animal experimentation, according to the 3R's principle of replacement, reduction, and refinement. Furthermore, the associated costs of animal purchasing, housing, and handling cannot be ignored, as well [57]. Under this perspective, 3D cultures (such as organoids) and innovative microfluidic devices (such as "organs-on-chip") represent useful platforms to perform significant investigations *in vitro* on multiple topics, including the pathogenesis of COPD, IPF, or other lung diseases. They grant simultaneous multicellular culture and cell–cell interactions that overcome the limitations of standard monolayer cell cultures, allowing a step forward towards reproducing the complexity of tissues. Moreover, specific protocols and setups make it possible to simulate many more elaborated pathogenetic features, such as ontogenetic-like mechanisms, tunable biomechanical cues, altered gas/liquid interfaces, as well as immune cells recruitment and activation (**Figure 2**). Physiologically relevant *in vitro* systems are also suitable to discover and test new drugs and therapeutics, supporting the clinical translation of novel protocols in a "personalized medicine" perspective [1].

Three-dimensional culture systems offer multiple advantages for *in vitro* phenotype control in order to obtain physiologically relevant settings [58]. The simplest 3D culture system is represented by spheroids, which can be obtained

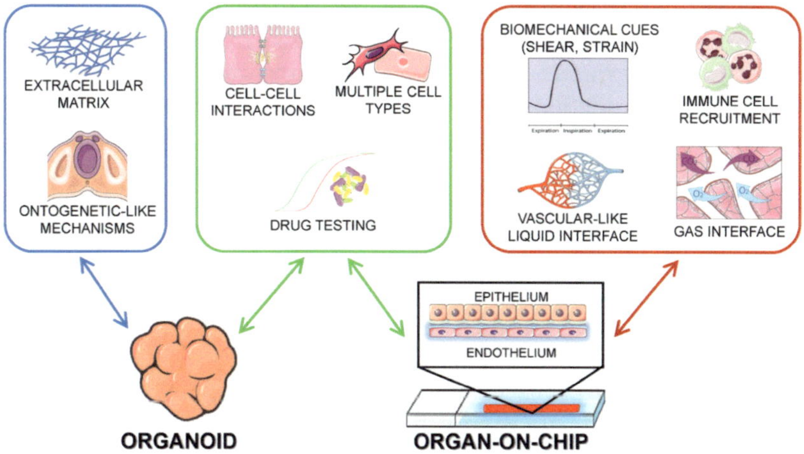

Figure 2.
In vitro models of lung pathology. Both organoids and lungs-on-chip allow the creation of 3D systems where complex cell–cell interactions and multi-cellular cultures are possible. Moreover, drug discovery and testing in these settings can provide important preliminary results in vitro. Several features, though, are better reproduced in specific 3D systems, for example the inclusion of an extracellular matrix, or the modelling of ontogenetic-like mechanisms better fit in organoid cultures. Conversely, biomechanical cues, gas and liquid interfaces, and immune cells response are more finely tunable with organs-on-chip technology. Figure was prepared using images from Servier Medical Art by Servier (https://smart.servier.com), which are licensed under a Creative Commons Attribution 3.0 Unported License.

from embryonic-like stem cells or several resident lung cell types, particularly those with a facultative stemness potential (e.g. pneumocytes, Clara cells [59]). Lung cell spheroids, despite their simplicity, can provide useful preliminary models even for the study of complex pathological issues. Alveoli-like structures obtained from distal airway stem cells [60] or Oct-4+ progenitor cells [61] have been used for the study of viral infections (e.g. H1N1 influenza virus, or SARS-CoV), the pathogenesis of tissue damage, and subsequent mechanisms of tissue repair. As another example, lung spheroids from stromal primitive cells [62] have significantly contributed to the elucidation of novel pathogenetic mechanisms during organ reconditioning procedures, in particular during ex vivo lung perfusion (EVLP) protocols before lung transplantation. In fact, OXS strongly contributes to tissue damage during EVLP. It has been shown that inhibition of NOX2 activity during thermic stress and starvation (mimicking EVLP conditions) can reduce ROS production, thus being protective for lung epithelial cells [63, 64]. Finally, specific interference of cigarette smoke with Wnt/β-catenin signaling has been described in human fibroblasts, impairing their capacity to support spheroid growth of lung epithelial cells, which can be considered in this case as a stemness assay linked to the activation of a repair mechanism [65].

The more complex example of organotypic 3D cultures is represented by organoids. Lung organoids are self-assembling structures of lung cell types that replicate cell–cell interaction, cell-ECM interaction, and organ structure and function at the microscale, as similar as possible to in vivo histological architecture. They can be used as models of both physiological and pathological settings. Strikoudis et al. have modelled pulmonary fibrosis in lung organoids to study Hermansky-Pudlak syndrome (HSP) [66]. IPF and HSP both are characterized by lung fibrosis, and are now considered as similar clinical entities, albeit with distinct etiology. Lung organoids were generated from embryonic stem cells (ESCs) with specific mutations that strongly predispose to HSP. The resulting organoids displayed a fibrotic phenotype, with an enhanced number of mesenchymal cells, and increased deposition of fibronectin and collagen. Interestingly, HSP organoids share a strong signature with lung samples from IPF patients, including the overexpression of interlukin-11 (IL-11), a key driver of the fibrotic process that is stimulated also from OXS [67]. This finding validates HSP lung organoids as a tool to study IPF and other lung diseases characterized by fibrosis [66]. Similarly, Wilkinson et al. have developed an organoid from induced pluripotent stem cell (iPSC)-derived fibroblasts functionalized with hydrogel beads, that acts as a 3D alveolar template within a rotating bioreactor [68]. Interestingly, they discovered that organoid formation was not possible in their conditions without the inclusion of fetal lung fibroblasts. Treatment of cultures with exogenous TGF-β1 consistently increased contraction, expression of Collagen 1 and α-SMA, and the emergence of fibroblastic foci within the treated organoid. This system showed features of tissue scarring similar to IPF, thus confirming the feasibility of organoid culture systems to model lung fibrosis. Moreover, these lung organoids can recapitulate even a more complex and representative lung microenvironment when cultured with endothelial and epithelial cells [68]. As an example, using lung organoids from patients with IPF, Surolia et al. described a 3D model to predict the invasive response of IPF fibroblasts to antifibrotic drugs therapy. They observed that inhibition of vimentin intermediate filaments assembly can reduce the invasiveness of lung fibroblasts derived from the majority of the IPF patients tested, uncovering a possible novel therapeutic target for pulmonary fibrosis [69].

Overall, these 3D self-assembled systems recapitulate numerous pathogenetic features of diseases, but nonetheless still show several limitations in their application as models, such as lack of vascular network, immune cells, and other supporting cells (**Figure 2**). These features need to be implemented to reach higher levels of physiological relevance for lung disease modelling [69].

4. Organs-on-chip for the study of lung diseases

In the last decade, the integration of advanced bioengineering approaches (e.g. 3D multicellular cultures) with microfluidic and microfabricated substrates has led to the development of devices called "organs-on-chip" [70]. These bioengineered tools allow fine control and tuning of the microenvironment architecture, media composition, and cell–cell interactions. The combination of lung cells and micro/nanoengineering devices gave rise to new *in vitro* models for the study of therapeutic approaches in pulmonary diseases. In fact, lungs-on-chip can recapitulate typical features of the parenchymal structure, and primary physiological or pathological conditions of the human lung microenvironment, such as liquid and gas interfaces [71] (**Figure 2**). In 2010, Hu et al. for the first time created a lung-on-chip using a soft lithography technique. Soft lithography offers the advantage to control the molecular structure of surfaces, the pattern of complex molecules relevant to biology, and to fabricate channel structures appropriate for microfluidics [72]. They produced a biomimetic microdevice that recapitulates the crucial alveolar-capillary interface of the human lung. This device is a 2.5D system since it contains monolayers of epithelial and endothelial cells that mimic the alveolar-capillary barrier, and permits investigation under dynamic conditions, with biomechanical cues in the form of SS due to perfusion, and strain similar to breathing [71]. However, ECM components are lacking in this model, and this significantly limits the relevance of this device, in particular concerning the study of pulmonary fibrosis. To address these limitations, other groups have designed arrays of 3D microtissue that are suspended over multiple flexible poly-dimethylsiloxane (PMDS) micropillars [73–75]. In particular, Sellgren et al. produced an advanced model by co-culturing interstitial fibroblasts with epithelial and endothelial cells [75]. They demonstrated the feasibility of including a stromal layer within lung-on-chip devices. Similarly, Asmani et al.

have developed a human lung device to model key biomechanical events occurring during lung fibrogenesis, which include progressive stiffening and contraction of alveolar tissue. They used this system for predicting the efficacy of anti-fibrotic drugs for IPF patients, demonstrating that preventative treatments with these drugs can reduce tissue contractility, and counteract tissue stiffening and decline in tissue compliance [73]. Overall, these new approaches will give a better understanding of the complex pathogenesis of IPF.

As discussed above, COPD is a syndrome defined by progressive and chronic airflow limitation, due to the fact that lungs become inflamed, damaged, and narrowed. The main cause is smoking, but others exist such as long-term exposure to harmful fumes or dust, and rare genetic conditions [43]. As for IPF, the animal models of COPD present some limitations. For example, modelling cigarette smoke exposure fails to recapitulate some major airway phenotypes of COPD, such as hyperplasia of basal and mucin-producing cells, and mucus plugging of the airways [76]. Before the advent of lung-on-chip technology, the best-established *in vitro* model to study COPD disease and to address cigarette smoke-induced damage on human airway epithelial cells was the air-liquid-interface (ALI) culture system [77]. The defining feature of ALI cultures is that the basal surface of the cell is in contact with a liquid culture medium, whereas the apical surface is exposed to air [78]. These systems mimic the conditions found in the human airway, and drive differentiation towards different phenotypes [79]. One major limitation of conventional ALI models is that these static culture systems make dynamic processes, such as nutrient exchange and immune cell migration [80], difficult to study.

In this regard, innovative approaches, such as microfluidic lungs-on-chip, have been developed in the last years and helped filling this gap. In 2016, Benam et al. developed the human lung "small airway-on-a-chip", a microfluidic device that supports and drives full differentiation of a columnar, pseudostratified, mucociliary bronchiolar epithelium, composed of cells isolated from healthy individuals or people with COPD, underlined by a functional microvascular endothelium [81]. They demonstrated that COPD small airway chips recapitulate important features of the disease, such as selective cytokine hypersecretion and neutrophil recruitment from the vascular flow in response to epithelial activation by pathogen-like stimuli. Moreover, exposure of the healthy epithelium to interleukin-13 (IL-13) reconstituted the asthmatic phenotype that involves goblet cells hyperplasia, cytokine hypersecretion, and decreased ciliary function [82]. The same group improved this system by developing a "Breathing-Smoking Human Lung-on-Chip", a novel device that consists of four components: a small airway on-chip, a smoke generating robot, a micro-respirator, and a control software that mimics human smoking and breathing. This smoking airway-on-a-chip system effectively recapitulated several key smoke-triggered molecular changes that are known to occur in lung epithelial cells, including increased OXS [83]. When human airway chips fabricated using cells from healthy donors were exposed to whole cigarette smoke, the authors observed a significant increase in the expression of the anti-oxidant gene heme oxygenase 1 (HMOX1), and increased phosphorylation of the transcription factor nuclear factor-like 2 (Nrf2). The latter induces expression of cytoprotective genes, including HMOX1, protecting cells from OXS and chemical toxicity. Furthermore, they identified new smoke-induced dysfunction, such as reduced ciliary beating, a novel biomarker of COPD disease, and studied the epithelial responses to smoke generated by electronic cigarettes [84]. However, the main limitation of this system is the absence of cellular stromal components.

As mentioned before, COPD represents a group of lung diseases that also include refractory severe asthma. In this regard, Nesmith et al. have designed a human airway musculature-on-a-chip with bronchiolar smooth muscle cells on an elastomeric thin film. To recapitulate asthmatic inflammation *in vitro*, they exposed this biomimetic tissue to IL-13, which resulted in hypercontractility and altered relaxation. Interestingly, the authors were able to show reverse asthmatic hypercontraction of smooth muscle cells using a muscarinic antagonist and a β-agonist, which are used clinically to relax constricted airway [85]. Similarly, Villenave et al. developed a model of severe asthma-on-chip containing a fully differentiated mucociliary bronchiolar epithelium underlined by a microvascular endothelium with fluid flow [86]. They infected the engineered tissue with human Rhinovirus (HRV), a leading cause of asthma exacerbation in children and adults; this led to a pro-inflammatory response characterized by ciliated cells death, goblet cells hyperplasia, release of cytokines, recruitment from the fluid flow and extravasation of human neutrophils across the endothelium. Infection of IL-13-treated Airway Chips with HRV to mimic the molecular response observed in severe asthma patients, induced upregulation of adhesion molecules (E- and P-Selectin, ICAM-1) in endothelial cells, and increase of neutrophil recruitment when compared with IL-13 or HRV stimulation alone [87]. The same group implemented this device to study the integrity of epithelial monolayers-on-chip, measuring trans-epithelial electrical resistance (TERT). They designed a new microfluidic device within a human lung airway chip that contains embedded electrodes, and demonstrated its utility for the assessment of airway barrier function, formation, and disruption in response to relevant external stimuli [88]. These studies suggest that Airway Chips may provide unique opportunities to explore lung pathogenesis, including responses to drug treatments for the evaluation of safety and efficacy of new drugs. Moreover, the possibility of studying the involvement and activation of immune cells certainly brings added value to these systems, allowing the study of physiologically relevant issues within an integrated model.

5. Conclusions

Basic and translational research on lung biology and pathology can greatly benefit from the development of 3D *in vitro* models that can maintain cell phenotypes and functions in a physiologically relevant way. Lung organoids and lungs-on-chip allow the creation of different kinds of *in vitro* microenvironments (**Figure 2**), that can be useful for the study of specific diseases, and for the elucidation of novel pathogenetic pathways. They represent in fact important translational models for the study of clinically relevant issues, for the identification of novel therapeutic targets, and for preliminary testing of new drugs. The main challenge in future developments is represented by the standardization of integrated protocols for the simultaneous inclusion of extracellular matrix, stromal components, immune cells, and biomechanical cues within 3D *in vitro* models. This step forward would provide a clinically relevant system for lung research, which would include all the actors involved in endogenous responses occurring *in vivo*. Nonetheless, despite several limitations still existing, the complexity of these models has been rapidly increasing in the past decade, and they must be considered as complementary in all respects to *in vivo* studies carried on in animal models.

Acknowledgements

This work was supported by grant # 2015BN82FK from the Italian Ministry of Education, University and Research (MIUR) to Isotta Chimenti.

Conflict of interest

The authors declare no conflict of interest.

Author details

Vittorio Picchio[1], Vittoria Cammisotto[1], Francesca Pagano[2], Roberto Carnevale[1,3] and Isotta Chimenti[1,3]*

1 Department of Medical Surgical Sciences and Biotechnologies, Sapienza University, Italy

2 Institute of Biochemistry and Cell Biology, National Council of Research (IBBC-CNR), Italy

3 Mediterranea Cardiocentro, Napoli, Italy

*Address all correspondence to: isotta.chimenti@uniroma1.it

References

[1] Evans K V., Lee JH. Alveolar wars: The rise of *in vitro* models to understand human lung alveolar maintenance, regeneration, and disease. Vol. 9, Stem Cells Translational Medicine. John Wiley and Sons Ltd.; 2020. p. 867-81.

[2] Raghu G, Remy-Jardin M, Myers JL, Richeldi L, Ryerson CJ, Lederer DJ, *et al.* Diagnosis of idiopathic pulmonary fibrosis An Official ATS/ERS/JRS/ALAT Clinical practice guideline. Am J Respir Crit Care Med. 2018 Sep 1;198(5):e44-68.

[3] Bocchino M, Agnese S, Fagone E, Svegliati S, Grieco D, Vancheri C, *et al.* Reactive Oxygen Species Are Required for Maintenance and Differentiation of Primary Lung Fibroblasts in Idiopathic Pulmonary Fibrosis. Koenigshoff M, editor. PLoS One. 2010 Nov 16;5(11):e14003.

[4] King TE, Pardo A, Selman M. Idiopathic pulmonary fibrosis. In: The Lancet. Lancet; 2011. p. 1949-61.

[5] Zaman T, Lee JS. Risk Factors for the Development of Idiopathic Pulmonary Fibrosis: a Review. Curr Pulmonol Reports. 2018 Dec;7(4):118-25.

[6] Kropski JA, Blackwell TS, Loyd JE. The genetic basis of idiopathic pulmonary fibrosis. Eur Respir J. 2015 Jun 1;45(6):1717-27.

[7] Raghu G, Chen SY, Hou Q, Yeh WS, Collard HR. Incidence and prevalence of idiopathic pulmonary fibrosis in US adults 18-64 years old. Eur Respir J. 2016 Jul 1;48(1):179-86.

[8] Jo HE, Glaspole I, Grainge C, Goh N, Hopkins PMA, Moodley Y, *et al.* Baseline characteristics of idiopathic pulmonary fibrosis: Analysis from the Australian Idiopathic Pulmonary Fibrosis Registry. Eur Respir J. 2017 Feb 1;49(2).

[9] Molyneaux PL, Cox MJ, Willis-Owen SAG, Mallia P, Russell KE, Russell AM, *et al.* The role of bacteria in the pathogenesis and progression of idiopathic pulmonary fibrosis. Am J Respir Crit Care Med. 2014 Oct 15;190(8):906-13.

[10] Oh CK, Murray LA, Molfino NA. Smoking and idiopathic pulmonary fibrosis. Vol. 2012, Pulmonary Medicine. Hindawi Limited; 2012.

[11] Conti S, Harari S, Caminati A, Zanobetti A, Schwartz JD, Bertazzi PA, *et al.* The association between air pollution and the incidence of idiopathic pulmonary fibrosis in Northern Italy. Eur Respir J. 2018 Jan 1;51(1).

[12] Kaur A, Mathai SK, Schwartz DA. Genetics in idiopathic pulmonary fibrosis pathogenesis, prognosis, and treatment. Vol. 4, Frontiers in Medicine. Frontiers Media S.A.; 2017.

[13] Fingerlin TE, Murphy E, Zhang W, Peljto AL, Brown KK, Steele MP, *et al.* Genome-wide association study identifies multiple susceptibility loci for pulmonary fibrosis. Nat Genet. 2013 Jun;45(6):613-20.

[14] Noth I, Zhang Y, Ma SF, Flores C, Barber M, Huang Y, *et al.* Genetic variants associated with idiopathic pulmonary fibrosis susceptibility and mortality: A genome-wide association study. Lancet Respir Med. 2013 Jun;1(4):309-17.

[15] Seibold MA, Wise AL, Speer MC, Steele MP, Brown KK, Loyd JE, *et al.* A common MUC5B promoter polymorphism and pulmonary fibrosis. N Engl J Med. 2011 Apr 21;364(16):1503-12.

[16] Lozano R, Naghavi M, Foreman K, Lim S, Shibuya K, Aboyans V, *et al.* Global and regional mortality from

235 causes of death for 20 age groups in 1990 and 2010: A systematic analysis for the Global Burden of Disease Study 2010. Lancet. 2012 Dec 1;380(9859):2095-128.

[17] Decramer M, Janssens W, Miravitlles M. Chronic obstructive pulmonary disease. In: The Lancet. Lancet Publishing Group; 2012. p. 1341-51.

[18] Vos T, Allen C, Arora M, Barber RM, Brown A, Carter A, *et al.* Global, regional, and national incidence, prevalence, and years lived with disability for 310 diseases and injuries, 1990-2015: a systematic analysis for the Global Burden of Disease Study 2015. Lancet. 2016 Oct 8;388(10053):1545-602.

[19] Barnes PJ. Chronic obstructive pulmonary disease. Vol. 343, New England Journal of Medicine. N Engl J Med; 2000. p. 269-80.

[20] Gooptu B, Ekeowa UI, Lomas DA. Mechanisms of emphysema in α1-antitrypsin deficiency: Molecular and cellular insights. Vol. 34, European Respiratory Journal. Eur Respir J; 2008. p. 475-88.

[21] Mannino DM, Buist AS. Global burden of COPD: risk factors, prevalence, and future trends. Vol. 370, Lancet. Lancet; 2007. p. 765-73.

[22] Eisner MD, Anthonisen N, Coultas D, Kuenzli N, Perez-Padilla R, Postma D, *et al.* An official American Thoracic Society public policy statement: Novel risk factors and the global burden of chronic obstructive pulmonary disease. Vol. 182, American Journal of Respiratory and Critical Care Medicine. Am J Respir Crit Care Med; 2010. p. 693-718.

[23] PRYOR WA, STONE K. Oxidants in Cigarette Smoke Radicals, Hydrogen Peroxide, Peroxynitrate, and Peroxynitrite. Ann N Y Acad Sci. 1993;686(1):12-27.

[24] Zinellu E, Zinellu A, Fois AG, Carru C, Pirina P. Circulating biomarkers of oxidative stress in chronic obstructive pulmonary disease: A systematic review. Vol. 17, Respiratory Research. BioMed Central Ltd.; 2016. p. 150.

[25] Raghu G, Collard HR, Egan JJ, Martinez FJ, Behr J, Brown KK, *et al.* An Official ATS/ERS/JRS/ALAT Statement: Idiopathic pulmonary fibrosis: Evidence-based guidelines for diagnosis and management. Am J Respir Crit Care Med. 2011 Mar 15;183(6):788-824.

[26] Reczek CR, Chandel NS. ROS-dependent signal transduction. Vol. 33, Current Opinion in Cell Biology. Elsevier Ltd; 2015. p. 8-13.

[27] Lee J, Cho YS, Jung H, Choi I. Pharmacological regulation of oxidative stress in stem cells. Vol. 2018, Oxidative Medicine and Cellular Longevity. Hindawi Limited; 2018.

[28] Carnevale R, Sciarretta S, Violi F, Nocella C, Loffredo L, Perri L, *et al.* Acute Impact of Tobacco vs Electronic Cigarette Smoking on Oxidative Stress and Vascular Function. Chest. 2016 Sep 1;150(3):606-12.

[29] Frati G, Carnevale R, Nocella C, Peruzzi M, Marullo AGM, De Falco E, *et al.* Profiling the Acute Effects of Modified Risk Products: Evidence from the SUR-VAPES (Sapienza University of Rome-Vascular Assessment of Proatherosclerotic Effects of Smoking) Cluster Study. Vol. 22, Current Atherosclerosis Reports. Springer; 2020. p. 1-11.

[30] Kuwano K, Nakashima N, Inoshima I, Hagimoto N, Fujita M, Yoshimi M, *et al.* Oxidative stress in lung epithelial cells from patients with idiopathic interstitial pneumonias. Eur Respir J. 2003 Feb 1;21(2):232-40.

[31] Barreiro E, Peinado VI, Galdiz JB, Ferrer E, Marin-Corral J, Sánchez F, *et al.*

Cigarette smoke-induced oxidative stress: A role in chronic obstructive pulmonary disease skeletal muscle dysfunction. Am J Respir Crit Care Med. 2010 Aug 15;182(4):477-88.

[32] Zemskov EA, Lu Q, Ornatowski W, Klinger CN, Desai AA, Maltepe E, *et al.* Biomechanical Forces and Oxidative Stress: Implications for Pulmonary Vascular Disease. Antioxid Redox Signal. 2019 Oct 20 [cited 2020 Sep 8]; 31(12):819-42.

[33] Davies PF. Flow-mediated endothelial mechanotransduction. Vol. 75, Physiological Reviews. American Physiological Society; 1995. p. 519-60.

[34] Mohan S, Koyoma K, Thangasamy A, Nakano H, Glickman RD, Mohan N. Low shear stress preferentially enhances IKK activity through selective sources of ROS for persistent activation of NF-κB in endothelial cells. Am J Physiol Physiol. 2007 Jan;292(1):C362-71.

[35] White SJ, Hayes EM, Lehoux S, Jeremy JY, Horrevoets AJG, Newby AC. Characterization of the differential response of endothelial cells exposed to normal and elevated laminar shear stress. J Cell Physiol. 2011 Nov;226(11):2841-8.

[36] Lu X, Kassab GS. Nitric oxide is significantly reduced in ex vivo porcine arteries during reverse flow because of increased superoxide production. J Physiol. 2004 Dec 1;561(2):575-82.

[37] Kubes P, Suzuki M, Granger DN. Nitric oxide: An endogenous modulator of leukocyte adhesion. Proc Natl Acad Sci U S A. 1991 Jun 1;88(11):4651-5.

[38] Marginean C, Popescu MS, Vladaia M, Tudorascu D, Pirvu DC, Petrescu F. Involvement of Oxidative Stress in COPD. Curr Heal Sci J. 2018; 44(1):48-55.

[39] Murray CJL, Lopez AD. Measuring the global burden of disease. Vol. 369, New England Journal of Medicine. Massachussetts Medical Society; 2013. p. 448-57.

[40] Repine JE, Bast A, Lankhorst I. Oxidative stress in chronic obstructive pulmonary disease. Vol. 156, American Journal of Respiratory and Critical Care Medicine. American Thoracic Society; 1997. p. 341-57.

[41] Kinnula VL, Crapo JD. Superoxide dismutases in the lung and human lung diseases. Vol. 167, American Journal of Respiratory and Critical Care Medicine. Am J Respir Crit Care Med; 2003. p. 1600-19.

[42] Pastori D, Andreozzi P, Carnevale R, Bartimoccia S, Limaj S, Melandri S, *et al.* Does the Coexistence of Chronic Obstructive Pulmonary Disease and Atrial Fibrillation Affect Nox2 Activity and Urinary Isoprostanes Excretion? Vol. 31, Antioxidants and Redox Signaling. Mary Ann Liebert Inc.; 2019. p. 786-90.

[43] Birben E, Sahiner UM, Sackesen C, Erzurum S, Kalayci O. Oxidative stress and antioxidant defense. Vol. 5, World Allergy Organization Journal. BioMed Central Ltd.; 2012. p. 9-19.

[44] Cheresh P, Kim SJ, Tulasiram S, Kamp DW. Oxidative stress and pulmonary fibrosis. Vol. 1832, Biochimica et Biophysica Acta - Molecular Basis of Disease. Biochim Biophys Acta; 2013. p. 1028-40.

[45] Jones DP. Extracellular redox state: Refining the definition of oxidative stress in aging. In: Rejuvenation Research. Rejuvenation Res; 2006. p. 169-81.

[46] Osborn-Heaford HL, Ryan AJ, Murthy S, Racila AM, He C, Sieren JC, *et al.* Mitochondrial Rac1 GTPase import and electron transfer from cytochrome c are required for pulmonary fibrosis. J Biol Chem. 2012 Jan 27;287(5):3301-12.

[47] Murthy S, Adamcakova-Dodd A, Perry SS, Tephly LA, Keller RM, Metwali N, *et al.* Modulation of reactive oxygen species by Rac1 or catalase prevents asbestos-induced pulmonary fibrosis. Am J Physiol - Lung Cell Mol Physiol. 2009 Nov;297(5).

[48] He C, Murthy S, McCormick ML, Spitz DR, Ryan AJ, Carter AB. Mitochondrial Cu,Zn-superoxide dismutase mediates pulmonary fibrosis by augmenting H2O2 generation. J Biol Chem. 2011 Apr 29;286(17):15597-607.

[49] Liu G, Cheresh P, Kamp DW. Molecular basis of asbestos-induced lung disease. Vol. 8, Annual Review of Pathology: Mechanisms of Disease. Annual Reviews Inc.; 2013. p. 161-87.

[50] Galluzzi L, Kepp O, Trojel-Hansen C, Kroemer G. Non-apoptotic functions of apoptosis-regulatory proteins. Vol. 13, EMBO Reports. EMBO Rep; 2012. p. 322-30.

[51] Crestani B, Besnard V, Boczkowski J. Signalling pathways from NADPH oxidase-4 to idiopathic pulmonary fibrosis. Vol. 43, International Journal of Biochemistry and Cell Biology. Elsevier Ltd; 2011. p. 1086-9.

[52] Carnesecchi S, Deffert C, Donati Y, Basset O, Hinz B, Preynat-Seauve O, *et al.* A key role for NOX4 in epithelial cell death during development of lung fibrosis. Antioxidants Redox Signal. 2011 Aug 1;15(3):607-19.

[53] Manoury B, Nennan S, Leclerc O, Guenon I, Boichot E, Planquois JM, *et al.* The absence of reactive oxygen species production protects mice against bleomycin-induced pulmonary fibrosis. Respir Res. 2005 Jan 21;6(1).

[54] Liu RM, Gaston Pravia KA. Oxidative stress and glutathione in TGF-β-mediated fibrogenesis. Vol. 48, Free Radical Biology and Medicine. Free Radic Biol Med; 2010. p. 1-15.

[55] Todd NW, Luzina IG, Atamas SP. Molecular and cellular mechanisms of pulmonary fibrosis. Vol. 5, Fibrogenesis and Tissue Repair. BioMed Central; 2012. p. 11.

[56] Moeller A, Ask K, Warburton D, Gauldie J, Kolb M. The bleomycin animal model: A useful tool to investigate treatment options for idiopathic pulmonary fibrosis? Vol. 40, International Journal of Biochemistry and Cell Biology. Pergamon; 2008. p. 362-82.

[57] Sneddon LU, Halsey LG, Bury NR. Considering aspects of the 3Rs principles within experimental animal biology. Vol. 220, Journal of Experimental Biology. Company of Biologists Ltd; 2017. p. 3007-16.

[58] Shamir ER, Ewald AJ. Three-dimensional organotypic culture: Experimental models of mammalian biology and disease. Vol. 15, Nature Reviews Molecular Cell Biology. Nature Publishing Group; 2014. p. 647-64.

[59] Kotton DN, Morrisey EE. Lung regeneration: Mechanisms, applications and emerging stem cell populations. Vol. 20, Nature Medicine. Nature Publishing Group; 2014. p. 822-32.

[60] Kumar PA, Hu Y, Yamamoto Y, Hoe NB, Wei TS, Mu D, *et al.* Distal airway stem cells yield alveoli *in vitro* and during lung regeneration following H1N1 influenza infection. Cell. 2011 Oct 28;147(3):525-38.

[61] Ling TY, Kuo M Der, Li CL, Yu AL, Huang YH, Wu TJ, *et al.* Identification of pulmonary Oct-4+ stem/progenitor cells and demonstration of their susceptibility to SARS coronavirus (SARS-CoV) infection *in vitro*. Proc Natl Acad Sci U S A. 2006 Jun 20;103(25):9530-5.

[62] Chimenti I, Pagano F, Angelini F, Siciliano C, Mangino G, Picchio V, *et al.*

Human lung spheroids as *in vitro* niches of lung progenitor cells with distinctive paracrine and plasticity properties. Stem Cells Transl Med. 2017;6(3).

[63] Pagano F, Nocella C, Sciarretta S, Fianchini L, Siciliano C, Mangino G, *et al.* Cytoprotective and Antioxidant Effects of Steen Solution on Human Lung Spheroids and Human Endothelial Cells. Am J Transplant. 2017 Jul 1;17(7): 1885-94.

[64] Carnevale R, Biondi-Zoccai G, Peruzzi M, De Falco E, Chimenti I, Venuta F, *et al.* New insights into the steen solution properties: Breakthrough in antioxidant effects via NOX2 downregulation. Oxid Med Cell Longev; 2014.

[65] Khedoe PPSJ, Ng-Blichfeldt J-P, Van Schadewijk A, Marciniak SJ, Koenigshoff M, Gosens R, *et al.* Impairment of lung organoid formation by cigarette smoke treatment of mesenchymal cells. In: European Respiratory Journal. European Respiratory Society (ERS); 2018. p. PA4274.

[66] Strikoudis A, Cieślak A, Loffredo L, Chen YW, Patel N, Saqi A, *et al.* Modeling of Fibrotic Lung Disease Using 3D Organoids Derived from Human Pluripotent Stem Cells. Cell Rep. 2019 Jun 18;27(12):3709-3723.e5.

[67] Nishina T, Komazawa-Sakon S, Yanaka S, Piao X, Zheng D-M, Piao J-H, *et al.* Interleukin-11 Links Oxidative Stress and Compensatory Proliferation. Sci Signal. 2012 Jan 17;5(207):ra5–ra5.

[68] Wilkinson DC, Alva-Ornelas JA, Sucre JMS, Vijayaraj P, Durra A, Richardson W, *et al.* Development of a Three-Dimensional Bioengineering Technology to Generate Lung Tissue for Personalized Disease Modeling. Stem Cells Transl Med. 2017 Feb;6(2):622-33.

[69] Surolia R, Li FJ, Wang Z, Li H, Dsouza K, Thomas V, *et al.* Vimentin intermediate filament assembly regulates fibroblast invasion in fibrogenic lung injury. JCI Insight. 2019 Apr 4;4(7).

[70] Wu Q, Liu J, Wang X, Feng L, Wu J, Zhu X, *et al.* Organ-on-a-chip: Recent breakthroughs and future prospects. Vol. 19, BioMedical Engineering Online. BioMed Central Ltd.; 2020. p. 1-19.

[71] Huh D, Matthews BD, Mammoto A, Montoya-Zavala M, Hsin HY, Science DEI. Reconstituting Organ-Level Lung Functions on a Chip. 2010.

[72] Whitesides GM, Ostuni E, Takayama S, Jiang X, Ingber DE. SOFT LITHOGRAPHY IN BIOLOGY AND BIOCHEMISTRY. 2001.

[73] Asmani M, Velumani S, Li Y, Wawrzyniak N, Hsia I, Chen Z, *et al.* Fibrotic microtissue array to predict anti-fibrosis drug efficacy. Nat Commun. 2018 Dec 1;9(1):1-12.

[74] Zhao R, Boudou T, Wang WG, Chen CS, Reich DH. Decoupling cell and matrix mechanics in engineered microtissues using magnetically actuated microcantilevers. Adv Mater. 2013 Mar 25;25(12):1699-705.

[75] Sellgren KL, Butala EJ, Gilmour BP, Randell SH, Grego S. A biomimetic multicellular model of the airways using primary human cells. Lab Chip. 2014 Sep 7;14(17):3349-58.

[76] Mortaz E, Adcock IA. Limitation of COPD Studies in Animal Modeling. Tanaffos. 2012;11(3):7-8.

[77] Adamson J, E L, Phillips G, D M. *In vitro* Models of Chronic Obstructive Pulmonary Disease (COPD). In: Bronchitis. InTech; 2011.

[78] Bals R, Beisswenger C, Blouquit S, Chinet T. Isolation and air-liquid interface culture of human large airway and bronchiolar epithelial cells. J Cyst Fibros. 2004 Aug 1;3(SUPPL. 2):49-51.

[79] van Riet S, Ninaber DK, Mikkers HMM, Tetley TD, Jost CR, Mulder AA, *et al. In vitro* modelling of alveolar repair at the air-liquid interface using alveolar epithelial cells derived from human induced pluripotent stem cells. Sci Rep. 2020 Dec 1;10(1):1-12.

[80] Shrestha J, Razavi Bazaz S, Aboulkheyr Es H, Yaghobian Azari D, Thierry B, Ebrahimi Warkiani M, *et al.* Lung-on-a-chip: the future of respiratory disease models and pharmacological studies. Crit Rev Biotechnol. 2020 Feb 17;40(2):213-30.

[81] Benam KH, Mazur M, Choe Y, Ferrante TC, Novak R, Ingber DE. Human lung small airway-on-a-chip protocol. In: Methods in Molecular Biology. Humana Press Inc.; 2017. p. 345-65.

[82] Benam KH, Villenave R, Lucchesi C, Varone A, Hubeau C, Lee HH, *et al.* Small airway-on-a chip enables analysis of human lung inflammation and drug responses *in vitro*. Nat Methods. 2016 Feb 1;13(2):151-7.

[83] Benam KH, Novak R, Ferrante TC, Choe Y, Ingber DE. Biomimetic smoking robot for *in vitro* inhalation exposure compatible with microfluidic organ chips. Nat Protoc. 2020 Feb 1;15(2):183-206.

[84] Benam KH, Novak R, Nawroth J, Hirano-Kobayashi M, Ferrante TC, Choe Y, *et al.* Matched-Comparative Modeling of Normal and Diseased Human Airway Responses Using a Microengineered Breathing Lung Chip. Cell Syst. 2016 Nov 23;3(5):456-466.e4.

[85] Nesmith AP, Agarwal A, Mccain ML, Parker KK. Human Airway Musculature on a Chip: An *In vitro* Model of Allergic Asthmatic Bronchoconstriction and Bronchodilation.

[86] Nawroth JC, Lucchesi C, Cheng D, Shukla A, Ngyuen J, Shroff T, *et al.* A Micro-Engineered Airway Lung-Chip Models Key Features of Viral-Induced Exacerbation of Asthma. Am J Respir Cell Mol Biol. 2020 Jul 24;

[87] Villenave R, Lucchesi C, Lee H-H, Nguyen J, Varone A, Karalis K, *et al.* Severe Asthma on-Chip: a novel platform enables study of viral-induced exacerbations in asthma and drug response *In vitro*. In: European Respiratory Journal. European Respiratory Society (ERS); 2017. p. PA4136.

[88] Henry OYF, Villenave R, Cronce MJ, Leineweber WD, Benz MA, Ingber DE. Organs-on-chips with integrated electrodes for trans-epithelial electrical resistance (TEER) measurements of human epithelial barrier function. Lab Chip. 2017 Jul 7;17(13):2264-71.

Section 2

Immuno-Molecular Strategies for Understanding Diseases and Their Management

Chapter 4

Immune System Modulations in Cancer Treatment: Nanoparticles in Immunotherapy

Kadriye Kızılbey, Nelisa Türkoğlu and Fatma Ceren Kırmızıtaş

Abstract

Cancer immunotherapy is based on the idea of overcoming the main problems in the traditional cancer treatments and enhancing the patient's long-term survival and quality of life. Immunotherapy methods aimed to influence the immune system, to detect and eradicate the tumors site and predict the potential results. Nowadays, nanomaterials-based immunotherapy approaches are gaining interest due to numerous advantages like their ability to target cells and tissues directly and reduce the off-target toxicity. Therefore, topics about immune system components, nanomaterials, their usage in immunotherapy and the benefits they provide will be discussed in this presented book chapter. Immunotherapy can be divided into two groups mainly; active and passive immunotherapy including their subtitles such as immune checkpoint inhibitors, adoptive immunotherapy, CAR-T therapies, vaccines, and monoclonal antibodies. Main classification and the methods will be evaluated. Furthermore, state-of-art nanocarriers based immunotherapy methods will be mentioned in detail. The terms of size, charge, material type and surface modifications of the nanoparticles will be reviewed to understand the interference of immune system and nanoparticles and their advantages/disadvantages in immunotherapy systems.

Keywords: tumor, cancer immunotherapy, vaccination, immunomodulation, antigen receptors, nanoparticles, bottom-up method, top-down method

1. Introduction

Understanding the immune system and its components may enlighten future potential treatments to generate disease progression such as cancer. For almost 30 years, by targeting the immune system by therapeutics brought a totally new point of view in the field of cancer treatment. Accordingly, besides the commonly preferred cancer treatments, the treatments developed specifically for the patient and the diseases come to the forefront. To date, immunotherapy is a method developed as an alternative to conventional cancer treatments [1, 2]. The immune system which is an awareness system based on distinguishing between "self" and "non-self" works in harmony with cells, related tissues, and organs respectively to protect the organisms. The main goal of the immune system is to defense to battle against "enemies". There are two types of immune responses; humoral and cellular immunity. Humoral immunity is primarily mediated by B and T lymphocytes and their products. It is also characterized by a weak response and a strong immunological memory. Cellular immunity components are natural killer (NK) cells, eosinophils, macrophages, and lymphocytes (B and T cells). Cellular immunity works faster than humoral immunity via activation and proliferation of B cells and activation of antigen-presenting cells (APCs). Cellular immunity can recognize tumors immediately, but it does not provide long-term immunity. Specifically, B and T cells primarily mediate the antitumor response. The CD4+ T cells pretend as "helper cells" and excrete cytokines relying on their profile either Th1 or Th2. Humoral and cellular immunity plays a crucial role in antitumor response [3, 4].

2. Cancer

Today, non-communicable diseases are held accountable as the leading cause of death worldwide. Among these diseases, cancer is predicted as one of the most important disease in the world that causes deaths and reduces the life quality [5]. Soon, it is thought that the number of cancer patients and cancer-related deaths will increase [6]. The definition of cancer for the first time in 3000 BC was used in inscriptions called the Edwin Smith Papyrus, the part of an ancient Egyptian textbook on trauma surgery. Cancer is generally characterized by the growth of abnormal cells beyond their normal limits. Cancer disease can affect almost any part of the body and has many anatomical and molecular subtypes, each of which requires specific treatment strategies. The main factors causing cancer are as follows; ionizing radiation, ultraviolet rays, age, inadequate physical activity, smoking and alcohol consumption, nutrition and diet, chemicals, microorganisms, and genetic factors. It is known that environmental factors are much more effective in the formation of the disease than hereditary factors. The most important reason is stated as mutations that occur in genes. Most cancers are caused by a series of mutations that allow cells to divide faster, escape internal and external controls, and prevent programmed cell death. As the cells continue to divide under the influence of mutations in solid tissue such as organ, bone or muscle the resulting mass is called tumor. Solid tumors are classified as; benign (noncancerous) and malignant (cancerous). Benign tumors do not have the ability to metastasize; they can only grow where they are located. On the other hand, malignant tumors have the ability to spread to neighboring tissues and organs from where they are formed. Many types of cancer initially show no symptoms. The main symptoms observed can be given as; unexplained, and rapid weight loss, fever, malaise, pain, swelling and bleeding. However, each type of cancer has its own specific symptoms, so the treatment method of each cancer type differs. **Figure 1** shows a schematic representation of tumor cells progression.

2.1 Cancer treatment methods

Cancer is an individual disease; hence treatment methods vary from patient to patient. The method of treatment should be chosen by considering the degree and course of the disease, age and health situation of the patient. Generally, most of the patients have the combination of treatment methods. Surgical intervention, radiation therapy, chemotherapy, and hormone therapy are defined as traditional cancer treatments in the literature [2]. In recent years, immunotherapy has also been the increasingly used method in cancer treatment.

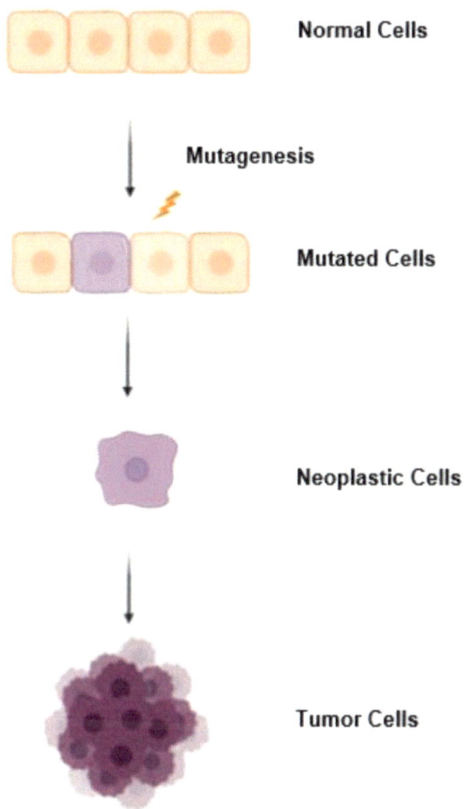

Figure 1.
Schematic representation of tumor cells progression (created with BioRender.com).

2.1.1 Surgical intervention

Surgical intervention, a local treatment method, can also be used in the combination of other treatment methods. It is applied in tumors without metastasis that only exist in one area such as solid tumors; but not effective in leukemia and cancer types that spread. Also, surgical intervention is preferred when the tumor is in an untreatable part of the body by other treatment methods such as radiation therapy or chemotherapy (cannot reach the brain). In order to remove the tumor without damaging the neighboring healthy cells, the size of the tumor can be reduced by other methods. The surgical intervention method works against cancer in three ways; eradicating the entire tumor, debulking a tumor, and palliate the disease symptoms. Only eradicating the entire tumor may cure the patients if the cancer cells are located in small area at one place. Debulking a tumor is used to reduce the tumors size while surgery is combined with other treatment methods. Palliate the symptoms, the last way in surgical intervention, is to remove the tumor to reduce the pain or pressure caused by the tumor. There are some disadvantages of surgical intervention such as the possibility of leaving microscopic residues around the tumor after surgery, the health status of the patients, and the success of the surgery [7, 8].

2.1.2 Radiation therapy

Radiation therapy or radiotherapy (RT) is based on the principle of using a fairly high dose of radiation to shrink the tumor by killing the cancer cells. There are various types of radiotherapy depends on the general state of the patient and disease. The principle of radiation therapy is to destroy cancer cells as possible without damaging healthy cells. Because, in the late 20th century, scientist discovered that radiation therapy not only cures cancer cells but also may be the cause of cancer itself. The notable side effect, it can kill and harm healthy cells significantly. Thereby it has side effects such as hair loss, vomiting, and loss of appetite that will affect your daily life. The choice of the exact type of radiation therapy relies on several circumstances such as the type, stage, size and location of cancer, and medical history of the patients. Reducing the tumor mass by radiation therapy is helpful to decrease the pressure of tumor on the nearest healthy cells. Additionally, it is used before surgical intervention to shrink tumor mass to make it suitable for surgery and after surgery; the microscopic residues on the edge of the tumor can be removed much more easily. Also, this method of therapy is very suitable for making systemic therapy [9, 10].

2.1.3 Chemotherapy

Chemotherapy (CT), also as chemo, is the most commonly used method in cancer treatment. The aim is to kill cancer cells using chemotherapeutic agents. This method is developed in the late 20th century and combined with surgery and/or radiation therapy. Over the years, many chemotherapeutic drugs showed great impact and gained success for the treatment of many types of cancer. The aim of the treatment can be stated as reducing the size of the tumor, reducing the effects of the symptoms seen in the patient, preventing metastasis, and reducing the total number of tumor cells in the body. The drugs used in chemotherapy direct the cell to death by stopping or decelerating the cancer cell proliferation. Some of these drugs are natural and some of them are synthetic. Hair loss, vomiting, loss of appetite, fever, diarrhea and fatigue are temporary side effects of the drugs that end after the treatment [11, 12].

2.1.4 Hormone therapy

Hormones, in the classical sense, are organic compounds that are synthesized in ductless glands such as the pituitary gland, adrenal gland, thyroid gland, and parathyroid gland, which are known as endocrine organs, and act on certain target tissue that is carried by the blood. All cells communicate with each other *via* hormones. In the human body, hormones either can be small proteins (insulin, etc.) or stimulator for a cell to generate new proteins or cease making products. One possible featured outcome is cell growth and proliferation. Even though cancer cells are abnormal, they still keep the ability to react to signals of hormones. The main idea of hormone-based treatments is to deprive cancer cells of hormone signals. Otherwise, they would be stimulating to continue dividing. The main theme of the drugs that are used in this method relies on preventing the activity of hormone within the target cell or blocking the production of the related hormone. Hormone therapy is often preferred for the treatment of prostate and breast cancer. Generally, hormone therapy is combined with other treatment methods depending on the cancer type. Hormone therapy is very suitable for adjuvant and neoadjuvant therapy to reduce tumor mass. The term adjuvant therapy is about reducing the risk of cancer recurrence after major cancer treatment. Hormone therapy is also appropriate for the removal of cancer cells that spread to different parts of the body. Like all other methods, hormone therapy has common side effects. But these effects depend on the body's response to the therapy and the type of hormone therapy. Side effects are influenced by different terms such as patients' sex and type of hormone that is used. Hot flashes, weakened bones, nausea, and fatigue are common side effects for men. Menstrual irregularities for women who are not menopausal and vaginal dryness are seen in addition to the common side effects. To date, there are several hormone based drugs based on the hormonal signals, but their principles are diverse from each other. They all attack different parts of the pathways to decelerate to the growth of cancer [13].

2.1.5 Immunotherapy

Nowadays, cancer treatment is moving from non-specific methods to specific methods. Although success is achieved in the destruction of tumors with surgery and radiotherapy, cancer may recur due to cancerous cell debris in the damaged area. Cancer immunotherapy, an individualized method, is referred to as the "fifth step" of the treatment following the traditional methods mentioned above [14]. The immunotherapy method; boost the immune system to fight against cancer, train the immune system component's to memory, attack the cancer cells, and heighten the immune response *via* biological substances. For the last decades, immunotherapy becomes a promising method to fight against cancer. Immunotherapy can be applied using either external substances or their body cells [4].

3. Historical background of cancer immunotherapy

It is common knowledge that many cases of regression of tumor growth after high fever attacks or infectious diseases have been reported throughout history from Ancient Egypt to the 18th century. However, the relationship between the immune system and cancer was noticed in the middle of the 18th century with the developing technology. In the mid-18th century, two German doctors, Busch and Fehleisen, independently reported cases of tumor regressions of patients after erysipelas infection (*Streptococcus pyogenes* infection). In the literature, the first systematic immunotherapy study for the treatment of malignant tumors was conducted in 1891 by William B. Coley, a surgical oncologist. Coley injected the heat-inactivated *Streptococcus pyogenes* and *Serratia marcescens* organisms into the patient to stimulate the patient's immune system. After the project that he initiated, Coley has seen a regression in the tumor in more than 1000 sarcoma patients who cannot undergo surgical intervention. In a very short time, humanity evaluated this mixture as a great invention, "Coley Toxins". However, the word "toxin" was an unfortunate choice; the more acceptable name for the treatment was "mixed bacteria vaccine". Although the bacteria had some side effects such as fever and malaise, it is not as toxic as chemotherapy or radiotherapy and does not destroy the immune system [15, 16]. Coley's life-long cancer immunotherapy studies that will spearhead for many scientists have started after this project. In 1900, Paul Ehrlich stated that the first findings of the treatment, which would later be called antibody-mediated passive immunotherapy, had an important place in the treatment of tumors. In 1975, George Köhler and Cesar Milstein developed hybridoma technology for monoclonal antibody production. This was followed by the first successful use of monoclonal adults in human neoplasia in 1982 and the FDA (US Food and Drug Administration) approval of muromonab-CD3 (Orthoclone OKT3) in 1986. In 1997, both the first humanized monoclonal antibody, daclizumab (Zenapax), and the first monoclonal antibody for malignancy, rituximab (Rituxan), were approved by the FDA. This was followed by the FDA approval of gemtuzumab ozogamicin (Mylotarg) in 2000, the first toxin-bound monoclonal antibody, and ibritumomab tiuxetan (Zevalin) in 2002, the first radionuclide-bound monoclonal antibody [17].

Another area that cancer immunotherapy has advanced was using the patient's body cells. In the 1960s, the tumor immune surveillance hypothesis was put forward by Burnet. Since 1995, persuasive studies on effective tumor-specific immunity have attracted great interest. In particular, many studies show the ability of dendritic cells to elicit tumor-specific T cell immunity has led to this situation. Following preclinical researches, many studies involving various types of cancer have been conducted in patients. Recent studies have also made the immunosurveillance hypothesis quite popular [18, 19]. Immunotherapy studies have increased their importance in the 21st century with the licensing of clinical studies carried out with developing technology and methods [20]. Immunotherapy was declared as "breakthrough of the year" by Science magazine in 2013 after the clinical success achieved and has become even more prominent. Also, in 2018 James Allison and Tasuku Honjo received the Nobel Prize in Physiology and Medicine for their work based on the use of the immune system to destroy cancer cells. In the past two decades, great strides have been made in cancer immunotherapy. With all these spectacular developments, the number of cancer immunotherapy studies is increasing day by day [21, 22]. There are certain categories in cancer immunotherapy applications. These are the mechanism of innate and acquired immune resistance, internal and external resistance to immunotherapy, self-neutralization of tumor cells and antigen- presenting cells, inhibition of immunity by exosome release mechanisms, the response of tumor cells to therapy. Like all other methods, cancer immunotherapy has several advantages and disadvantages. Higher precision and specificity, long-term survival rate, fewer side effects than traditional treatment methods, removing residual tumor cells and microscopic lesions that remain in the body after treatment and improving the body's immune function are the advantages of immunotherapy. Also, it can control and kill more than one tumor type and it uses the body's immune system to increase immune response. Higher treatment costs, various non-specific toxic side effects after treatment are the disadvantages of immunotherapy. There is a high selectivity for patients in treatment. When the tumor type is "immunosuppressant type" or "immune exclusion type", the effect of immunotherapy treatment is considerably weak. Additionally, in particular, the use of immune checkpoint inhibitors can have adverse consequences leading to autoimmune diseases and even death [23].

4. Classification of immunotherapy

Cancer immunotherapy is generally classified in three ways; passive, active and combination immunotherapy depending on the mechanism of the therapeutic agent and the state of the patient's immune system. Classification of passive and active cancer immunotherapy studies is shown in **Table 1**.

4.1 Passive immunotherapy

The main purpose of passive immunotherapy is to increase the current anti-tumor response by using therapeutics that can be produced under laboratory conditions. It is preferred to use the treatment in patients with weak or dysfunctional immune systems. It is designed to attack tumor cells independently by modifying the components of the immune system in the laboratory. Monoclonal antibodies and adoptive cell therapy are frequently used passive immunotherapy methods [4, 20, 24].

	Passive immunotherapy	Active immunotherapy
NON-SPECIFIC	Adoptive Cell Therapy	Cytokines Immune Checkpoint Inhibitors
SPECIFIC	Monoclonal Antibodies	Cancer Vaccines Oncolytic Viruses

Table 1.
Classification of immunotherapy.

4.1.1 Monoclonal antibodies

For the past 20 years, monoclonal antibodies are the most commonly used FDA approved treatment in clinical immunotherapy studies. They are large artificial proteins with high antigen specificity produced by particular B cells. Due to their antigen specificity, their capacity to bind to epitopes on the surface of the tumor cell is high [25]. So, antibodies specific to antigens of cancer cells are produced in ex vivo conditions and transferred to the patient to increase the immune response. Antibodies in these targeted therapies are guided directly to the antigen on the surface of cancer cells. Different signaling functions can be created by the interaction of monoclonal antibodies and receptors on the surface of malignant tumors. Antibodies are used in treatment can be classified as naked, conjugated, radiolabeled, chemically labeled, and bispecific monoclonal antibodies. Naked monoclonal antibodies are most commonly used in cancer immunotherapy and bind directly to the antigen without any radioactive markers or drugs. Conjugated monoclonal antibodies are used to transfer chemotherapeutic drugs or radiolabeled particles to cancer cells. Radiolabeled monoclonal antibodies are created by adding radioactive particles to naked antibodies. Chemically labeled antibodies are monoclonal antibodies with a high chemotherapeutic effect. Radioactive or chemically labeled monoclonal antibodies aim to destroy the target cell with the toxins they contain or the radiation they emit. Bispecific antibodies carry two types of antibodies in their structure and can bind to two different antigens that are receptors for these two antibodies at the same time [18, 26, 27]. The first drug including monoclonal antibodies approved by the FDA was rituximab (Rituxan, Genentech) was used in the clinic at 1997. Today, with developing technology, many new drugs have emerged for the treatment of different types of cancer [25].

4.1.2 Adoptive cell therapy

It gathered speed with the studies carried out in the 20th century about the discovery of tumor-specific antigens located not on healthy cells but just on the tumor cells. Thus the importance of adoptive T cell transfer has been understood. Adoptive cell therapy is the transfer of natural or genetically modified T cells to patients in *ex vivo* conditions instead of stimulating the immune system. The transferred cells can be autologous or allogeneic targeted to a particular antigen in the host cell. It was pointed out that the stage of an immune response in the host is skipped directly by this step. To create a targeted immune response, autologous cells can recognize tumor antigens, move towards the tumor and exit the circulation. The transfer of T cells to destroy tumor cells is carried out in two ways; the infiltrating (TIL) of tumor specific T cells from existing tumor cells and the use of genetically modified T cells to specifically identify tumor cells. In both methods, the T cell is processed *ex vivo* and then transferred back to the patient [28]. The first successful cellular therapy in history was performed on an advanced melanoma patient with autologous TIL. The specific T cell receptor (TCR) is obtained by genetically modifying T cells. T cells and tumor-specific antigens are matched with HLA recognition by TCR technology. A minimal cytotoxic effect occurs by this natural pairing. TCRs also have disadvantages such as the low expression on the surface and short lifespan of T cells *in vivo*. Although the first studies ended up with disappointment, today, the other genetically modified T cell is chimeric antigen receptors, CAR. Many studies are conducted around the world on CAR-T technology and it is believed that positive results will be achieved in the near future [29, 30].

4.2 Active immunotherapy

Active immunotherapy aims to destroy cancer cells by stimulating the immune system by vaccination, immunomodulation, or targeting specific antigen receptors. The method is carried out employing cancer vaccines, oncolytic viruses, immune checkpoint inhibitors, and cytokines [20].

4.2.1 Cancer vaccines

The purpose of vaccination is to create an immune response to detect and destroy cancer cells. Cancer vaccines, containing whole, part, or purified antigens of tumor cells, can be peptide-based, immune cell or dendritic cell-based, or tumor cell-based. After the tumor cells are removed from the body, the patient is vaccinated and an immune response is created against the tumor cells that may remain in the body. Variable antigen expression, low immune response, diminishing the immune response in the tumor microenvironment and a decrease in activity over time are the restrictions of the cancer vaccine applications [4, 25].

- **Peptide-based vaccines** are designed to create an immune response against tumor antigens that interact with HLA molecules on the surface of tumor cells. Their toxic effects on healthy cells are low due to their antigen-specific design, but tumor antigen peptides and the patient's HLA type should be well characterized [31].

- **Immune or dendritic cell-based vaccines** consist of the use of tumor-associated antigens or autologous tumor cells and dendritic cells (DC) obtained from monocyte cells in early-stage cancer vaccines. In 2010, the drug called Sipuleucel-T (Provenge, Dendreon Corp.) is the first DC-based cancer vaccine was approved by the FDA for the treatment of prostate cancer. DC-based vaccines today use innovative *in vitro* culturing techniques enriched with cytokines, enhancing immunogenicity and improving DC function. DC-based cancer vaccines can be designed differently for both *ex vivo* and *in vivo* applications for various cancer types [4].

- **Tumor cell-based cancer vaccines** use the entire tumor cell to create an immune response. Unlike peptide-based vaccines, tumor cells are not specific to antigens on their surface, but the range of epitopes to which they can bind is wider. These vaccines can be prepared using the patient's cells (autologous) or using another patient's tumor cells (allogeneic). Tumor cell-based vaccines such as M-Vax (AVAX Technologies) can be used in the treatment of many different types of cancer in clinical studies [25].

4.2.2 Oncolytic viruses

These are called genetically altered viruses that can naturally penetrate only cancer cells and kill them. Talimogene laherparepvec (T-Vec) is the first oncolytic virus-based drug, approved by the FDA, that the protection mechanisms developed against viral infections are impaired in most cancer cells. By taking advantage of this degradation, viruses can reproduce more intensely in cancer cells than healthy cells. Recently, replication specific to cancer cells was obtained and a reovirus variant called Reolysin (exhibiting oncolytic behavior in cells with activated Ras signaling pathway) has been developed. In 1991, positive results were gained in the treatment of brain cancer with a mutation in a genetically modified type 1 herpes simplex virus [32].

4.2.3 Immune checkpoint inhibitors

Several inhibitory receptors and ligands expressed on T cells, antigen-presenting cells, and tumor cells have recently been important elements of immunosuppression in the tumor microenvironment. Because of their biological role as regulators of T cell activation, these receptor/ligand pairs have been termed "immune checkpoints". Immune checkpoints are cell membrane proteins involved in the regulation of the immune response. Multiple controls or "checkpoints" are present or activated to ensure that the immune-inflammatory response is not continuously activated after tumor antigens have generated a response. Immune checkpoints are signals that can halt an existing immune response. The over -expression of these signals by tumor cells affects tumor cell-specific T-cell immunity in the cancer microenvironment. The aim of treatments involving inhibition of the immune checkpoint is to use and strengthen the immune system by disrupting the negative immune system. In 2011, the drug called Ipilimumab was used in clinical use for melanoma patients by using immune control point drugs. As of March 2019, 7 immunotherapy drugs based on checkpoints are used in clinical practice. Monoclonal antibodies that bind to immune checkpoints bind with cytotoxic T lymphocyte-associated molecule-4 (CTLA-4), programmed cell death protein 1 (PD-1), and programmed cell death ligand (PDL-1) [33].

- **PD-1/PDL-1**, under normal circumstances, PD-1 has two ligands; PD-L1 and PD-L2. Blocking the interaction between PD-1 and PDL-1 with antibodies enhances the immune response against cancer cells, "releases the brakes" in the immune system and allows for the attack of tumor cells that express PDL-1. Nivolumab and Pembrolizumab are the first two drugs approved by the FDA in 2014 [34].

- **CTLA-4** inhibition increases the activation of cytotoxic T cells. Thus, immune blockade due to Treg cells is inhibited and antitumor activation is observed. Ipilimumab is the first drug approved by the FDA for CTLA-4 treatment in 2011 [35].

4.2.4 Cytokines

They are the main regulators of innate and adaptive immune systems that allow cells of the immune system to communicate in paracrine and autocrine systems over short distances. Unlike other therapeutic agents, these molecules directly stimulate immune cells, for example Interleukin-21 (IL-21) can act as agents involved in active immunotherapy [36]. The use of cytokines in cancer immunotherapy showed tumor regression, prevention of metastasis formation, improvement of immunological memory and decrease in risk of disease recurrence with increased survival. The use of cytokine (IL-2, GM-CSF, IFN-α) -based biological therapy in combination with conventional therapies is under clinical development [37]. In 1986, IFN-α became the first FDA approved cytokine for the treatment of leukemia.

Subsequently, IL-2 was approved by the FDA in 1992 for metastatic kidney cancer and 1998 for advanced melanoma treatment [36].

4.3 Combinational immunotherapy

Combinational immunotherapy refers the use of a different anticancer agent for treatments of cancer. Conjugation of IL-2 and HER-2 monoclonal antibody proved to be a very forceful combination in immunotherapy. Lately, PD-1 and CTLA-4 conjugation has been examined. The results revealed that the combined system was safe and had no significant toxic effect [38].

5. Nanoparticles in cancer immunotherapy

Nanoparticle-based biomaterials have a critical role in cancer immunotherapy compared with conventional drugs [39]. Immunotherapy often targets tumor cells, immune and stromal cells in the tumor microenvironment [40]. Additionally, side reactions occurring due to the interactions between nanoparticles (NPs) and cells can be adjusted by modifications of nanoparticles [41]. Nanoparticle-based drug delivery systems can improve the solubility, *in vivo* stability, and pharmacokinetic profile. Also, they protect drugs from premature release and degradation in the living system. These systems can be designed according to the microenvironment of the target such as pH, redox potential or enzymes, and external dynamics such as light, electrical and magnetic fields. Targeted delivery with NPs can also reduce toxicity and immune-related side effects [2]. The size and the shape of the NP are very effective in therapeutic efficacy by changing its pharmacokinetics, transportation, and cellular uptake [42]. Recent advances in nanoparticle formulations have generated a wide range of other shapes like rods, prisms, cubes, stars, and discs out of spherical. It is considered as non-spherical particles have higher blood circulation periods, prolonged margination effects, and higher penetration capacities within solid tissues and tumors [43]. The charge of NP has great priority in the transition of it into cells. Besides, NP-ligand coupling conditions and the elasticity of NP upgrade transportation and accumulation of NP in the living system [44, 45]. Generally, it is well known that cationic NPs create a higher immune response than neutral or anionic NPs [43]. The size, shape, elasticity, optical, magnetic, and electrical properties of nanoparticles can be modified to increase the usage of NPs in cancer therapy as a carrier [2, 41, 46]. High specificity, efficacy, diagnosing, imaging, and therapeutic properties make NPs candidates in immunotherapy for effective cancer treatment. Liposomes, micelles, polymeric, metallic, and inorganic NPs have a wide range of usage in cancer immunotherapy [44].

5.1 Classification of nanoparticles

The nanoparticles are generally categorized into tree class as organic, inorganic, and carbon-based. Dendrimers, micelles, and liposomes are the most widely known organic nanoparticles. These biodegradable, non-toxic, and capsule-shaped nanoparticles appear to be an ideal choice for drug delivery due to their sensitivity to thermal and electromagnetic radiation. Inorganic nanoparticles, metal, and metal oxide-based NPs do not contain carbon in their structure. Aluminum, cadmium, cobalt, copper, gold, iron, lead, silver, and zinc can be used to fabricate metallic NPs in 10 to 100 nm size range. Carbon-based nanoparticles, fullerenes, graphene, carbon nanotubes (CNT), and carbon nanofibers, are build up from carbon in nanosize [47].

5.2 Preparation methods of nanoparticles

It can be viewed two different ways to synthesize nanoparticles; bottom-up and top-down methods (**Table 2**). These techniques also can be divided as chemical and physical methods. Although both methods have positive and negative features, the chemical one has more disadvantages due to the wet reaction steps it has [48].

5.2.1 Bottom-up method

It is also known as a constructive method like building-up of material from an atom. Sol–gel, spinning, chemical vapor deposition (CVD), pyrolysis, and bio-synthesis are the foremost methods in this technique. Nanoparticles, nanoshells, and nanotubes with narrow size distribution can be synthesized by this approach. Besides, in this method deposition parameters can be controlled. However large scale production is difficult and chemical purification is needed.

- **Sol–gel method:** It is a simple, wet chemical process based on hydrolysis and polycondensation reactions [49]. This process indicates the chemical transformation of a system from a "sol" phase, a colloidal solution of solids suspended in a liquid phase, into a "gel" phase, a solid macromolecule submerged in a solvent [50]. The chemical and physical properties of the materials as high surface area and the stability can be obtained by the method via modifying experimental conditions. Metal oxide and chloride precursors are used in sol–gel process, and then a liquid and a solid phase separation occur after removing precursors either by shaking, stirring, or sonication. Nanoparticles are acquired in this phase separation by sedimentation, filtration, or centrifugation [47].

- **Spinning disc processing (SDP):** The method consists of a rotating disc inside a reactor generally filled with nitrogen or other inert gases to remove oxygen inside and avoid chemical reactions. The purpose of spinning is to merge atoms or molecules. The parameters of this process such as the liquid flow rate, disc rotation speed, liquid/precursor ratio, location of feed, and disc surface may vary for different systems and determine the characteristics of NPs [51].

Top-down approach		Bottom-up approach	
Physical processing methods		Physical and chemical processing methods	
		Physical techniques	
Mechanical methods:	Cutting, etching, grinding	Physical Vapor Deposition (PVD):	Evaporation (thermal, e-beam)
	Ball milling		Sputtering
Lithographic techniques:	Photo Lithography		Plasma Arching
	Electron Beam Lithography		Laser Ablation
		Chemical techniques	
		Chemical Vapor Deposition (CVD):	Plasma enhanced CVD (PECVD)
		Self Assembling	Electronic deposition, sol-gel method, emulsion, pyrolysis

Table 2.
Techniques in Top-Down and Bottom-Up approaches.

- **Chemical vapor deposition (CVD):** It is the deposition technique of thin films of gaseous reactants onto a substrate. Gaseous reactants can be elemental and compound semiconductors, metal alloys, and amorphous or crystalline compounds. In the CVD process, a volatile material (chemically reactive) is coming together with other gases to produce a nonvolatile solid material that deposit at the atomic level on a suitable substrate. This is a well-organized process that some kind of reactors should be used depending on the type of precursors, deposition conditions, and the forms of the energy introduced to the system to stimulate the planned chemical reaction. Metal–organic, plasma-enhanced, low-pressure, laser-assisted, and aerosol-assisted CVDs are the most accepted methods [52]. The deposition is carried out in a reaction chamber at the temperature suitable for the reaction, the substrate is heated and the chemical reaction occurs when the heated substrate contact with the combined gas. The substrate temperature is an important parameter for this method to gain pure, uniform, hard, and strong nanoparticles [47, 53].

- **Spray pyrolysis**: It is a method often used in industry for large scale production of NPs. Generally, nanometals and metal oxides are produced by this simple, reproducible, size controllable and low-cost method [54]. This process consists of a precursor with flame where the precursor solution is sprayed or injected using a nanoporous nebulizer onto the hot substrate into the furnace at high pressure to form a droplet. The precursor can be either liquid or vapor. After evaporation, the precursor decomposes to recover nanoparticles or films on the substrate. Some of the furnaces have laser or plasma to produce high temperature to facilitate evaporation [55].

- **Biosynthesis:** It is an alternative to conventional physical and chemical nanoparticle synthesizing methods. Plants are preferred in this green and environmentally friendly cost-effective technique to prepare non-toxic and biodegradable nanoparticles [56]. In this method, several microorganisms as bacteria, fungus, and yeasts, etc. are used along with the precursors to produce nanoparticle for bioreduction and capping purposes. The biosynthesized nanoparticles have unique and enhanced properties that find a wide range of applications in drug delivery systems [57].

5.2.2 Top-down method

This method is also known as a destructive method due to the reduction of bulk material to nanometric scale particles. Contrary to bottom-up, large-scale production is possible and chemical purification is unnecessary in the top-down method. Broad size distribution (10–1000 nm), varied particle shapes, control over deposition parameters and reaction costs are disadvantages of this method. There are many techniques in this method, but mechanical milling, nanolithography, laser ablation and sputtering are among the most frequently used ones.

- **Mechanical milling**: This process has been used for a long time in mineral, ceramic processing, and powder metallurgy industries. The aim of mechanical milling of materials consists of minimizing particle size, blending, changing particle shapes, and synthesizing nanoparticles in a high energy mill with a convenient medium. At nanoparticle synthesis elements are granulated in an inert atmosphere. Mechanical milling is an economical method for nanosize production of large quantities [58]. The dynamics of mechanical milling vary according to energy transfer to the material from the balls [59]. Type of mill, the powder supplied to drive the milling chamber, milling speed, size and size distribution of the balls, dry or wet milling, the temperature of milling and the duration of milling are the factors that affect the energy transfer. Also, deformations, fractures, and the type of welding cause variations in particle shape and size [58].

- **Nanolithography**: This is the fabrication of molecules in a nanometric size range of 1 to 100 nm. Lithography is a combination of deposition and etching to have high-resolution topography. There are two main methods called as masked and maskless lithography. These 2 methods contain many techniques inside. While a mask or a mold is needed in masked lithography to fabricate patterns, maskless lithography produces unstable patterns without the use of mask. Photolithography, soft lithography, and nanoimprint lithography are the main techniques in masked lithography. Maskless lithography consists of electron beam lithography, ion beam lithography, and scanning probe lithography [60].

  The process is about printing material in a required shape or structure on a light- sensitive material. The main advantage of nanolithography is to make several copies with the desired shape and size from a single nanoparticle. On the other hand, the necessity of some equipment and their costs are the disadvantages of nanolithography [61].

- **Laser ablation (LA):** The laser irradiates the surface of the sample with a changeable wavelength of the laser and the refractive index of the solid or liquid target material in this complex PVD process. The laser removes electrons from the target material in a high electric field and those scattered electrons meet with the atoms of the bulk sample, where the energy transfer occurs. This leads to the heating of the surface and vaporization. The material is converted to a plasma state at high laser flux. There are some different applications in this method such as welding, cladding, cutting, cleaning, and generation of nanoparticles. During applications environmental conditions such as vacuum, air, gas and liquid can be changed. Pulsed-laser ablation types of solid target materials have great potential in the fields of laser-material microprocessing, nanotechnology and device fabrication. Besides, Laser Ablation Synthesis in Solution (LASiS) is a common and reliable top-down method that provides an alternative solution to the conventional chemical synthesis of metal-based nanoparticles. Also, organic solvents and water can be used in LASiS for NP synthesis and the method can be called as a 'green' process [62, 63].

- **Sputtering:** The principle of this physical process is to use the energy of plasma on the surface of a material, to arrange the atoms of the material and deposit them on the substrate with energetic ions. After the bombardment with ions, the ejection of atoms from the target occurs and then they deposit onto a substrate in the vacuum sputtering chamber. This high vacuum-based coating technique is included in the group of PVD processes. The shape, size, and composition of the nanoparticles vary with the layer thickness, temperature, and annealing time and substrate type [64].

6. Conclusion

The application of polymeric NPs in cancer therapy has been studied for decades. Poly(lactic-co-glycolic acid) (PLGA), chitosan, and polyethylene glycol

(PEG) are the most common, FDA-approved polymeric carriers for drug and bio-agent delivery. PLGA and chitosan contain hydrophobic domains which also capable of activating immune cells by their adjuvant character. In general, PLGA-based NPs for cancer immunotherapy is based on targeting dendritic cells. Micelles and liposomes are also convenient for the delivery of therapeutics and antigens. Recently, immunomodulatory nanoliposomes with 100 nm size were designed to deliver cancer antigens. The researches continued until today has indicated the importance of NPs in cancer immunotherapy. The antigen-NP conjugated systems help to introduce the immune-therapeutic agent to antigen-presenting cells efficiently. A high immune effect occurs with the presence of immunotherapeutic agent-loaded nano delivery systems in comparison to free immunotherapeutic agents. Prolongation, antigenicity, adjuvant selection, and inflammation are the most critical parameters for designing and engineering NPs.

On the other hand, there are still some issues to be solved in cancer immunotherapy. In some cases, insufficient information about cancer cells causes drugs not to present the expected effect. Scientists are unable to have precise information about the behavior of nanoparticles in the living system. In addition to these, there are difficulties in adjusting the toxicity, characterization, and monitoring behavior of nanomaterials in biochemical pathways. Moreover, failure to comply with the rules in drug use in such practices makes the work of the researchers even more difficult.

Besides, nanotechnology is promising for oncological applications for precise diagnoses and struggles with cancer cells. In light of the information mentioned in the literature, it is seen that interdisciplinary approaches and researches about the design and development of nanoparticle-based cancer immunotherapy are promising. Nanotechnology-based studies enable a therapeutic efficacy with a low dose of therapeutics, avoid cytotoxicity, and not to destroy the healthy cells of the patient. The quality and duration of cancer patients' lives can be improved by developing new methodologies in cancer immunotherapy based on nanoparticles.

Author details

Kadriye Kızılbey[1*], Nelisa Türkoğlu[2] and Fatma Ceren Kırmızıtaş[2]

1 İstanbul Yeni Yüzyıl University, Biomedical Engineering Department, İstanbul, Turkey

2 Yıldız Technical University, Department of Molecular Biology and Genetic, İstanbul, Turkey

*Address all correspondence to: kadriyekizilbey@gmail.com

References

[1] P. Huang *et al.*, "Nano-, micro-, and macroscale drug delivery systems for cancer immunotherapy," *Acta Biomaterialia*. 2019, doi: 10.1016/j.actbio.2018.12.028.

[2] J. Nam, S. Son, K. S. Park, W. Zou, L. D. Shea, and J. J. Moon, "Cancer nanomedicine for combination cancer immunotherapy," *Nature Reviews Materials*. 2019, doi: 10.1038/s41578-019-0108-1.

[3] H. Zhang and J. Chen, "Current status and future directions of cancer immunotherapy," *Journal of Cancer*. 2018, doi: 10.7150/jca.24577.

[4] R. Kokate, "A Systematic Overview of Cancer Immunotherapy: An Emerging Therapy," *Pharm. Pharmacol. Int. J.*, 2017, doi: 10.15406/ppij.2017.05.00112.

[5] F. Bray, J. Ferlay, I. Soerjomataram, R. L. Siegel, L. A. Torre, and A. Jemal, "Global cancer statistics 2018: GLOBOCAN estimates of incidence and mortality worldwide for 36 cancers in 185 countries," *CA. Cancer J. Clin.*, 2018, doi: 10.3322/caac.21492.

[6] V. K. Chaturvedi, A. Singh, V. K. Singh, and M. P. Singh, "Cancer Nanotechnology: A New Revolution for Cancer Diagnosis and Therapy," *Curr. Drug Metab.*, 2018, doi: 10.2174/1389200219666180918111528.

[7] A. Sudhakar, "History of Cancer, Ancient and Modern Treatment Methods," *J. Cancer Sci. Ther.*, 2009, doi: 10.4172/1948-5956.100000e2.

[8] O. Baykara, "Kanser Tedavisinde Güncel Yaklaşımlar," *Balıkesır Sağlık Bilim. Derg.*, 2017, doi: 10.5505/bsbd.2016.93823.

[9] J. R. Robbins, J. M. Longo, and M. Straza, "Radiation therapy," in *Cancer Regional Therapy: HAI, HIPEC, HILP, ILI, PIPAC and Beyond*, 2019.

[10] K. E. Burton, "Radiotherapy," in *Management of Adult Glioma in Nursing Practice*, 2019.

[11] V. T. DeVita and E. Chu, "A history of cancer chemotherapy," *Cancer Research*. 2008, doi: 10.1158/0008-5472.CAN-07-6611.

[12] B. A. Chabner and T. G. Roberts, "Chemotherapy and the war on cancer," *Nature Reviews Cancer*. 2005, doi: 10.1038/nrc1529.

[13] A. Nicolini, *G. Rossi*, P. Ferrari, R. Morganti, and A. Carpi, "A new immunotherapy schedule in addition to first-line hormone therapy for metastatic breast cancer patients in a state of clinical benefit during hormone therapy," *J. Mol. Med.*, 2020, doi: 10.1007/s00109-020-01881-3.

[14] N. Zaidi and E. M. Jaffee, "Immunotherapy transforms cancer treatment," *Journal of Clinical Investigation*. 2019, doi: 10.1172/JCI126046.

[15] E. F. McCarthy, "The toxins of William B. Coley and the treatment of bone and soft-tissue sarcomas.," *Iowa Orthop. J.*, 2006.

[16] S. J. Oiseth and M. S. Aziz, "Cancer immunotherapy: a brief review of the history, possibilities, and challenges ahead," *J. Cancer Metastasis Treat.*, 2017, doi: 10.20517/2394-4722.2017.41.

[17] T. A. Waldmann, "Immunotherapy: Past, present and future," *Nature Medicine*. 2003, doi: 10.1038/nm0303-269.

[18] M. Barbaros, Burak; Dikmen, "Kanser İmmünoterapisi," *Erciyes Üniversitesi Fen Bilim. Enst. Derg.*, 2015.

[19] C. R. Parish, "Cancer immunotherapy: The past, the present and the future," *Immunology and Cell Biology*.

2003, doi: 10.1046/j.0818-9641.2003.01151.x.

[20] N. E. Papaioannou, O. V. Beniata, P. Vitsos, O. Tsitsilonis, and P. Samara, "Harnessing the immune system to improve cancer therapy," *Annals of Translational Medicine*. 2016, doi: 10.21037/atm.2016.04.01.

[21] M. Sambi, L. Bagheri, and M. R. Szewczuk, "Current challenges in cancer immunotherapy: Multimodal approaches to improve efficacy and patient response rates," *Journal of Oncology*. 2019, doi: 10.1155/2019/4508794.

[22] M. J. Smyth and M. W. L. Teng, "2018 Nobel Prize in physiology or medicine," *Clinical and Translational Immunology*. 2018, doi: 10.1002/cti2.1041.

[23] S. Tan, D. Li, and X. Zhu, "Cancer immunotherapy: Pros, cons and beyond," *Biomedicine and Pharmacotherapy*. 2020, doi: 10.1016/j.biopha.2020.109821.

[24] D. V. Yuzhakova, M. V. Shirmanova, T. F. Sergeeva, E. V. Zagaynova, and K. A. Lukyanov, "Immunotherapy of cancer (Review)," *Sovremennye Tehnologii v Medicine*. 2016, doi: 10.17691/stm2016.8.1.23.

[25] C. Lee Ventola, "Cancer immuno-therapy, part 1: Current strategies and agents," *P T*, 2017.

[26] A. A. Özlük, M. G. Oytun, and D. Günenç, "Kanser immünoterapisi," *İstanbul Bilim Üniversitesi Florence Nightingale Transplant. Derg.*, 2017, doi: 10.5606/FNG.TRANSPLANTASYON.2017.004.

[27] I. Kimiz-Gebologlu, S. Gulce-Iz, and C. Biray-Avci, "Monoclonal antibodies in cancer immunotherapy," *Molecular Biology Reports*. 2018, doi: 10.1007/s11033-018-4427-x.

[28] Ö. Met, K. M. Jensen, C. A. Chamberlain, M. Donia, and I. M. Svane, "Principles of adoptive T cell therapy in cancer," *Seminars in Immunopathology*. 2019, doi: 10.1007/s00281-018-0703-z.

[29] P. Ataca and O. Arslan, "Chimeric Antigen Receptor T Cell (Car T Cell) Therapy In Hematology," *Turkish J. Hematol.*, 2015, doi: 10.4274/tjh.2015.0049.

[30] E. Wrona and P. Potemski, "A novel immunotherapy — The history of CAR T-cell therapy," *Oncology in Clinical Practice*. 2019, doi: 10.5603/OCP.2019.0016.

[31] K. Kaur and G. L. Khatik, "Cancer Immunotherapy: An Effective Tool in Cancer Control and Treatment," *Curr. Cancer Ther. Rev.*, vol. 16, no. 1, pp. 62-69, 2020.

[32] G. Marelli, A. Howells, N. R. Lemoine, and Y. Wang, "Oncolytic viral therapy and the immune system: A double-edged sword against cancer," *Frontiers in Immunology*. 2018, doi: 10.3389/fimmu.2018.00866.

[33] S. Khan and D. E. Gerber, "Autoimmunity, checkpoint inhibitor therapy and immune-related adverse events: A review," *Seminars in Cancer Biology*. 2019, doi: 10.1016/j.semcancer.2019.06.012.

[34] K. KAHVECİ and M. Türkoğlu, "İmmün Kontrol Noktası İnhibitörleri Ctla-4 ve Pd-1/Pd-l1'in İmmünoterapideki Yeri," *Mehmet Akif Ersoy Üniversitesi Fen Bilim. Enstitüsü Derg.*, vol. 10, no. 2, pp. 210-218.

[35] J. A. Seidel, A. Otsuka, and K. Kabashima, "Anti-PD-1 and anti-CTLA-4 therapies in cancer: Mechanisms of action, efficacy, and limitations," *Frontiers in Oncology*. 2018, doi: 10.3389/fonc.2018.00086.

[36] T. A. Waldmann, "Cytokines in cancer immunotherapy," *Cold Spring*

Harb. Perspect. Biol., 2018, doi: 10.1101/cshperspect.a028472.

[37] S. Lee and K. Margolin, "Cytokines in cancer immunotherapy," *Cancers*. 2011, doi: 10.3390/cancers3043856.

[38] J. Ciccolini, D. Barbolosi, N. André, S. Benzekry, and F. Barlesi, "Combinatorial immunotherapy strategies: most gods throw dice, but fate plays chess," *Annals of Oncology*. 2019, doi: 10.1093/annonc/mdz297.

[39] C. Wang, Y. Ye, Q. Hu, A. Bellotti, and Z. Gu, "Tailoring Biomaterials for Cancer Immunotherapy: Emerging Trends and Future Outlook," *Advanced Materials*. 2017, doi: 10.1002/adma.201606036.

[40] S. K. Rajendrakumar *et al.*, "Self-assembled, adjuvant/antigen-based nanovaccine mediates anti-tumor immune response against melanoma tumor," *Polymers (Basel).*, vol. 10, no. 10, p. 1063, 2018.

[41] Y. Yu and J. Cui, "Present and future of cancer immunotherapy: A tumor microenvironmental perspective," *Oncology Letters*. 2018, doi: 10.3892/ol.2018.9219.

[42] R. Agarwal, V. Singh, P. Jurney, L. Shi, S. V. Sreenivasan, and K. Roy, "Mammalian cells preferentially internalize hydrogel nanodiscs over nanorods and use shape-specific uptake mechanisms," *Proc. Natl. Acad. Sci. U. S. A.*, 2013, doi: 10.1073/pnas.1305000110.

[43] W. Park, Y.-J. Heo, and D. K. Han, "New opportunities for nanoparticles in cancer immunotherapy," *Biomater. Res.*, 2018, doi: 10.1186/s40824-018-0133-y.

[44] S. P. Surendran, M. J. Moon, R. Park, and Y. Y. Jeong, "Bioactive nanoparticles for cancer immunotherapy," *International Journal of Molecular Sciences*. 2018, doi: 10.3390/ijms19123877.

[45] V. Manolova, A. Flace, M. Bauer, K. Schwarz, P. Saudan, and M. F. Bachmann, "Nanoparticles target distinct dendritic cell populations according to their size," *Eur. J. Immunol.*, 2008, doi: 10.1002/eji.200737984.

[46] B. D. Chithrani, A. A. Ghazani, and W. C. W. Chan, "Determining the size and shape dependence of gold nanoparticle uptake into mammalian cells," *Nano Lett.*, 2006, doi: 10.1021/nl052396o.

[47] A. M. Ealias and M. P. Saravanakumar, "A review on the classification, characterisation, synthesis of nanoparticles and their application," in *IOP Conference Series: Materials Science and Engineering*, 2017, doi: 10.1088/1757-899X/263/3/032019.

[48] H. R. Ghaffarian, M. Saiedi, M. A. Sayyadnejad, and M. A. Rashidi, "Synthesis of ZnO nanoparticles by spray pyrolysis method," *Iran. J. Chem. Chem. Eng.*, 2011.

[49] L. Baraket and A. Ghorbel, "Control preparation of aluminium chromium mixed oxides by sol-gel process," *Stud. Surf. Sci. Catal.*, 1998, doi: 10.1016/s0167-2991(98)80233-4.

[50] S. Prasad, V. Kumar, S. Kirubanandam, and A. Barhoum, "Engineered nanomaterials: nanofabrication and surface functionalization," in *Emerging Applications of Nanoparticles and Architecture Nanostructures*, Elsevier, 2018, pp. 305-340.

[51] H. Bagheri Farahani, M. Shahrokhi, and A. Molaei Dehkordi, "Experimental investigation and process intensification of barium sulfate nanoparticles synthesis via a new double coaxial spinning disks reactor," *Chem. Eng. Process. - Process Intensif.*, 2017, doi: 10.1016/j.cep.2017.02.007.

[52] M. Benelmekki and A. Erbe, "Nanostructured thin films–background, preparation and relation

to the technological revolution of the 21st century," in *Frontiers of Nanoscience*, 2019.

[53] A. Kumar and D. Nanda, "Methods and fabrication techniques of superhydrophobic surfaces," in *Superhydrophobic Polymer Coatings*, 2019.

[54] K. Okuyama and W. W. Lenggoro, "Preparation of nanoparticles via spray route," *Chem. Eng. Sci.*, 2003, doi: 10.1016/S0009-2509(02)00578-X.

[55] D. S. Kumar, B. J. Kumar, and H. M. Mahesh, "Quantum nanostructures (QDs): an overview," in *Synthesis of Inorganic Nanomaterials*, Elsevier, 2018, pp. 59-88.

[56] T. M. Abdelghany *et al.*, "Recent advances in green synthesis of silver nanoparticles and their applications: about future directions. A review," *Bionanoscience*, vol. 8, no. 1, pp. 5-16, 2018.

[57] N. I. Hulkoti and T. C. Taranath, "Biosynthesis of nanoparticles using microbes—a review," *Colloids Surfaces B Biointerfaces*, vol. 121, pp. 474-483, 2014.

[58] T. P. Yadav, R. M. Yadav, and D. P. Singh, "Mechanical milling: a top down approach for the synthesis of nanomaterials and nanocomposites," *Nanosci. Nanotechnol.*, vol. 2, no. 3, pp. 22-48, 2012.

[59] S. C. Tjong and H. Chen, "Nanocrystalline materials and coatings," *Mater. Sci. Eng. R Reports*, vol. 45, no. 1-2, pp. 1-88, 2004.

[60] R. J. Varghese, S. Parani, S. Thomas, O. S. Oluwafemi, and J. Wu, "Introduction to nanomaterials: synthesis and applications," in *Nanomaterials for Solar Cell Applications*, Elsevier, 2019, pp. 75-95.

[61] G. Venugopal, M.-H. Jung, M. Suemitsu, and S.-J. Kim, "Fabrication of nanoscale three-dimensional graphite stacked-junctions by focused-ion-beam and observation of anomalous transport characteristics," *Carbon N. Y.*, vol. 49, no. 8, pp. 2766-2772, 2011.

[62] A. H. Hamad, K. S. Khashan, and A. A. Hadi, "Laser ablation in different environments and generation of nanoparticles," *Appl. Laser Ablation Thin Film Depos. Nanomater. Synth. Surf. Modif.*, p. 177, 2016.

[63] G. Mansoureh and V. Parisa, "Synthesis of metal nanoparticles using laser ablation technique," in *Emerging Applications of Nanoparticles and Architecture Nanostructures*, Elsevier, 2018, pp. 575-596.

[64] M. T. Nguyen and T. Yonezawa, "Sputtering onto a liquid: interesting physical preparation method for multi-metallic nanoparticles," *Sci. Technol. Adv. Mater.*, vol. 19, no. 1, pp. 883-898, 2018.

Chapter 5

Immunological Basis for the Development of Allergic Diseases-Prevalence, Diagnosis and Treatment Strategies

Siddanakoppalu N. Pramod

Abstract

Allergy is an immune disorder due to over responsiveness of immune system to a relatively normal and harmless antigen; derived from environmental and dietary substances commonly referred as allergens. Allergy is an IgE mediated type I hypersensitivity which is characterized by the degranulation of specialized white blood cells known as mast cells and basophils. Majority of characterized allergens are proteinaceous in nature and induce Th2 response. Specific Th2 cytokines elicit the induction of allergen specific IgE antibodies in sensitive individuals. The IgE binds to Fc epsilon receptor on basophil/mast cells and on exposure, allergens cross links the IgE and induce release of hypersensitivity mediators that result in allergic symptoms. The symptoms varies from mild allergies like hay fever, itchiness, rashes, rhinatisis, conjunctivitis to a severe condition such as Asthma and some time life threatening anaphylaxis. At present a various blood based test exist to diagnose allergies which include skin prick, patch test and Specific IgE tests. The best treatment available is to avoid exposure to allergens alternatively use of anti-histamines, steroids or other symptom reducing medications are in practice. Immunotherapy to desensitize the response to allergen and targeted therapy are promising for allergy in future.

Keywords: allergy, Th2 cells, Immunoglobulin E, Basophils, Mast cells, Histamine

1. Introduction

Allergies are among the most common chronic conditions worldwide. Symptoms of allergies range from making miserable to putting at risk for life-threatening reactions. According to the leading experts in allergy, an allergic reaction begins with the hyper sensitization of immune system to a relatively harmless antigen [1, 2]. The function of the immune system is to protect the body from invading pathogens that are responsible for inducing various diseases. In case of an allergic reaction, the immune systems wrongly recognize a common foreign antigen which are otherwise harmless substance as a pathogen. This foreign antigen is referred as an allergen. The immune components hyper react to the allergen and induce adverse immune response by production of Immunoglobulin E (IgE) antibodies [3, 4]. The IgE antibody has affinity to Fc epsilon R1 receptor present on the surface of cells that release inflammatory

IntechOpen

mediators on allergen stimulations. The released mediators like histamine and other chemicals cause allergic inflammation that result in allergic hypersensitivity. An inflammatory reaction typically triggers allergic symptoms in the throat and nose with itching and mucous discharge, sinus blockage and irritation in lining of stomach or itching on skin [2, 3]. In certain sensitive population, allergies can obstruct lung function and trigger bronchial hyper responsiveness and can induce asthmatic symptoms. In some people, a life-threatening serious reaction can occur called anaphylaxis [3–5]. A variety of proteins derived from diverse sources from different environment can act as allergens which were responsible for allergic reactions. The most common allergens were derived from plant flower pollens, dust mite, food sources, insect stings, animal hair and dander, mold, drugs, latex and medications.

The concept of “allergy” was originally introduced in 1906 by the Viennese pediatrician Clemens von Pirquet, after he noted that some of his patients were hypersensitive to normally innocuous entities such as dust, pollen, or certain foods [6, 7]. Earlier, all types of inappropriate hyper immune sensitive inflammatory reactions were termed as allergies. It was believed that, most were caused due to an improper increased activation of certain cells of the immune system that induce inflammation. Later, allergic IgE mediated mechanism was established that disproportionately activate certain cells of immune system to induce the release of inflammatory mediators [7]. A new classification system was proposed by Philip Gell and Robin Coombs in 1963 that described Type I to IV hypersensitivity reactions based on the immune mechanism and involvement of immune components [8]. In this system of classification, the allergic reactions or “allergy” was referred and restricted to immediate IgE mediated type I hypersensitivity. This is characterized by rapid onset of developing reactions and appearance of allergic or hypersensitivity symptoms in less than 20 mins after exposure to allergen. The landmark discovery for unrevealing the mechanism of allergy was through isolation and description of the importance of immunoglobulin E (IgE). In 1960, Kimishige Ishizaka and co-workers were first to report the antibody class IgE provided proof that is vital in mediating type I allergic hypersensitivity [9, 10]. The IgE which was now referred as allergic antibody or regenic antibody was primary immune component that can induce atopy or allergy among immune sensitive individuals [10].

2. Epidemiology

Prevalence of allergy or allergic disease fundamentally depends on various factors that govern the susceptibility of population to develop atopic condition. Predominantly genetic and environmental predisposition frames the basis for occurrence of allergy in an individual. Globally 8–10% of the population suffers from one or the other type of allergic disease which range from mild rhinatisis to sever asthma or anaphylaxis. At present a steady increase in the atopy was observed due to change in lifestyle, food habits and environment (**Table 1**). Several hypothesis and study provide evidence of genetic change in the population due to increased immune sensitivity and reduced antigen tolerance. Some report had identified an increase in allergic prevalence due to increase in perennial allergens exposure that happened by housing changes, increase in indoor environment with reduced ventilation and change in hygiene approach that decreased activation of immune regulatory control. The change in dietary habit, increased obesity, reduced physical exercise adds to hyper immune sensitization that increase atopy [11]. The reduced exposure to wild and native environmental antigens and high hygienic living standards expose fewer infections. It is reported that reduced infection at early

Allergy type	Allergic Prevalence and statistics
Allergic rhinatisis	• Worldwide allergic rhinatisis affects 10–30% of the population • Worldwide, sensitization (IgE antibodies) to foreign proteins in the environment is present up to 40% of the population • 7.5% adults and 9% children reported hay fever in an year
Asthma	• About 3 to 9% of the population suffer from allergic asthma. • Incidence had increased from 9.4 to about 18–20% in last five years in some European countries.
Drug allergy	• Worldwide adverse drug reactions may affect upto 10% of the world's population and affect 20% in hospitalized patients • Worldwide drugs may be responsible for up to 20% of fatalities due to anaphylaxis
Food allergy	• Around 8% of the population suffer from various food allergy • 6% at aged 0–2 years, 9% at aged 3–5 years, 8% at 6–18 yrs. and around 3–6% in adults. • 30–38% food allergic children have history of severe reaction and have multiple food allergies.
Insect allergy	• Worldwide, many allergic severe cases were reported with insect bite but lack systemic report. • In upto 50% of individuals who experience fatal reaction there is no documented history of previous systemic reaction.
Skin allergy	• Worldwide urticaria occurs with lifetime prevalence above 20% • Black children in US wwere likely to have had skin allergies (17%) than white (12%) or Asian (10%) children.
General allergy	• Worldwide, the rise in prevalence of allergic disease has continued in the industrialized world for more than 50 years • Worldwide sensitization rates to one or more common allergens among school children are currently approaching 40–50%.

Table 1.
Allergic conditions: Statistics and epidemiology [11, 12].

childhood age direct and polarize the developing immune system from Th1 type Th2, that makes the normal harmless antigen to a dangerous hypersensitive allergen that allow an increase in allergic disease.

Decreased rate of exposure to infection is not only increase immune sensitivity it also polarize immune response towards atopic mechanism. The hygienic hypothesis alone was unable to explain the increased prevalence of allergic disease. The recent evidences provide, importance of gastrointestinal microbial environment in development of atopy. Gut health, food and fecal-oral pathogens substantiate greater role to decide the risk for development of atopy. In some studies it was observed that an increased parasitic infection has been shown to associate with decreased prevalence of asthma [12]. This indicates the infection can exert the effect on Th1/Th2 regulation and it speculate that the dominance in Th1 response decrease atopy and Th2 link to IgE induction and elevates allergic diseases.

3. Signs and symptoms

Allergens are proteinaceous molecules that can be found in diverse substances in various forms. These can be inhaled, ingested and can also be exposed through contact to skin. Many allergens present in dust and pollens are airborne particles. These can be easily exposed through air and can induce symptoms in areas that contact

Affected Organ	Symptoms
Nose	Swelling of the nasal mucosa, runny nose and nasal irritation (Allergic rhinitisis)
Sinuses	Allergic sinusitis
Eyes	Redness and itching of the conjunctiva (Allergic conjunctivitis)
Airways	Sneezing, coughing, bronchoconstriction, wheezing and dyspnea. Sometimes outright attack of asthma, in severe cases the airway constructs due to swelling known as laryngeal edema.
Ears	Feeling of fullness, possibly pain and impaired hearing due to the lack of eusta chain tube drainage.
Skin	Rashes such as eczema and hives (Urticaria)
Gastrointestinal tract	Abdominal pain, bloating, vomiting and diarrhea.

Table 2.
Common symptoms of allergic reactions [13].

with allergen such as eyes, nose and lungs. Most common symptoms like hay fever also known as allergic rhinatisis cause runny nose, mucosal irritation and sneezing [13]. Some can also swollen eyes with itching and redness. Inhaled allergic particles can get into lungs and lead to bronchial hyper responsiveness. Particulate allergens inhaled through air can enter the lungs and cause asthmatic symptoms. Narrowing of the airways induce sneezing, coughing and through bronchoconstriction. The increased production of mucus thickens the airways and restricts the airflow to lungs that cause shortness of breath (dyspnea, bronchial hyper responsiveness and wheezing. Apart from these, the allergic reaction can be encounter through contact of allergens, ingestion through food and medications, insect bites and drug administration [13, 14]. Symptoms of contact and food allergy include itchy and swelling of the skin found during hives, gastrointestinal upset, edema, vomiting and diarrhea. Food allergies rarely cause respiratory (asthmatic) reactions, or rhinitis (**Table 2**).

Insect bites, drugs, medications and contact to insect stings with venom produce systemic allergic response affecting multiple organs. The exaggerated hyper immune response which is acute, life threatening and serious is called anaphylaxis and if not attended may induce death. The allergens effect multi organ system including digestive, respiratory, circulatory and cardiac system. Based on the severity and rate of sensitization the allergens can cause cutaneous reactions, edema, hypotension, bronchoconstriction, coma and sometime death [13]. Many allergenic substances such as latex can induce contact dermatitis and angioedema through skin rashes and irritations. The nature and source of allergens are diverse and they cause both cutaneous and systemic symptoms which range from very mild to severe depending on route of exposure and sensitization mechanism.

4. Causes

Causative agents for allergy or hypersensitivity reactions were allergens which present in many diverse sources in the environment. These allergy inducing factors have been placed in two categories (i) host factors and (ii) environmental factors [15]. The human host for the allergic reactions has different immune sensitivity due to various host factors that include gender, race, heredity and age. The genetic makeup and hereditary predisposition forms the basis for the increased incidence of allergic at certain population. However, there is insufficient evidence to explain the increase in allergic disorders with genetic factors alone. The change in food habits, living style and environmental pollutions and microbial exposure make huge contribution in allergic incidence. There were major environmental factors that alter the immune sensitization to induce atopy. To mention, the exposed inhalant and ingested allergen levels, exposure to infection diseases during early childhood and dietary changes. The alteration in environment certainly induces immune modulation that favors the development of allergic disease in susceptible population. The major class of allergens that predominantly cause allergic reactions belong to one of the following categories.

4.1 Food proteins

- One of the most common food allergies is sensitivity to peanuts. Tree nuts, including pecans, pistachios, pine nuts, and walnuts, are another common allergen.

- Egg allergies affect one to two percent of children. Milk, from cows, goats, or sheep, is another common allergy-causing food. Other foods containing allergenic proteins include soy, wheat, fish, shellfish, fruits, vegetables, spices, synthetic and natural colors, chicken, and chemical additives.

4.2 Non-food proteins

- Pollens, animal dander and dust can trigger an IgE-mediated cutaneous, respiratory, and systemic reaction. There is high prevalence of these allergies in the general population.

- The latex on contact induces delayed type hypersensitivity reaction which appears dry, crusted lesions called contact dermatitis. The delayed allergic response lasts 48 to96 hours. Rubbing the allergic lesions aggravates the reaction and it can lead to ulcerations. For the same latex the anaphylaxis reaction may occur in some sensitive individuals.

4.3 Toxins interacting with proteins

- Some food toxins on contact may induce delayed type of hypersensitivity with red rashes, blisters and edema.

- Another non-food protein reaction, urushiol-induced contact dermatitis, originates after contact with poison ivy, poison oak or sumac.

4.4 Genetic basis

- Allergic disease can be hereditary and there is a strong genetic basis for the development of allergic diseases. It has been reported that among homozygous twins, same allergic diseases were find 70% of the time; and in about 40% of non-identical twins same allergy has been reported [16].

- The allergic individuals are reported to have children with similar allergic diseases and with severe symptoms. The immune sensitivity is observed more with allergic lineage compared to non allergic parents. It was observed that the most common allergic disease are familial. It seems that the likelihood of developing allergies is inherited and related to an irregularity in the immune system.

4.5 Hygiene hypothesis

- Allergy is the result of a disproportional activation of immunological response to a relatively non-harmful antigen. An allergen induces the production of regenic IgE antibody through polarization of Th2 mediated immune response from natural Th1 response [17].

- Regular microbial infection elicits Th1 cytokines mediated immune response, which produce neutralizing IgM and IgG against infection agents like bacteria and viruses. This also downregulate Th2 mediated immune response.

- The proposed mechanism of hygienic hypothesis states, insufficient activation of Th1 mediated cytokine can lead to over reactive hyper stimulation of Th2 cytokines which polarize the immune response that leads to allergic diseases.

- This depicts that an individual spending early life in a clean, sterile and hygienic environment had less exposure to true microbial pathogens. The deficit in the development and activation of immune system in early life makes it more sensitive. These create hyper activation and over stimulation of immune components against harmless antigens and turn them into dangerous allergens.

4.6 Other environmental factors

- Geographical variations, climatic conditions, diet habits and lifestyles have considerable association with the incidence of allergic diseases.

- Globally, differences have been exhibited with the number of allergic individuals with in a population that reports allergic diseases [17]. The incidence of allergy is increasing in developed and industrialized countries compared to those which are more traditional or agricultural developing countries. The rate of allergic diseases were higher in urban population versus rural populations which substantiate hygiene hypothesis.

5. Hypersensitivity or inflammatory allergic reactions

The damaging immunologic reactions are called as hypersensitivity reactions. Although, current understanding of allergic diseases has grown vastly since then, this classification system remains useful even today. The 4 types of hypersensitivity reactions are; immediate hypersensitivity (type I) reactions, cytotoxic (type II), Ag-Ab complex mediated (type III), and T cell-mediated (type IV) delayed hypersensitivity [11]. The IgE mediated acute and immediate hypersensitivity reaction is the dominant out of the four types and forms the basis of allergic reactions that trigger and responsible for all allergic symptoms.

5.1 Immediate type IgE-mediated hypersensitivity

IgE mediated hypersensitivity is acute and the inflammation occurs immediately (within 30 mins) after exposure to an antigen (allergen). Allergen specific IgE antibodies binds to the FcεRI receptors present on basophils and mast cells. Exposure to allergen, specifically recognize the FCεRI bound IgE and cross link the adjacent IgEs and activate the signal cascade to trigger mast cell or basophils degranulation. The energy dependent degranulation process releases the non-cytotoxic, preformed inflammatory mediators that are responsible for induction of allergic symptoms within few minutes [18, 19]. Cellular degranulation releases two types of allergic mediators. Histamine, serotonin and tryptase are preformed mediators that are released by granular exocytosis. Other mediators like prostaglandins, leukotrienes were immediately synthesized de novo and released which act as pro-inflammatory signaling molecules. The two-phase mediators cause effect on glandular secretion, vascular permeability and smooth muscle contraction. These increase the immune cellular infiltration to the site of the inflammation within few minutes to hours and induce allergic reactions and referred as immediate hypersensitivity [20]. The allergic mediators, manifest and cause inflammation in many tissue sand organs (gastrointestinal system, respiratory system or generalized) either locally or systematically based on site of response (**Figure 1**). The symptoms range from mild atopic hay fever, rhinatisis, eczema to a chronic asthma and severe life threatening anaphylaxis.

5.2 Allergic sensitization and reaction

In an allergen sensitized subject with atopy, exposure of skin, nose, or airways to a single dose of allergen produces cutaneous wheal-and-flare reaction, sneezing and runny nose, or wheezing within minutes (**Table 3**), respectively [21, 22]. Depending upon duration and amount of allergen exposure the severity of allergic reaction may occur. Most IgE mediated hypersensitive reactions were immediate and express the clinical symptoms within an hour time. This may reach peak with late phase reactions in about 6 to 9 hours and after subsidizes slowly and resolves. In skin (**Figure 2**), the immediate reaction was characterized by itching and swelling and the late phase reaction by edematous erythoma which is read and forms blisters. In lungs it is noticed with nasal blockage, bronchial hyper responsiveness and further wheezing [22].

The type I hypersensitivity reaction has two stages, the earlier sensitization phase and the later effector phase. During the sensitization, the body encounter the antigen (allergen) for the first time and was recognized by the antigen presenting cells (APCs) as foreign antigen. The cells phagocytosed the antigen and present on the surface through MHC-II molecules. The naïve T helper lymphocyte recognize the presented antigen on MHC II and polarize the response towards Th2 by producing cytokines like interleukin −4 (IL-4) and interleukin −10 (IL-10). These interleukins interact with other type of lymphocytes known as B cells through specific receptors and instruct them through signal transaction that modulate gene transcription resulting in production of Immunoglobulin E (IgE) antibodies [23]. B cell turns into plasma cells and secrete large amount of IgE which circulates in the blood and on reaching basophils and mast cells, they recognize the specific receptors and binds the cell surface. The FcεR1 receptor has high affinity to IgE Fc portion and this referred as allergen sensitization. There was no observed inflammation or appearance of allergic symptoms during the sensitization phase.

5.3 Acute response

After the sensitization to allergen, the body had synthesized the IgE antibodies which occupied the surface of granulocytes; mast cell and basophils. The second exposure to the allergen, directly encounters the specific IgE antibodies present on the surface of allergic mast cells and basophils. The cross linking of two adjacent IgE molecules through multivalent allergen initiate the degranulation of these cells by activation of signaling cascade which results in exocytosis of preformed granular contents into the cellular space. The released histamine and other inflammatory chemical mediators (cytokines, interleukins, prostaglandins and leukotrienes)

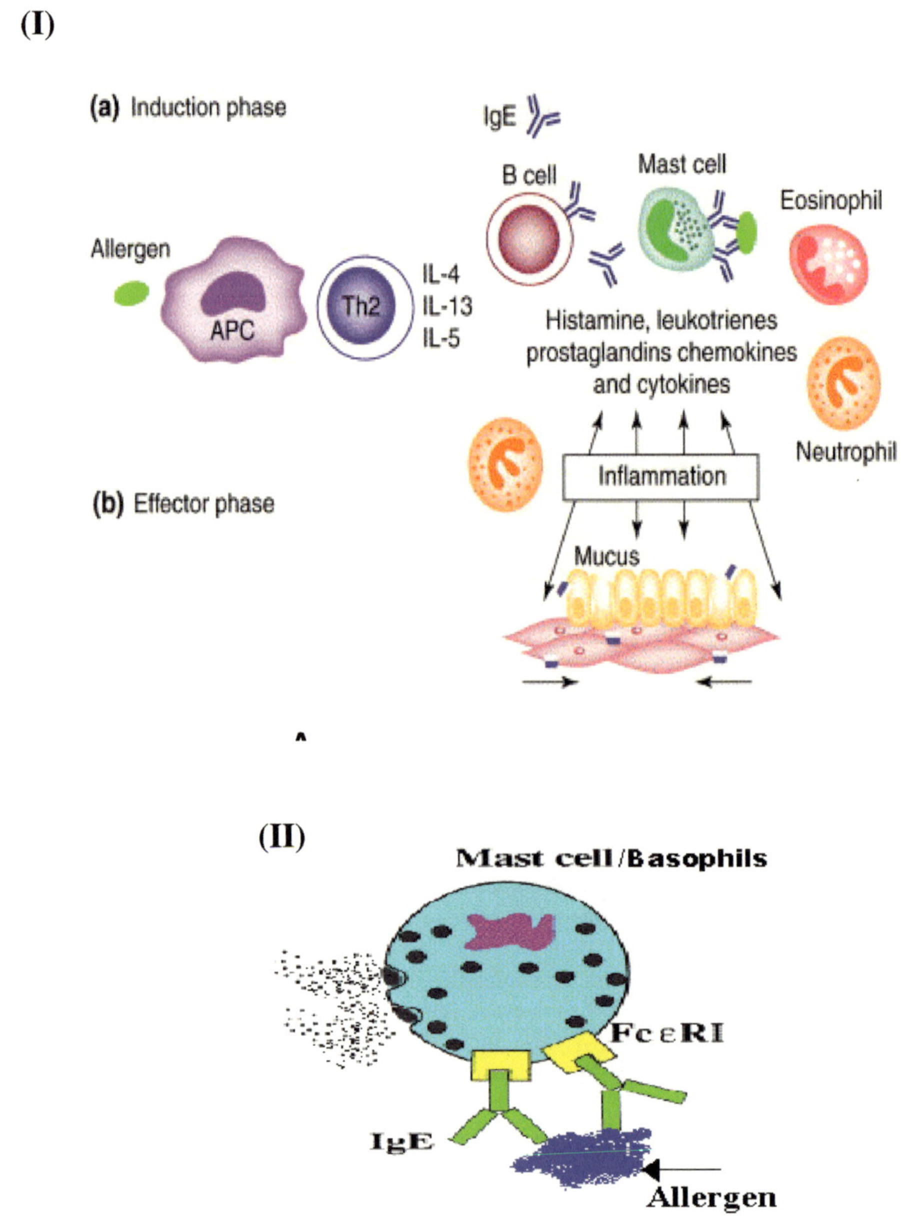

Figure 1.
*(I) Enhanced IgE dependent effector function and potential immunoregulatory function in mast cells or basophils after IgE-dependent upregulation of FcεRI surface expression (***Wedemeyer et al., 2000). (II). Mechanism that triggers the degranulation of mast cell or basophils. Crosslinkage can be mediated by: (A) the allergen that initiated the IgE response.*

induce systemic effects such as mucous secretion, smooth muscle contraction and vasodilatations [24, 25]. This results in the exacerbations of allergic symptoms like rhinorrhea, itchiness, dyspnea and anaphylaxis. Depending on the immune sensitivity of individual and mode and duration of exposure to allergen, the symptoms can be localized (organ or tissue specific); as asthma is localized to respiratory system and eczema to dermis or system-wide (classical anaphylaxis) where the whole body response with systemic effects.

Cell properties	Mast Cells	Basophils
Cell diameter	10–15 μm	5–7 μm
Nucleus	Bilobed or multi-lobed	Round or oval; eccentric
Cell surface contour	Smooth with occasional short, broad projections	Numerous narrow projections
Predominant localization	Connective tissues	Blood
Life span	Weeks or months	Days
Terminally differentiated	No	Yes
Major granule contents	Histamine, chondroitin sulphate, neutral proteinases, tryptase, heparin, TNFα	Histamine, chondroitin sulfate, neutral proteinases, major basic protein, Charcot-leyden protein
Mediators that are synthesized and released after degranulation	TNFα, PAF, LTC_4, PGD_2, IL-4	LTC_4

Abbreviations: TNF-tissue necrosis factor; PAF-platelet activating factor; LTC_4-leukotriene C_4; PGD_2-prostaglandin D_2.

Table 3.
Properties of human mast cells and basophils.

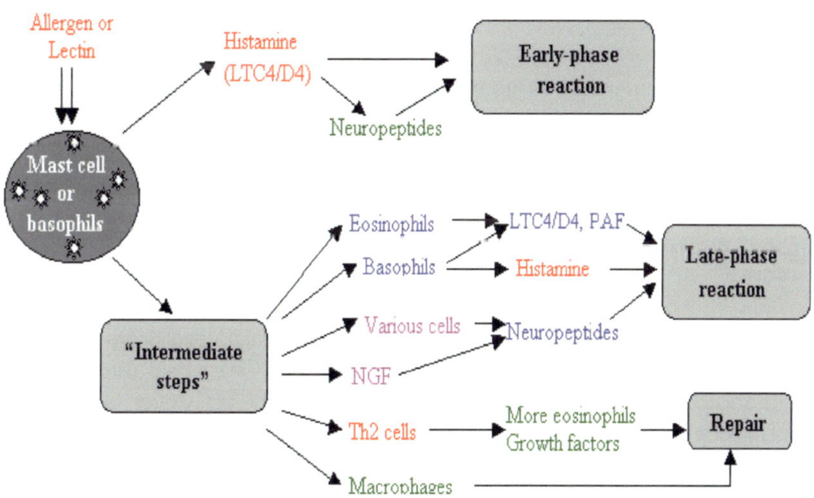

Figure 2.
The progression of allergic inflammation (e.g. in skin on testing), shows the schematic representation and components involved in the development of early-phase and late-phase reactions.

5.4 Late-phase response

Acute chemical mediators induce immediate allergic response. Once these acute response subsidies, often other leukocytes such as neutrophils, eosinophils and macrophages migrate to the site of inflammation to phagocytose and clear the damaged or inflamed tissues and cells. This results in allergic late phase response that usually lapse for 2 to 24 hours depending upon the site of inflammation and kind of allergic reaction [26]. Some time, the cytokines released from the degranulated mast cells play a role in inducing long term late phase allergic reactions that extend the symptoms for long duration. In case of allergic asthma, the late phase response persist longer that results in bronchoconstriction, impairing the lung function and cause wheezing.

Acute and late phase allergic response and their specific symptomatic disease [25, 26].

a. Immediate (early-phase reaction)

- Gastrointestinal
- Hives, angioedema
- Rhinitis, asthma
- Anaphylaxis

b. Immediate (late-phase reaction)

- Eczema/atopic dermatitis
- Eosinophilic gastroenteritis
- Urticaria

A strict relationship between genetic, skin behavior, immunological factors and trigger events such as environmental, psychological, and infections may be elicited and considered to be involved in the development and severity of allergy.

6. Immune signaling mechanism of IgE-mediated hypersensitivity (allergy)

Mast cells and basophils degranulation process is considered to be a prime signaling event for the development of allergic disease. Cross bridging of mast cell bound IgE molecules by allergen is thought to initiate the activation through Fc epsilon R1 receptor bound G protein coupled GTPase. This in turn causes the activation of phospholipase C and release phosphatidyl inositol bisphosphate (PIP2) and diacyl glycerol (DAG) from membrane lipids. The Inositol triphosphate (IP3) produced induce the release of calcium (Ca^{2+}) from endoplasmic reticulum and increase intracellular calcium levels [27, 28]. Increased Ca^{2+} in cytoplasm activates certain enzymes such as myosin light chain kinase and calmodulin. Calcium combined with DAG activates protein kinase C (PKC), these intracellular events trigger the migration of preformed granules in mast cells and basophils to their periphery. The preformed granules fuse with plasma membrane and release granular contents through exocytosis process [29]. In the same time these events also promote generation of lipid mediators like prostaglandins and leukotrienes resulting in the induction of allergic inflammation (**Figure 3**).

6.1 Modes of activation of mast cells/basophils

Cross linking of adjacent IgE by an allergen induces activation and degranulation of granulocytes (mast cells and basophils). The binding of IgE to cell surface is possible due to the molecule that has ability to bind IgE with high affinity and keeps it attached to cells surface. The Fc epsilon R1 receptor has strong affinity to Fc portion

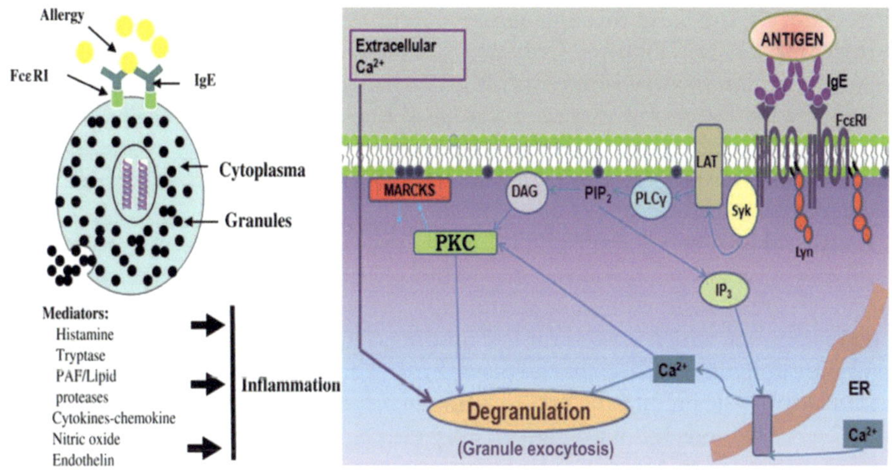

Figure 3.
Biochemical events of mast cell or basophil activation. There is a sequence of biochemical events following the bridging of cell bound IgE by an antigen. The climax of the whole process is the non-cytotoxic secretion of various chemical mediators such as histamine and other pharmacological substances.

of IgE which is made up of epsilon class of constant region. The allergen interacts with specific epitope to fragment of antigen binding (Fab) portion of the IgE. The allergen should be multivalent and with higher size so that it cross-link the two adjacent IgE to induce activation of mast cells for degranulation [30] to induce allergic reaction.

6.2 Cells and components of immune system involved in allergic reactions

The components of immune system are responsible for hyper immune response against allergen. Many cells and antibodies, cytokines are involved in various immune function that results in allergy or atopic reaction. The granulocytes like mast cells, basophils and eosinophils, lymphocytes such as Th2 cells and B cells play a prime role in development of allergic reactions. Inter molecular complex formed between allergen, IgE and FcεRI on the surface of mast cell or basophils are essential for activation of degranulation process to release mediators. That forms the basis for the immune activation to induce allergic diseases.

Atopic allergens: The allergens responsible for atopic diseases are derived principally from natural and airborne organic particles, especially plant pollens, fungal spores, and animal or insect debris, and to lesser extent from ingested food [31]. The ability of different pollens, molds, or foods to sensitize for IgE allergy varies, so that some of these environmental allergens are intrinsically more sensitizing than others, irrespective of the amount of exposure. Some lectins have been identified as allergens peanut agglutinin [32], soybean agglutinin [33] and wheat germ agglutinin [34] and are in general recognized as minor allergens in comparison with other common major allergens.

Mast cells and basophils: Mast cells are mononuclear cells with densely stained metachromatic granules while basophils are polymorphonuclear and are smaller in size (**Table 3**), approximately 5–7 microns versus 10–15 microns [35].

All the circulatory and connective tissues of the human system are susceptible for allergic response. It is due to the distribution of mast cells and basophils almost in all parts of the body. Mast cells are distributed essentially in all connective body parts and are often find adjacent to epidermal and microvasculature. The development, maturation and differentiation of mast cells influenced by cytokines and cellular growth factors like stem cell factor (c-kit ligand). Whereas basophils found in circulatory system and through hematopoietic cell linage precursor cells which are differentiated and matured from myeloid progenitor cells into specialized granulocytes along with eosinophils [36]. The basophils are circulatory and move through the blood and represent around 1% of the leukocytes. Mast cells are static and are found adhered to connective tissues across the body. These two cells contain preformed allergic pro-inflammatory mediators in the cellular granules which on degranulation cause allergic inflammations.

FcεRI and immunoglobulin E (IgE): The mast cells and basophils have high affinity receptors which has specificity to bind Fc portion of IgE antibody called Fc epsilon R1. This consist of four subunits ($\alpha\beta\gamma_2$) which represent one extracellular alpha (α) domain which is need for IgE binding. The beta (β) subunit is a transmembrane domain which spans the plasma membrane and the gamma (γ) subunits present as intrinsic membrane protein and are responsible for signal transaction [37]. The FcεRI binds IgE with high affinity (10^9 to 10^{10}/mole) and that is important phenomena for allergic response and development of allergy.

The Immunoglobulin (IgE) is referred as regenic antibody and it play a important role in allergic hypersensitivity reaction. It is a glycoprotein and belongs to one of the class of antibody with molecular weight of 190 kD and has 12% carbohydrate by weight which is present in the heavy chain at Fc portion. The serum concentration of IgE ranges from ng/mL to μg/mL with an atopic serum half-life of 2–3 days [38]. IgE recognizes mast cells and basophils through FcεR1 receptor and gets inactivated by heating at 56°C for at least 30 min. The cytokines IL-4, IL-13 and IL-10 induce synthesis of IgE by plasma cells. The detection allergen-specific IgE antibodies in the individual sera are considered as prominent diagnostic parameter and represent the allergic sensitivity.

Th2 or CD4+ cells: T helper cells are the immune responsive cells that have special interest in humoral immunity through induction of antibody production. The atopic individuals have high circulating allergen specific IgE antibodies. TH cells are circulatory lymphocytes which are characterized as CD4+ cells. There are two subsets of T-helper cells based on the antigen recognition and cytokine secretion. The Th1 cytokines direct the B cell to induce IgM and IgG. In contrast the Th2 type of response produce cytokines IL-4, IL-5, IL-10 and IL-13 [39, 40] these direct the B lymphocytes (B-cells) to produce allergic immunoglobulin IgE. This differentiates the function of Th2 from Th1-type cytokine (IFNγ and IL-2) response. The immunopathological hallmark of allergic disease is the infiltration of affected tissues by cells with a Th2-type cytokine profile [41, 42] that increase IgE production and allergic reactions.

Mediators released by mast cells and basophils: The pathophysiology of allergic reaction is exhibited with the inflammatory symptoms which are initiated by various allergic mediators released through the degranulation of mast cells and basophils. These cells synthesize and prestore granular mediators and instantly generate the lipid mediators [41]. The granular preformed mediators are rapidly released following activation; these represents; histamine, tryptase, serotonin, and other inflammatory cytokines. The others are synthesized de nova following mast cell activation and are release slowly. These include prostaglandin and leukotrienes that are metabolites of membrane lipids [43]. The complete list of mediators from mast cell and basophils are quite extensive and are undoubtedly account for multiple possible pathological consequences of allergic reactions.

7. *In vivo* and *in vitro* diagnosis of allergic reactions

Allergic disease needs diagnosis and prognosis for constant monitoring and treatment of symptoms. When allergic reaction is suspected in an individual based on the symptoms, the systemic diagnosis is essential for detection of causative allergen. A detailed case history of exposure and duration for appearance of symptoms with possible repetitive incidence will provide an idea of the type of allergy. The allergy diagnosis varies form case to case and it needs extensive inspection to identify the possible causative agent; the allergen. Based on the case history and information some can be identified rapidly however, the complex, obscure cases need repetitive interceptions to find the allergen. History, physical examination, onset of disease, duration of symptoms, time for resolving symptoms and kind and organ affected are required for initial assessment. This is followed by specific laboratory test which are required for the diagnosis and conformation of the onset of allergic reaction like eosinophils counts, total serum IgE levels, serum histamine levels and related medical examinations (**Figure 4**). It is important to correlate the detailed case history with the laboratory tests and that provide evidence of allergy [44].

Allergic disease is often episodic and that depends on the exposure to allergen to which the individual is sensitive. The case specific objective signs of the allergic symptoms can only be identified during the allergic incidence with proper physical examination. The observed symptoms have to correlate with subjective signs provided in the case history for the proper identification of allergic disease. Allergy diagnosis requires thorough examination to rule out other illness of the subject. A variety of *in vitro* and in vivo laboratory tests are available to supplement the history and physical examinations. There are qualitative and quantitative tests that predict the allergic reaction using sampling fluids and immune cellular responses through immunochemical techniques.

Many of the allergic symptoms shares common pathological behaviors with other illness and that need confirmative cross examination before planning

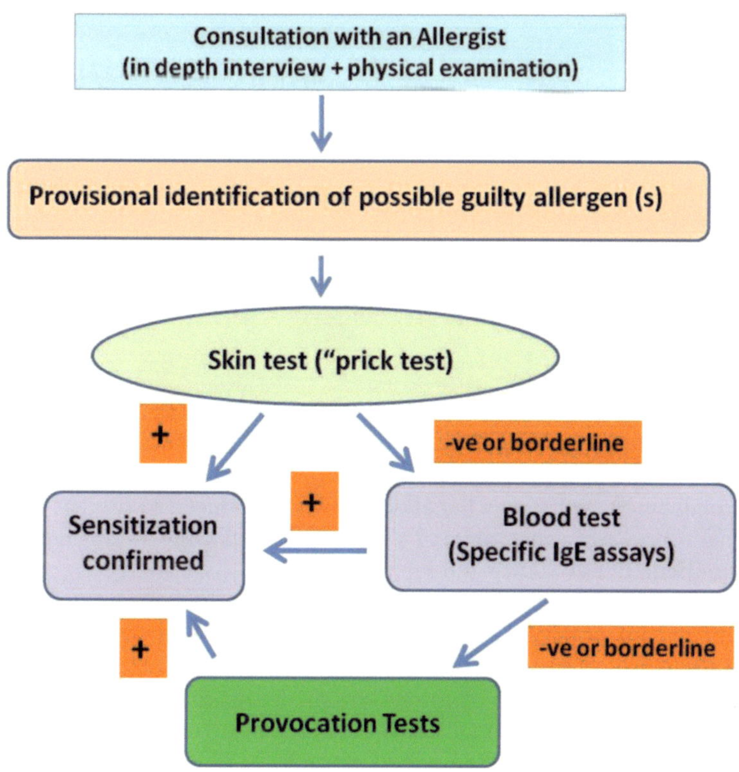

Figure 4.
Schematic diagram showing the identification and assessment of atopic status using clinical history, physical examination and laboratory tests.

treatment regime. For instance, the viral flu induce rhinitis and nasal conjunction which also the same with pollen or dust allergy. The food toxicity or certain diet can cause gastrointestinal upset which is quite same as that of food allergy. The common cold or viral flu airway infections induce bronchoconstriction that results in wheezing which exactly mimics the symptoms of allergic asthma. Henceforth, careful diagnosis is a prerequisite for the identification and treatment of allergic disorders.

7.1 Skin testing

Skin testing is the in vivo mimicking of allergic reaction that demonstrates the allergic sensitivity to specific allergen. The skin testing predict and confirm the presence of allergen specific IgE antibodies in the individual. These were most preferred over the blood testing during allergy diagnosis. Skin testing is also known as prick test and puncture testing. The most two types of allergic tests, which are commonly in use at clinical level for diagnosis are skin prick test (SPT) and prick by prick test (PPT). In the earlier one the suspected allergen sample was placed and was pricked with small needle and allowed for erythoma formation for 30 mins [45]. In later the sample was pricked initially with the needle and then the same was pricked to skin and the pricked area was observed for the development of reaction. The histamine was used as positive control and PBs as negative. The wheel and flare diameter was measured and was compared for the prediction of positive allergic reaction. Some time a similar intradermal test on the skin can also be used for assessment of allergic reaction to certain medication and drugs. The skin testing is widely used in allergic clinic with standard available panel of allergen samples to identify causative allergen and provide proper treatment for allergic symptoms.

7.2 Blood testing

Blood is the primary biological sample for diagnosis of illness in clinics. The blood sample contains various immune components that are related to allergic reactions [46]. Various blood allergy testing parameters and methods are available which can detect and diagnose allergy and identify allergens. The most often used are serum total IgE level; that estimate the IgE content in the subject serum per mL. The other is allergen specific IgE level which predicts the confirmative diagnosis of elicitor. Both are measured through radiometric (RAST) or colorimetric (ELISA) immune assays.

7.3 Other methods of testing

Allergen challenge testing: During allergen challenge test, the subject was monitored and the whole procedure was done in the presence of a expert clinician. In this, a small amount of suspected allergen was introduced to subject through oral or other routes and appearance of allergic reactions were monitored. This test provides confirmative evidence and identifies the causative allergen.

Elimination/Challenge tests: In this procedure, subject was instructed to avid coming in contact with allergen prior to test. During asymptomatic time, few suspected allergens were added with food or medicines and were given to subject and the appearance of allergic symptoms was recorded. Based on this a true allergen can be identified for planning treatment.

Patch testing: Patch testing is much in practice for identifying the contact dermatisis or delayed type of allergic reactions. In this case an allergen is placed on the patch and that is stick to the back of the subject. The symptoms will be observed after 24 hrs for the appearance of symptoms.

Unreliable tests: There are some allergic tests which are not considered for practice by International allergy council and those does not provide proper scientific evidence to identify allergy or allergens. Some of them are cytotoxicity testing, provocative tests, subcutaneous or sublingual testing. In future with substantiate research and technical improvement can be used for diagnosis of some of the allergic diseases.

8. Treatments

Advancement in allergy research had made enormous contribution for the treatment of mild to severe allergic diseases. There are many treatments available to treat various symptoms of allergic diseases and several medications are available and are effectively treat and manage atopic conditions. For anaphylaxis epinephrine shots are available which can be carried with the patients and for others, anti histamines and anti inflammatory drugs are routinely recommended to cope up with the symptoms [46, 47]. Depending on the source of allergens various diagnosis methods have been devised and based on those therapeutic methods have evolved to address problems associated with allergic reactions. The following are some commonly followed approaches to treat allergic diseases.

8.1 Avoidance

Avoiding exposure to allergen is the best and valid recommendation for limiting allergic reactions in sensitized individuals. It is one of the simple and traditional approach for treatment of allergy. However, it becomes difficult to avoid certain environmental allergen which are dispersed in the air and can be easily inhaled without any notice or have any control. In such cases the avoidance becomes difficult and need alternative therapeutic methods to address problems.

8.2 Pharmacotherapy

When allergen tracking and avoidance is not possible and exposure to allergen becomes inevitable then the pharmacotherapy can provide protection to ease of allergen induced symptoms. Many drugs have been designed which act as antagonistic to the allergic mediators and block their actions. Some common drug targets are anti histamines and anti leukotrienes which prevent the action of inflammatory mediators and block the appearance of allergic symptoms [46]. The FDA approve drugs that include antihistamines, adrenaline (epinephrine), theophylline and Glucocorticosteroids which acts primarily as anti inflammatory molecules. The anti-leukotrienes such as Montelukast (Singulair) or Zafirlukast (Accolate) are in common use along with mast cell stabilizer, decongestants and eosinophil chemotoxins are used as drugs to prevent and monitor acute and chronic allergic disorders [47].

8.3 Immunotherapy

In case the allergen has been identified and sensitization process is well established with the subject, in that case the desensitization or hyposensitization is adopted as treatment to vaccinate the allergic subject with small doses of allergen over a long period. During this the subject tolerates the allergen dose and reduces its sensitivity and increase IgG production over IgE that avoid allergic reactions. Studies have demonstrated the efficiency of this type of immune therapy and the long term practice had shown preventive effect of immunotherapy in reducing the development of atopy. A second form of immunotherapy involves the intravenous injection of monoclonal anti-IgE antibodies. These bind to free and B-cell associated IgE; signaling and induce their destruction [48]. A third type, Sublingual immunotherapy, is an orally-administered therapy that takes advantage of oral immune tolerance to non-pathogenic antigens such as foods and resident bacteria [49]. Allergen shot treatment may appear as future closest therapy to cure for allergy. This therapy requires close monitoring and long-term commitment for the efficient treatment by the subject.

8.4 Unproven and ineffective treatments

In some of the recent studies, an enzyme potentiated desensitization (EPD), experimental treatment has been tried and had not produced any promising results. Many hypoallergic food preparation now follow the same strategy. The treatment approach but failed to convince and had not accepted as effective [50]. EPD uses dilutions of allergen with an enzyme, beta-glucuronidase, that changes the allergen nature and polarize T-regulatory lymphocytes which favor desensitization, or down-regulate IgE induction and prevent allergic reactions.

9. Conclusion

Allergies may cause symptoms ranging from mild abdominal discomfort to life-threatening anaphylaxis. Avoiding offending allergen exposure may not be easy if the causative is uncommon or not identified. However, successful avoidance of exposure to allergens may reduce the symptoms in the allergic individuals. Presently most of the treatment and diagnosis methods available are only to reduce the symptoms but the medications methods will not provide any permanent relief from the allergic diseases. New studies and investigations are in progress from the researcher to provide solutions for the allergy treatment. New invention in the field of immunology provides methods and techniques to find new horizons to understand allergic disorders. Since allergies are associated with disorder of immune function, scientists working to find therapy will benefit from new developments in immunology research. Understanding the regulatory mechanism of Allergic disease and designing suitable method for rapid allergen detection and revisiting for the success of immunotherapy and pharmacotherapy will advance the allergy research and provide suitable bench to treat and prevent allergies. The advances in molecular biology that lead to understanding of immune molecular networks will certainly provide promising hope for the preventive and therapeutic solution to allergies in near future.

Conflict of interest

The author declare no conflict of interest.

Author details

Siddanakoppalu N. Pramod
Department of Studies in Biochemistry, Davangere University, Shivagangotri, Davangere, Karnataka, India

*Address all correspondence to: pramodsn@davangereuniversity.ac.in

References

[1] Kay AB. Overview of Allegy and allergic diseases: with a view to the future. British Medical Bulletin 2000;56:843-866.

[2] Douglass JA, Hehir RO, Diagnosis, treatment and prevention of allergic disease: the basics. The Medical journal of Australia 2006; 185: 228-233.

[3] Maurer D, Stingl G. Immunoglobulin E-binding structures on antigen-presenting cells present in skin and blood. J Invest Dermatol 1995;104:707-710.

[4] Dullaers M, De Bruyne R, Ramadani F, Gould HJ, Gevaert P, Lambrecht BN. The Who, where and when of IgE in allergic airway disease. J Allergy Clin Immunology 2012; 129:635-645.

[5] Romagnani S. Human TH1 and TH2 subsets: doubt no more. Immunol Today 1991; 12: 256-257

[6] Basu S and Banik K. Hypersensitivity: an overview. Immunol Cur Res. 2017;2: 1000105.

[7] Combs RRA. The hypersensitivity reaction: some personal reflections. Clin Exp Allergy 1992; 22: 673-680.

[8] Coombs RRA, Gell PGH. Classification of allergic reactions responsible for clinical hypersensitivity and disease. In: Gell PDH, Coombs RRA, Lachmann PJ (eds), *Clinical aspects of Immunology* 3rd Edition, Blackwell: Scientific Publications, Oxford, Chapter 25, 1975.

[9] Platts-Mills TAE. The continuing effect of the discovery of IgE by Kimishige Ishizaka. J Allergy Clin Immunol 2018;142: 788-789.

[10] Amarasekera M. Immunoglobulin E in health and disease 2011;1:12-15.

[11] Ebert Jr CS, Pillsbury HC. Epidemiology of Allergy. Otolaryngol Clin North Am. 2011;44:537-548.

[12] Constant SL, Bottomly K. Induction of Thl and Th2 CD4* T cell responses: the alternative approaches. Atmu Rev Immunol 1997; 15: 297-322.

[13] Aberg N, Hesselmar B, Aberg B, Eriksson B. Increase of asthma, allergic rhinitis and eczema in Swedish schoolchildren. Clin Exp Allergy 1995; 25: 815-819

[14] Mekori YA. Introduction to allergic disease. Crit Rev In Food Sci Nutr 1996; 36:1-12.

[15] Pawankar R, Canonica GW, Holgate ST, et al: Allergic diseases and asthma: a major global health concern. Curr Opin Allergy Clin Immunol 2012; 12:39-41

[16] Holt PG, Macaubas C, Stumbles PA, Sly PD. The role of allergy in the development of asthma. Nature 1999; 402 (Suppl): B12–B17

[17] Rogers PR, Croft M. Peptide dose, affinity, and time of differentiation can contribute to the Thl/Th2 cytokine balance. /Immunol 1999; 163: 1205-1213

[18] Helm RM, Burks AW. Mechanism of food allergy. Curr Opin Immunol 2000; 12: 647-653.

[19] Kay AB. Allergy and allergic disease. First of two parts. N England J Med 2001;344:30-37.

[20] Turcanu V, Stephens AC, Chan SMH, Rancé F, Lack G. IgE-mediated facilitated antigen presentation underlies higher immune responses in peanut allergy. Allergy 2010;65:1274-1281.

[21] Kambayashi T, Baranski JD, Baker RG, et al. Indirect involvement of allergen-captured mast cells in antigen presentation. Blood 2008;111:1489-1496.

[22] Kay AB. Modulation of inflammation in allergic disease. Allergy Clin Immunol Int 2004; 16: 206-210.

[23] Qiao HL, Yang J, Zhang YW. Relationship between specific serum IgE, cytokines and polymorphisam in the IL-4Ralpha in patients with pencillin allergy. Allergy 2005; 60: 1053-1059.

[24] Clark S, Camargo CA Jr. Epidemiology of anaphylaxis. Im-munol Allergy Clin North Am. 2007;27(2):145-163.

[25] Kemp SF, Lockey RF. Anaphylaxis: a review of causes and mechanisms. J Allergy Clin Immunol. 2002;110(3): 341-348.

[26] Galli SJ, Tsai M, Piliponsky AM. The development of allergic inflammation. Nature. 2008;454:445-454.

[27] Yamaguchi M, Lantz CS, Oettgen HC, Katona IM, Fleming T, Miyajima I, kinet JP, Galli SJ. IgE enhances mouse mast cell FcεRI expression in vitro and invivo: evidence for a novel amplification mechanism in IgE-dependent reactions. J Exp Med 1997; 185: 663-672.

[28] Turner H, Kinet JP. Signalling through the high-affinity IgE receptor FcεRI. Nature 1999; 402: B24–B30.

[29] Rivera J and Gilfillan AM. Molecular regulation of mast cell activation. J Allergy Clin Immunol 2006; 117: 1214-1226.

[30] Kjaer TMR, Frokiaer H. Dietary lectins and the immune response. Rev Food Nutr Toxicity 2005; 4: 271-295

[31] Woodfolk JA, Commins SP, Schuyler AJ, Erwin EA, Platts-Mills TAE. Allergens, sources, particles and molecules: why do we make IgE responses? Allergology International 2015;64:295-303.

[32] Burks AW, Cockrell G, Connaughton C, Guin J, Allen W, Helm RM. Identification of peanut agglutinin and soybean trypsin inhibitor as minor legume allergens. Int Arch Allergy Immunol 1994; 105:143-149

[33] Baur X, Pau M, Czuppon A, Fruhmann G. Characterization of soybean allergens causing sensitization of occupationally exposed bakers. Allergy 1996; 51:326-330

[34] Weichel M, Vergoossen NJ, Bonomi S, Scibilia J, Ortolani C, Ballmer-Weber BK, et al. Screening the allergenic repertoires of wheat and maize with sera from double-blind, placebo-controlled food challenge positive patients. Allergy 2006; 61: 128-135.

[35] Amin K, The role of mast cells in allergic inflammation. Respiratory medicine. 2012;1:9-14.

[36] Fulkerson PC, Rothenberg ME. Targetting Eosinophils in Allergy, inflammation and beyond. Nat Rev Drug Discov 2013:12:1-20.

[37] Garman SC, Sechi S, Kinet JP, Jardetzky TS. The analysis of the human high-affinity IgE receptor Fc epsilon RI alpha from multiple crystal forms. J Mol Biol 2001; 311: 1049-1062.

[38] Gould HJ, Sutton BJ, Beavil AJ, Beavil RL, McCloskey N, Coker HA, Fear D, Smurthwaite L. The biology of IgE and the basis of allergic disease. Annu Rev Immunol 2003; 21: 579-628.

[39] Romagnani S. Human TH1 and TH2 subsets: doubt no more. Immunol Today 1991; 12: 256-257.

[40] Till S, Durham S, Dickason R. Huston D, Berngre J, Walker S, Robinson D, Kay AB. IL-13 production by allergen stimulated T-cells is increased in allergic disease and associated with IL-5 but not IFN-gamma expression. Immunology 1997; 91: 53-57.

[41] Kay AB, Ying S, Varney V et al., Messenger RNA expression of the cytokine gene cluster, IL-3, IL-4, IL-5 and GM-CSF in allergen induced late phase cutaneous reactions in atopic subjects. J Exp Med 1991; 173: 775-778.

[42] Robinson DS, Hamid Q, Ying S, Tsicopoulos A, Barkans J, Bentley AM, Corrigan C, Durham SR, Kay AB. Predominant T_{H2}-type branchoalveolar lavage T-lymphocyte population in atopic asthma. New Eng J Med 1992; 326: 298-304.

[43] Popescu FD, Mediators involved in the pathophysiology of allergic rhinitis. Romanian J Rhinology 2012;6: 104-108.

[44] Tanno LK, Calderon MA, Li J, Casale T, Demoly P. Updating Allergy/or hypersensitivity diagnostic procedures in the WHO ICD-11 revison. J Allergy Clin Immunol Pract 2016; 4: 650-657.

[45] Doyen V, Casset Divaret-Chauveau A, Khayath N, Peiffalphin JC, De Blay F. Diagnosis of allergy in asthma. Rev Mal respire 2020; 37: 243-256.

[46] Bonyadi MR, Hassanzadeh D, Seyfizadeh N, Borzoueisileh S. Assessmentt of allergen specific igE by immunoblotting method in atopic dermatitis. Eur Ann Allergy Clin Immunol 2017; 49:213-219.

[47] Pullerits T, Praks L, Ristioja V, Lotvall J. Comparison of a nasal glucocorticoid, antileukotriene, and a combination of antileukotriene and antihistamine in the treatment of seasonal allergic rhinitis. J Allergy Clin Immunol 2002;109, 949-955.

[48] Wong AG, Lomas JMllergy testing and Immunotherapy. Pediatric Rev 2019; 40:219-228.

[49] Scurlock AM. Oral and sublingual immunotherapy for treatment of IgE mediated food allergy. Clin Rev Allergy Immunol 2018; 55: 139-152.

[50] Astarita C, Scala, G, Sproviero S, Franzese A. Effects of enzyme-potentiated desensitisation in the treatment of pollinosis: a double-blind placebo-controlled trial. J Invest Allergol Clin Immunol 1996;6: 248-255.

Chapter 6

SARS-CoV-2 and Coronavirus Ancestors under a Molecular Scope

Maram Adel Abdelghany,
Sarah Abdullah Gozai Alghamdi and Jehane Ibrahim Eid

Abstract

The Pandemic of COVID-19 has been thoroughly followed and discussed on many levels due to the high level of attention that it has brought by its effect on the world. While this disease might seem like to arise out of the blue, we will shed light on COVID-19 disease which is caused by the virus SARS-CoV2 and belong to family of coronaviruses. We will discuss current knowledge about SARS-CoV2 emergence, diagnosis, its mode of action, and genomic information, For an antiviral treatment to be used, it should be preceded by a foundation of information about the virus genome and its family as discussed in this review.

Keywords: Covid-19, SARS-COV-2, genome, evolution, immunopathology, phylogenetic

1. Introduction

The Coronaviridae have a wide variety of host species, which infect many mammalian and avian species and result in high respiratory, hepatic, and central nervous system diseases. Coronaviruses in humans and fowl mainly cause infections in the upper respiratory tract and enteric infections are caused by pig and bovine Coronavirus [1]. Coronaviruses CoVs are divided into four genera and in 1937 the first coronavirus was identified [2, 3]. Coronaviruses are a family of helical nucleocapsid and extremely large genomes enveloped positive-stranded RNA viruses. Coronaviruses are composed of: 1) Nucleocapsid Protein (N): helical nucleocapsid protein component and is supposed to bind genomic RNA in a bead-on-string mode. 2) spike protein (S): Viral envelope component that mediates binding to the receptor and merging of cell membranes if the virus and host. 3) Membrane Protein (M): the most present component and gives its form to the virion envelope. 4) Envelope Protein (E): A small, only minor component of virions and a small polypeptide between 8.4 and 12 kDa (76–109 amino acids). 5) Accessory Proteins: "Extra" genes may be interspersed with a group of canonical genes, replicase, S, E, M, and N with additional ORFs, or embedded in a separate ORF or heavily overlapped with another gene [4]. (**Figure 1**) Coronaviruses are also one of the few genomically proofreading RNA viruses that avoid the virus developing mutations that could weaken it. Such capacity may have contributed to the failure of specific antivirals like ribavirin to subdue SARS-CoV-2 meanwhile, can thwart viruses like hepatitis C. Drugs kill viruses by mutations. However, the proofreader can eliminate these changes in coronaviruses. Coronaviruses have a special trick that is fatal: they often recombine,

IntechOpen

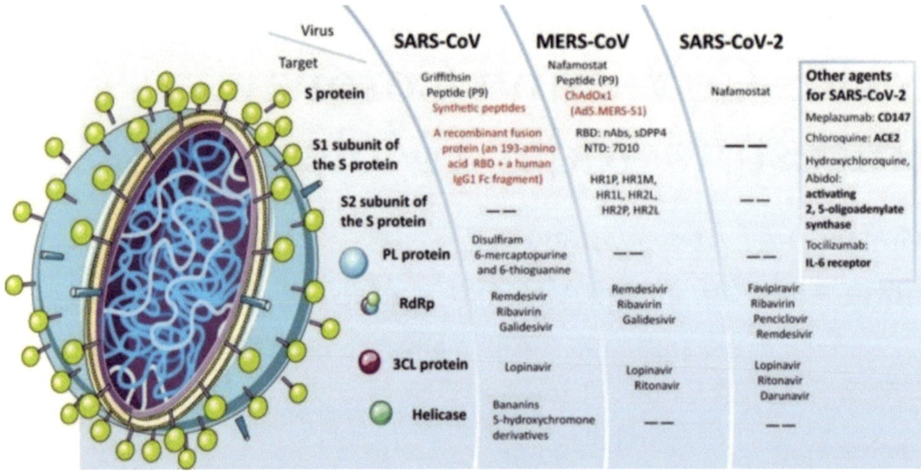

Figure 1.
Viral structure diagram showing the envelope, Centre and structure of the nucleoprotein. S, the spike protein and different drug candidates against the three coronaviruses [5].

exchange pieces of RNA and other coronaviruses. This is usually an insignificant trade between viruses like parts [6]. Both mammals are affected by alphacorona-viruses and beta-coronaviruses. Alpha-Coronaviruses and beta-Coronaviruses typically cause human breathing diseases and animal gastroenteritis. Gamma and delta coronaviruses infect birds, but some can infect mammals as well. The SARS-CoV, MERS-CoV viruses, and the other four human coronaviruses (HCoV-NL-63, HCoV-229E, HCoV-OC43 and HKU1) are responsible for severe respiratory syndromes in people with mild conditions in immunocompetent hosts, although some infections are severe in infants and elderly people [7]. Coronavirus transcription is characterized by the development of several mRNAs containing the sequences corresponding to the two ends of the genome. The production of subgenomic mRNA requires discontinuous transcripts. Transcription is known as the process by which subgenomical mRNAs are generated, and replications are the process by which genomic-sized RNA, which also acts as mRNA, is generated [8]. Human coronaviruses (HCoV) were first detected and developed in the nasal cavities of common cold patients in the 1960s. Two human coronaviruses-OC43 and 229E are responsible for about 30% of common colds [9, 10]. Middle East Coronavirus Respiratory Syndrome (MERS-CoV) has also been a global health concern. The initial report for MERS-CoV was in 2012. More than 2000 civilians have been infected in 27 countries in the Middle East and 4 subcontinents. During the SARS outbreak in 26 countries, more than 8000 cases were recorded in 2003 [11]. The ongoing coronavirus disease outbreak (COVID-19), first reported in December 2019 in Wuhan, China. As of 5th of April 2020, the world health organization (WHO) announced this disease as a global public health emergency to extend to 206 countries and territories across the world, with two international correspondence performed on 3,090,445 confirmed cases reported cases, including 217,769 deaths [12]. SARS-CoV-2 virus, the cause of COVID-19 disease that lead to an emergency outbreak that has been going for several months, now it may as well continue to its spread until the finding of new treatments along with the implementation of effective countermeasures. The newly evolving coronavirus (SARS-CoV 2) is becoming increasingly largescale. In the last few weeks, complete genomic sequences were released in order to understand the development and molecular characteristics of the virus by the global scientific community. In this review we will discuss the genomic structure of the virus, the possible relations between several viruses of the same family and the suspected origins and spill over that might have led to such epidemic and molecular diagnostics used to detect.

2. Emergence of a new virus

The life of people over the centuries has been influenced by Zoonotic diseases. Many of these situations are especially variable in complexity, dynamics and shifts over time, as they emerge, and reappear. Transmission of the pathogen from an animal to human, also known as zoonotic spillovers, is a global public health issue and remains an ambiguous phenomenon, while associated with multiple outbreaks [13]. A mixture of many factors is needed to fulfill a zoonotic spillover, including ecological, epidemiological and behavioral determinants of pathogen transmission and inherent human factors influencing susceptible infection, as well as dietary and societal factors linked with foodborne zoonotic spillover [14]. A new virus is a virus that mutated and went through an evolution process to adapt to new kinds of hosts by a process called spillover. Spillover can happen in wild animals' market as a virus can mutate and go on infecting a new host where it further mutates within new host until it adapts to this new host and become infectious [15]. Over the past two decades, many outbreaks of Zoonotic diseases such as SARS, the Hendra virus and the Nipah virus have been related to the bat-borne viruses. The most definitive proof was included from the separation of the CoV from bats in China, there was over 98% similarity in the genome sequence to SARS-CoV, and can use SARS-CoV-receptor ACE2 on cells of the human race. It is hard to evaluate the possibility for spillover of several similar SARS-CoV Bat CoVs as a result of infringing isolation of viruses, but it should be noted that a "consensus" virus developed through reverse genetics has high evidence of human infection it is clear that bats are the most likely original cause of the current 2019 CoV outbreak in Wuhan, China, which started in December 2019, continuing to spread to many city and province areas in China from a "wet market." The probability of food transmission of derived animal products was also suggested, as it has recently been pointed out to affect the present epidemic as well as the chance of common near contact with animals (a not unusual scenario in these types of markets). Their possible adaptations may lead to new and stable reservoirs, such as human hosts. Those are ideas and problems arisen from the emerged SARS-CoV2, that immediately compares SARS-CoV and MERS-CoV with other beta-coronaviruses with similar natural, intermediate animal hosts with also the possibility of human-to-human transmission in comparison [13, 16].

3. Genomic characteristics

During infection, the genome has many roles. It first functions as mRNA that is translated into a huge polyprotein called replicase that involves a ribosomal frame-shifting event for complete synthesis. The replicase is the only genome-derived translation product; all downstream ORFs are expressed by Subgenomic RNAs. Next, the genome is the replication and transcription template. Finally, the genome is involved in assembly, as progeny genomes are found in progeny viruses [4]. The genomic RNA for coronavirus of about 30 000 nucleotides encodes structural virus proteins, non-structural proteins with a key part in viral RNA synthesis (which is understood to be replicase transcriptase proteins) and non-structural proteins that are not necessary for viral replication in cell culture but which in vivo tend to be a selective advantage (which is referred to in vivo) [8]. Cis-acting sequence and structural elements involved in the replication, transcription, translation, and packaging are incorporated within RNA virus genomes. Some of these signals are intended to enable the interaction of selective viral RNAs with RNA synthesis machines while some allow or modify events that happen meanwhile the synthesis or assembly of viral protein [17, 18]. Coronaviruses contain the hugest genomes of any RNA virus, and this has hindered the production of full-length coronavirus cDNAs along with the discovery that certain cDNAs originating in the replicase areas of genes are unstable in bacteria. However, the assembly of long-lasting cDNAs in porcine coronavirus transmissible gastroenteritis viral genomic RNA (TGEV) has reportedly been identified with two methods. First, a TGEV full-length cDNA was installed on a bacterial artificial chromosome (BAC). Second, the TGEV total cDNA was installed in-vitro using a series of adjacent cDNAs within it engineered unique restriction sites, cDNA of the RNA transcripts derived from bacteriophage T7-RNA polymerase have been then used for infectious virus production [19]. SARS-CoV-2 (**Figure 2**) has a long genome with ORF1ab polyprotein, along with four main structural proteins, involving Spike surface glycoprotein, small envelope protein, matrix protein and nucleocapsid protein, which is also the case in other beta-coronaviruses (**Figure 3**; **Table 1**). In the ORF1ab polyprotein there were two deletions (three nucleotides and 24 nucleotides) and also one at the 3′ end of the genome (ten nucleotides) [21] (**Figure 4**).

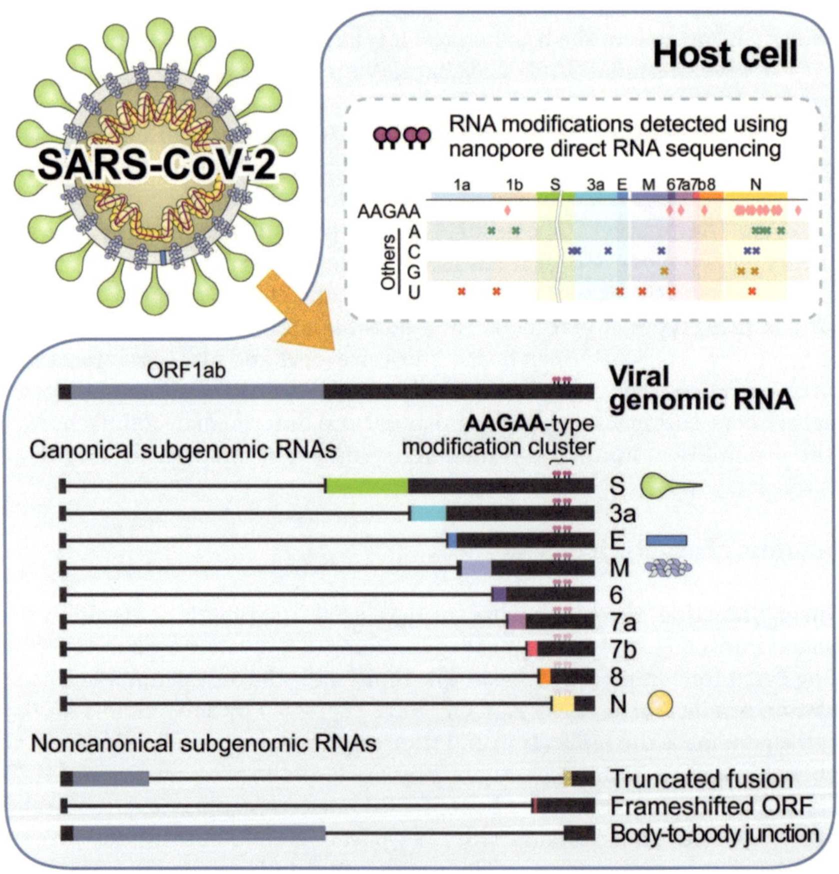

Figure 2.
The structure of SARS-CoV-2 transcriptome [20].

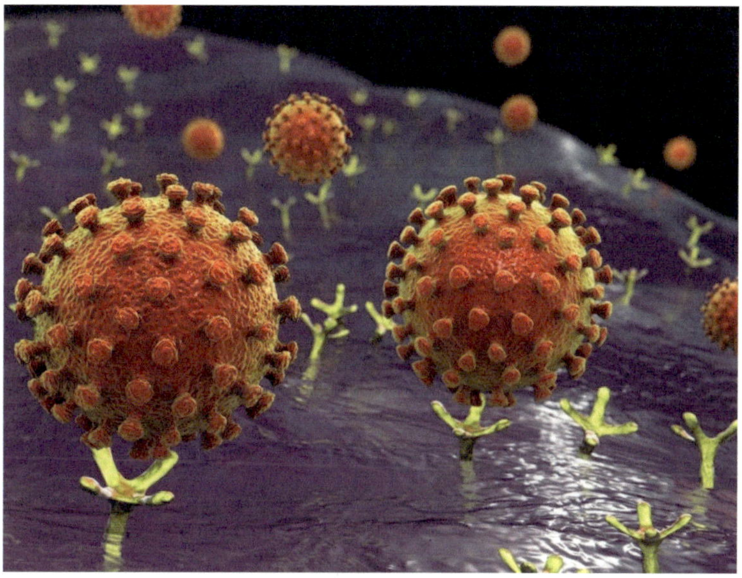

Figure 3.
SARS-CoV-2 binding by its spike protein to ACE2 receptor [12].

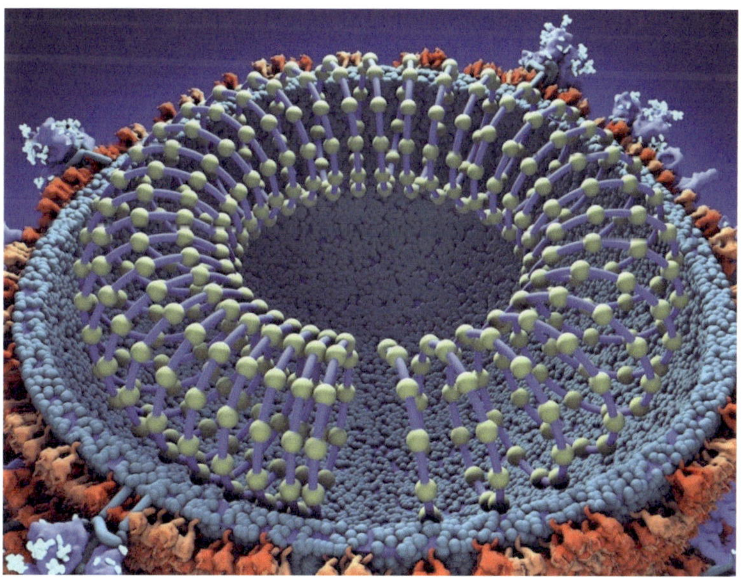

Figure 4.
SARS-CoV-2 inner proteins illustration [12].

3.1 Transfection

The unusual variations in host diversity and tissue tropism between coronaviruses are primarily due to differences in the spike glycoprotein. The S protein is a broad, glycoprotein type I membrane containing disruptive functional fields near the amino (S1) and carboxy (S2) Termini. Via their receptor specificity and probably by their membrane fusion activities in the cell entry of viral tropism, these spikes can be identified [1]. ACE2 is a primary determinant for the SARS-CoV Host range [23, 24]. The life cycle of COVID-19 starts with the binding of its Angiotensin Converting Enzyme (ACE2) receptor expressed in various

Group 1	
Human coronavirus 229E	HCoV-229E
Porcine enteric (transmisible gastroenteritis virus, TGEV; and porcine epidemic diarrhea virus, PEDV) and respiratory (PRCoV) coronavirus	PCoV
Canine coronavirus	CCoV
Feline coronavirus, including feline infectious peritonitis virus (FIPV)	FCoV
Group 2	
Human coronavirus OC43	HCoV-OC43
Bovine coronavirus	BCoV
Turkey coronavirus BCoV related	TCoV-B
Murine coronaviruses including mouse hepatitis virus (MHV)	MCoV
Porcine hemagglutinating encephalomyelitis virus	HEV
Rat coronavirus including sialodacryoadenitis virus (SDAV)	RtCoV
Group 3	
Avian coronavirus including infectious bronchitis virus (IBV)	ACoV
Turkey coronavirus IBV related	TCoV-I
Unclassified coronavirus	
Rabbit coronavirus	RbCoV

Table 1.
Some of coronaviridae family members [22].

cell types in the body and other susceptible cells throughout the body. (ACE2), the membrane-associated enzyme Carboxypeptidase, is a crucial regulator for cardiac function. Now, recognized and characterized with a sudden second role for ACE2 in mediating viral entry and cell fusion in the form of SARS-CoV spike glycoprotein partner. The coronaviridae family includes this zoonotic virus. The virus has a healthy ssRNA genome and little structural and non-structural protein. Different points of view have been identified with great similarity to SARS-CoV. The Approach of the virus is through S1 protein, which then integrates to the virus membrane with endosomal membranes, possibly by S2 mediation. Then the viral genome is released into the cytoplasm of the cell [25–30]. S-protein has two sub-units with one sub-unit directly binding to the receptor enabling the entrance of the virus into cells. The S-protein RNA binding domain in COVID-19 has a more advanced SARS-CoV homology. Although some of the residues essential to binding are not alike, the structural conformation was not changed in general by the non-identical residues [31]. CoV spike (S) is a key goal for vaccines, antibodies and diagnosis. A 3.5 angle-resolution cryo-electron microscopy structure for the SARS-CoV-2 S was developed cutting conformation in order to promote medical response. The prominent trimer 's state possesses rotation in a receptor-accessible conformation in one of the three receptor binding domains (RBDs). Biophysical and structural verification is also given that the SARS-CoV-2 S protein has more affinity than severe acute respiratory (SARS)-CoV S-binding enzyme 2, (ACE2) [32] (**Figure 5**).

3.2 Replication

Untranslated regions of RNA (UTRs) have 5′ and 3′ viruses that carry RNA-specific signals. The 5′ capped coronavirus genome compromise a 3' UTR

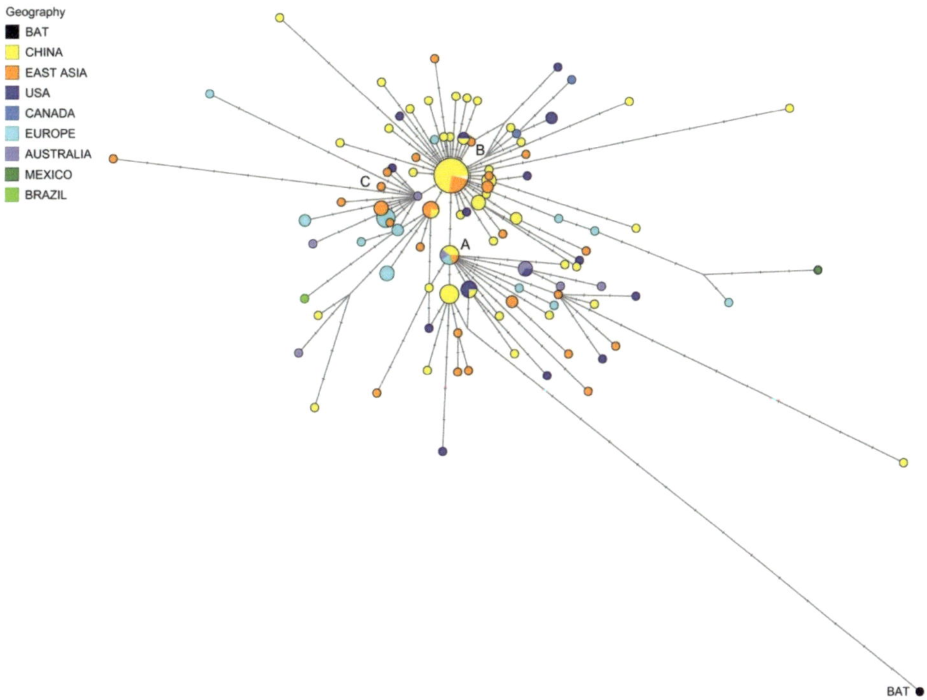

Figure 5.
Phylogenetic tree of 160 SARS-CoV-2 genomes [33].

consisting of 300 to 500 nucleotides) in addition to a poly(A) tail. Host-related factors involving two class-II viruses bovine coronavirus (BCV) and Mouse Hepatitis Coronavirus (MHV) were studied, in order to better understand coronavirus replication. Using gel mobility shift assays unique host protein interactions were identified with BCV 3' UTR [287 nt plus poly(A) tail]. The MHV 3′ -UTR [301 nt in addition to poly(A) tail] rivalry indicates that interactivity for the two viruses are preserved. UV cross-linking studies observed proteins with molecular masses of 99, 95 and 73 kDa. The ranges 40- to 50 and 30 kDa even contained less heavily labeled proteins. For binding the 73-kDa protein a poly(A) tail was needed. The 73 kDa proteins have been identified as cytoplasmic poly(A)-binding protein (PABP) by an Immuno-precipitation of UV-cross-linked proteins. To define the significance of the poly(A) tail, the replication of the impaired genomes BCV Drep and MHV MIDI-C was used alongside with several mutants. After transfection to the supporting virus-infected cells, the defect genomes with shortened, 5- or 10-A poly(A) tails have been replicated. BCV Drep RNA lacking a poly(A) tail did not replicate while MHV MIDI-C RNA replication was detected with a deleted tail after multiple mutations of the virus. The kinetics of replication is delayed in both mutants. Noticeable extension or addition of the poly(A) tail in mutants in the replication assay associated with the presence of these RNAs. RNAs exhibit less in vitro PABP binding in shorted Poly(A) tails, indicating decreased RNA replication interactions with protein. The data show strongly that the poly(A) tail is a significant indication for the replication of coronavirus [34]. The virus initiates replication and assembly of protein that is followed by the release of new infectious particles into novel target cells. These events are followed by proinflammation chemokines and cytokines producing and triggering which lead to significant pulmonary damage-causing atypical pneumonia with quick abnormalities and failure [35, 36].

3.3 Transcription

For differing coronaviruses, the number of mRNAs varies. The number of mRNAs in the coronavirus species is the number of functional genes. A few hours after infection with the virus in most viral cell systems, coronavirus mRNA synthesis can be identified and proceeded until cells have become invasive. Subgenomic mRNAs are found to be heterogeneous (M. [37]). A two-component support based on expression system was developed and individual genomes were created by selective recombination or by using infectious cDNA clones. Transcription sequences have mainly been characterized by helper-dependent expression systems and can now be validated via single genomes. The coronavirus genome was created through modification of infectious cDNA, resulting in efficient expression of the foreign gene (20 g ml – 1) and stable (20 passages) [22, 38]. In a Study, The creation of the full-length infectious cDNA clone and a functional duplicate of the Urbani strain as a bacterial artificial chromosome (BAC) of the extreme acute respiratory syndrome (SARS-CoV). Through this method, the viral RNA was expressed in the cytomegalovirus promoter's cell nucleus and further multiplied by viral replicas in the cytoplasm. The *Escherichia coli* infectious clone and duplicate have been completely stable. The use of the SARS-CoV replica has been shown to be important in efficient coronavirus-RNA synthesis for the recent identification of RNA-processes enzyme exoribonuclease, endo-ribonucleases and 2-line-O-ribose that are found to be essential [39]. The RNA-dependent RNA synthesis is used for coronaviral transcription. The result is that a nested range of 6 to 8 mRNAs of different sizes is produced, depending on the strain of the coronavirus. The mRNAs are five prime and three prime genome -co- terminals. The most significant mRNA is the genomic RNA (gRNA) for both rep1a and rep1b genes. A discontinuous transcription process fuses a lead sequence of 93 nucleotides) (originating from the 5 prime at the end of a genome to 5 prime of the mRNA coding sequence (body) [40]. The RNA virus genomes are comprised of a series of cis-acting and structural elements involved in viral replication. A bulky secondary loop structure was previously established at the upstream end of the 3-way untranslated region (3 tablets of UTR) of the Mouse Hepatitis Virus (MHV) coronavirus genome. This element has proved to be important for viral replication, beginning immediately downstream of the nucleocapsid gene stop codon. A 3 UTR pseudoknot of the corresponding downstream closely related to the bovine coronavirus BCoV. It is an essential pseudoknot for replication and has a preserved counterpart for each coronavirus in groups 1 and 2 [17]. More than one ORF is comprised of 5 'unique regions within multiple mRNA s. For example, mRNA 5 of MHV, which has two ORFs in the coding region which can encode two p 1 3 and pl0 proteins, respectively. A negative-stranded RNA template that is represented in an only very small percentage (1–2%) of the intracellular virus-specific RNAs is clearly mediated for Coronavirus RNA synthesis. This negative strand was synthesized by the virus-encoded RNA from the inbound virion. This is likely because the positive-sequenced RNA exceeds the negative-stranded RNA for several rounds of mRNA synthesis. Thus, negative-stranded RNA has more stability. This stability is attributed to the presence in the coronavirus-infected cells of all of the negatively-stranded RNA as a double-stranded RNA [41–43]. The transcriptome Structure was unknown despite the SARS-CoV-2 genome being recorded recently. a high-resolution map was presented of the SARS-CoV-2 transcriptome and epitranscriptome using two complementary sequencing techniques. DNA nanoball sequencing reveals that due to discontinuous transcription occurrences the transcriptome is highly complexSARS-CoV-2 yields transcripts that code unknown ORFs with fusion, deletion and/or frameshift in addition to the canonical genomic and 9 subgenomic RNAs. 41 sites for RNA modification on viral transcripts were also found with the most common motif being AAGAA with nanopore direct RNA sequencing [20].

3.4 Morphogenesis

Expression studies showed that coronavirus envelope protein E and the more present membrane glycoprotein M were required and adequate to assemble virus particles into cells. Clustered charged-to-alanine Mutagenesis of the gene E was carried, which integrated mutations in mouse hepatitis virus E (MHV) E protein, as a step forward in our understanding of the role of the mouse hepatitis virus E (MHV) E protein. One was apparently lethal and one was a wild-type phenotype of four probable clustered charged-to-alanine E gene mutants. The other two mutants were partly affected by temperature, developing tiny plaques at a nonpermissive temperature. Reverting analyses of these two mutants showed that each mutation was the reason for the temperature-sensitive phenotype and promoted probable interactions among E protein monomers. In permissive temperature, both temperature-sensitive mutants have been substantially thermolabile, indicating that their assembly fails. In the case of the electron microscopy, virions of one of the mutants were discovered to have remarkably aberrant morphology when compared with the wild type: most mutant virions had pinched and extended forms that were seen seldom in the wild [44–46]. Specific recombination of RNA was utilized to create mutants containing chimeric nucleocapsid (N) protein genes in mouse hepatitis virus (MHV) that replace bovine coronavirus N gene segments in place of the correct MHV sequences. This described portions of the two N proteins which were functionally equivalent, given evolutionary divergences. These regions included mostly the RNA binding domain centrally located and two putative spacers connecting the three N protein domains. On the other hand, a bovine coronavirus cannot be transferred from the amino terminus N, the acidic carboxy-terminal region and the central domain serine and arginine- rich section, probably because these parts of a molecule are engaged in protein–protein interactions that are unique to each virus (or possibly each host). The results show that the recombination of the coronavirus genome can be used to produce extensive substitutions and recombinants that cannot otherwise be produced between two viruses separated by species barrier [47].

4. Mutations

RNA viruses must establish an equilibrium between the adaptability to new environmental circumstances or the necessity to preserve the intact and replicative genome to ensure survival and propagation for the host cells. Various virus families with the biggest and most complex replicating RNA genomes identified, up to 32 kb of positive RNA, such as coronaviruses, can achieve these objectives. CoVs, including (MHV) and SARS-CoV, express 3 to 5′ of exoribonuclease (ExoN) activity in nsp14. The exoN genetic inactivation of alanine replacement with retained active DE-D Residues in Engineered SARS-CoV and MHV Genomes leads to viable mutants, which display 15 to 20 times higher mutation rates and up to 18 times higher than those endured for other RNA fidelity mutants. Nsp14-ExoN, therefore, is important for the fidelity of the replication and possibly acts as a direct mediator or regulator for a more complex RNA proof-reader, an exceptional process in RNA virus biology. The removal of nsp14-mediated proofreading mechanisms will have significant consequences for our interpretation of RNA virus evolution and will also provide a robust model to research the correlation between fidelity,

Genomic region	No. nt mutations	Missense mutation	SARV-CoV-2 strain
5′ UTR	8	N/A	
ORF1ab polyprotein	48	29	
		A (117) → T	USA/CA3/2020/EPI_ISL_408008 USA/CA4/2020/EPI_ISL_408009
		P (309) → S	France/IDF0515/2020/EPI ISL_408430
		S (428) → N	USA/CA1/2020/EPI_ISL_406034
		T (609) → I	USA/CA5/2020/EPI_ISL_408010
		A (1176) → V	Japan/TY-WK-012/2020/EPI_ISL_408665
		L (1599) → F	Korea/KCDC03/2020/EPI_ISL_407193
		I (1607) → V	USA/CA3/2020/EPI_ISL_408008 USA/CA4/2020/EPI_ISL_408009
		M (2194) → T	Shenzhen/SZTH-004/2020/EPI_ISL_406595
		L (2235) → I	Wuhan/WH01/2019/EPI_ISL_406798
		I (2244) → T	Wuhan/IPBCAMS-WH-03/2019/ EPI_ISL_403930
		G (2251) → S	Wuhan/WIV05/2019/EPI_ISL_402128
		A (2345) → V	Shandong/IVDC-SD-001/2020/EPI_ISL_408482
		G (2534) → V	Wuhan/IPBCAMS-WH-05/2020/ EPI_ISL_403928
		D (2579) → A	Wuhan/WIV07/2019/EPI_ISL_402130
		N (2708) → S	Wuhan/IPBCAMS-WH-01/2019/ EPI_ISL_402123
		F (2908) → I	Wuhan/IPBCAMS-WH-01/2019/ EPI_ISL_402123
		T (3058) → I	France/IDF0515/2020/EPI_ISL_408430
		S (3099) → L	Shenzhen/HKU-SZ-005/2020/EPI_ISL_405839
		L (3606) → F	Yunnan/IVDC-YN-003/2020/EPI_ISL_408480 Shandong/IVDC-SD-001/2020/EPI_ISL_408482 Chongqing/IVDC-CQ-001/2020/ EPI_ISL_408481 Singapore/3/2020/EPI_ISL_407988 France/IDF0515/2020/EPI_ISL_408430 USA/AZ1/2020/EPI_ISL_406223
		E (3764) → D	Japan/KY-V-029/2020/EPI_ISL_408669
		N (3833) → K	Wuhan/WH01/2019/EPI_ISL_406798
		W (5308) → C	Taiwan/2/2020/EPI_ISL_406031
		T (5579) → I	USA/CA2/2020/EPI_ISL_406036
		I (6075) → T	England/02/2020/EPI_ISL_407073 England/01/2020/EPI_ISL_407071
		P (6083) → L	Japan/AI/I-004/2020/EPI_ISL_407084
		F (6309) → Y	Sichuan/IVDC-SC-001/2020/EPI_ISL_408484
		E (6565) → D	Shenzhen/SZTH-004/2020/EPI_ISL_406595
		K (6958) → R	Wuhan/WIV05/2019/EPI_ISL_402128

Genomic region	No. nt mutations	Missense mutation	SARV-CoV-2 strain
		D (7018) → N	Wuhan/WIV02/2019/EPI_ISL_402127
Spike polyprotein	14	8	
		F (32) → I	Wuhan/HBCDC-HB-01/2019/EPI_ISL_402132
		H (49) → Y	Guangdong/20SF174/2020/EPI_ISL_406531 Guangdong/20SF040/2020/EPI_ISL_403937 Guangdong/20SF028/2020/EPI_ISL_403936
		S (247) → R	Australia/VIC01/2020/EPI_ISL_406844
		N (354) → D	Shenzhen/SZTH-004/2020/EPI_ISL_406595
		D (364) → Y	Shenzhen/SZTH-004/2020/EPI_ISL_406595
		V (367) → F	France/IDF0372/2020/EPI_ISL_406596 France/IDF0373/2020/EPI_ISL_406597
		D (614) → G	Germany/BavPat1/2020/EPI_ISL_406862
		P (1143) → L	Australia/QLD02/2020/EPI_ISL_407896
Intergenic region	5	N/A	
Envelope protein	0	0	
Matrix protein	2	1	
		D (209) → H	Singapore/2/2020/EPI_ISL_407987
Intergenic region	6	N/A	
Nucleocapsid protein	/	4	
		T (148) → I	Shenzhen/SZTH-004/2020/EPI_ISL_406595
		S (194) → L	Shenzhen/SZTH-003/2020/EPI_ISL_406594 Foshan/20SF207/2020/EPI_ISL_406534 USA/CA3/2020/EPI_ISL_408008 USA/CA4/2020/EPI_ISL_408009
		S (202) → N	Australia/QLD02/2020/EPI_ISL_407896
		P (344) → S	Guangzhou/20SF206/2020/EPI_ISL_406533
3'UTR	3	N/A	
Complete genome	93	42	

Table 2.
Mutations of SARS-CoV-2 strains found throughout the whole genome. The number in the parentheses shows where amino acid is found in its protein [21].

diversity and pathogenesis [48–52]. COVID-19 is very related to SARS-CoV Middle East Respiratory Syndrome (MERS). Yet another human attack by coronaviruses. A research attempted to explore potential changes/developments in the 'spike protein' element that enables the virus to bind to cell receptor(s) and in the silicon design and discovery of B epitopes in which antibody synthesis is used to neutralize and block this connection. The findings show that this protein varies constantly between

Accession	Location-date	Nucleotide variation	Gene	Amino acid change	Mutation type
MT240479	04-03-2020/Pakistan Gilgit	1 1497G > A	Orf1ab		Synonymous mutation
MN996527	30/Dec/2019-China Wuhan	21316G > A	Orf1ab	D7018N	Missense
MN996527	30/Dec/2019-China Wuhan	24292A > G	S		Synonymous mutation
LC528232	10/Feb/2020-Japan	11083 T > G	Orf1ab	L3606F	Missense
LC528232	10/Feb/2020-Japan	29642C > T	ORF10		Synonymous mutation
LR757995	05/Jan/2020-China Wuhan	28144 T > C	ORF8	L84S	Missense
LR757998	12/26/2019-China Wuhan	6968C > A	Orf1ab	L2235I	Missense
LR757998	12/26/2019-China Wuhan	11749 T > A	Orf1ab		Synonymous mutation
MN938384	1/10/2020-China Shenzhen	8782C > T	Orf1ab		Synonymous mutation
MN938384	1/10/2020-China Shenzhen	28144 T > C	ORF8	L84S	Missense
MN938384	1/10/2020-China Shenzhen	29095C > T	N		Synonymous mutation
MN975262	11/Jan/2020-China	8782C > T	Orf1ab		Synonymous mutation
MN975262	11/Jan/2020-China	9534C > T	Orf1ab	T3090I	Missense
MN975262	11/Jan/2020-China	29095C > T	N		Synonymous mutation
MN975262	11/Jan/2020-China	28144 T > C	ORF8	L84S	Missense
MN975262	11/Jan/2020-China	8782C > T	Orf1ab		Synonymous mutation
MN985325	19/Jan/2020-USA WA	28144 T > C	ORF8	L84S	Missense
MN994467	23/Jan/2020-USA CA	1548G > A	Orf1ab	S428N	Missense
MN994467	23/Jan/2020-USA CA	8782C > T	Orf1ab		Synonymous mutation
MN994467	23/Jan/2020-USA CA	26729 T > C	M		Synonymous mutation
MN994467	23/Jan/2020-USA CA	28077G > C	ORF8	V62L	Missense
MN994467	23/Jan/2020-USA CA	28144 T > C	ORF8	L84S	Missense
MN994467	23/Jan/2020-USA CA	28792A > C	N		Synonymous mutation
MN994467	23/Jan/2020-USA CA	1912C > T	Orf1ab		Synonymous mutation

Accession	Location-date	Nucleotide variation	Gene	Amino acid change	Mutation type
GWHABKF00000001	23/Dec/2019-China Wuhan	3778A > G	Orf1ab		Synonymous mutation
GWHABKF00000001	23/Dec/2019-China Wuhan	8388A > G	Orf1ab	N2708S	Missense
GWHABKF00000001	23/Dec/2019-China Wuhan	8987 T > A	Orf1ab	F2908I	Missense
GWHABKK00000001	30/Dec/2019-China Wuhan	24325A > G	S		Synonymous mutation
GWHABKK00000001	30/Dec/2019-China Wuhan	21316G > A	Orf1ab	D7018N	Missense
GWHABKH00000001	30/Dec/2019-China Wuhan	6996 T > C	Orf1ab	I2244T	Missense
GWHABKJ00000001	01/Jan/2019-China Wuhan	7866G > T	Orf1ab	G2534V	Missense
GWHABKM00000001	30/Dec/2019-China Wuhan	21137A > G	Orf1ab	K6958R	Missense
GWHABKM00000001	30/Dec/2019-China Wuhan	7016G > A	Orf1ab	G2251S	Missense
GWHABKO00000001	30/Dec/2019-China Wuhan	8001A > C	Orf1ab	D2579A	Missense
GWHABKO00000001	30/Dec/2019-China Wuhan	9534C > T	Orf1ab	T3090I	Missense
MT188341	05/Mar/2020-USA MN	6035A > G	Orf1ab		Synonymous mutation
MT188341	05/Mar/2020-USA MN	8782C > T	Orf1ab		Synonymous mutation
MT188341	05/Mar/2020-USA MN	16467A > G	Orf1ab		Synonymous mutation
MT188341	05/Mar/2020-USA MN	18060C > T	Orf1ab		Synonymous mutation
MT188341	05/Mar/2020-USA MN	21386insT	Orf1ab		Insertion
MT188341	05/Mar/2020-USA MN	21388-21390insTT	Orf1ab		Insertion
MT188341	05/Mar/2020-USA MN	23185C > T	S		Synonymous mutation
MT188341	05/Mar/2020-USA MN	28144 T > C	ORF8	L84S	Missense
MT188339	09/Mar/2020-USA MN	8782C > T	Orf1ab		Synonymous mutation
MT188339	09/Mar/2020-USA MN	17423A > G	Orf1ab	Y5720C	Missense
MT188339	09/Mar/2020-USA MN	18060C > T	Orf1ab		Synonymous mutation
MT188339	09/Mar/2020-USA MN	21386C > T	Orf1ab		Synonymous mutation

Accession	Location-date	Nucleotide variation	Gene	Amino acid change	Mutation type
MT188339	09/Mar/2020-USA MN	22432C > T	S		Synonymous mutation
MT188339	09/Mar/2020-USA MN	28144 T > C	ORF8	L84S	Missense
MT121215	02/Feb/2020-China Shanghai	6031C > T	Orf1ab		Synonymous mutation
MT123290	05/Feb/2020-China Guangzhou	15597 T > C	Orf1ab		Synonymous mutation
MT123290	05/Feb/2020-China Guangzhou	29095C > T	N		Synonymous mutation
MT126808	2/28/2020-Brazil	26144G > T	ORF3a	G251V	Missense
MT066175	31/Jan/2020-Taiwan	8782C > T	Orf1ab		Synonymous mutation
MT066175	31/Jan/2020-Taiwan	28144 T > C	ORF8	L84S	Missense
MT093571	07/Feb/2020-Sweden	13225C > G	Orf1ab		Synonymous mutation
MT093571	07/Feb/2020-Sweden	13226 T > C	Orf1ab		Synonymous mutation
MT093571	07/Feb/2020-Sweden	17423A > G	Orf1ab	Y5720C	Missense
MT093571	07/Feb/2020-Sweden	23952 T > G	S		Synonymous mutation
MT066156	30/Jan/2020-Italy	11083 T > G	Orf1ab	L3606F	Missense
MT066156	30/Jan/2020-Italy	26144G > T	ORF3a	G251V	Missense
LC522975	20/JAN/2020-JAPAN	8782C > T	Orf1ab		Synonymous mutation
LC522975	20/JAN/2020-JAPAN	29095C > T	N		Synonymous mutation
LC522975	20/JAN/2020-JAPAN	28144 T > C	ORF8	L84S	Missense
LC522975	20/JAN/2020-JAPAN	2662C > T	ORF1ab		Synonymous mutation
LC522974	20/JAN/2020-JAPAN	8782C > T	ORF1ab		Synonymous mutation
LC522974	20/JAN/2020-JAPAN	29095C > T	N		Synonymous mutation
LC522974	20/JAN/2020-JAPAN	28144 T > C	ORF8	L84S	Missense
LC522974	20/JAN/2020-JAPAN	2662C > T	ORF1ab		Synonymous mutation
LC522973	20/JAN/2020-JAPAN	8782C > T	ORF1ab		Synonymous mutation
LC522973	20/JAN/2020-JAPAN	29095C > T	N		Synonymous mutation
LC522973	20/JAN/2020-JAPAN	3792C > T	ORF1ab	A1176V	Missense
LC522973	20/JAN/2020-JAPAN	29095C > T	N		Synonymous mutation
LC522973	20/JAN/2020-JAPAN	2662C > T	ORF1ab		Synonymous mutation
LC522973	20/JAN/2020-JAPAN	28144 T > C	ORF8	L84S	Missense

Accession	Location-date	Nucleotide variation	Gene	Amino acid change	Mutation type
LC522972	20/JAN/2020-JAPAN	29303C > T	N	P344S	Missense
LC522972	20/JAN/2020-JAPAN	25810C > G	ORF3a	L140V	Missense
LC522972	20/JAN/2020-JAPAN	11557G > T	ORF1ab	E3764D	Missense
LC522972	20/JAN/2020-JAPAN	15324C > T	ORF1ab		Synonymous mutation
LC521925	21/JAN/2020-JAPAN	1912C > T	ORF1ab		Synonymous mutation
LC521925	21/JAN/2020-JAPAN	18512C > T	ORF1ab	P6083L	Missense
LC521925	21/JAN/2020-JAPAN	359_382del	ORF1ab	G32_ L39del	Deletion
MN988713	21/JAN/2020-USA Chicago	24034C > T	S		Synonymous mutation
MN988713	21/JAN/2020-USA Chicago	26729 T > C	M		Synonymous mutation
MN988713	21/JAN/2020-USA Chicago	8782C > T	ORF1ab		Synonymous mutation
MN988713	21/JAN/2020-USA Chicago	490 T > A	ORF1ab	D75E	Missense
MN988713	21/JAN/2020-USA Chicago	3177C > T	ORF1ab	P971L	Missense
MN988713	21/JAN/2020-USA Chicago	28854C > T	N	S194L	Missense
MN988713	21/JAN/2020-USA Chicago	28077G > C	ORF8	V62L	Missense
MN988713	21/JAN/2020-USA Chicago	28144 T > C	ORF8	L84S	Missense
MN997409	21/JAN/2020-USA Arizona	8782C > T	ORF1ab		Synonymous mutation
MN997409	21/JAN/2020-USA Arizona	29095C > T	N		Synonymous mutation
MN997409	21/JAN/2020-USA Arizona	11083G > T	ORF1ab	L3606F	Missense
MN997409	21/JAN/2020-USA Arizona	28144 T > C	ORF8	L84S	Missense
MT072688	26/JAN/2020-USA: Massachussetts	24034C > T	S		Synonymous mutation
NMDC60013002–09	01/JAN/2019-China Wuhan	27493C > T	ORF7a	P34S	Missense
NMDC60013002–09	01/JAN/2019-China Wuhan	28253C > T	ORF8		Synonymous mutation
NMDC60013002–10	30/Dec/2019-China Wuhan	20679G > A	ORF1ab		Synonymous mutation
NMDC60013002–01	30/Dec/2019-China Wuhan	11764 T > A	ORF1ab	N3833K	Missense
NMDC60013002–06	30/Dec/2019-China Wuhan	24325A > G	S		Synonymous mutation
NMDC60013002–04	05/Dec/2019-China Wuhan	28144 T > C	ORF8	L84S	Missense

Table 3.
Coding mutation list detected in SARS-CoV-2 genomes [57].

Accession	Location-date	Nucleotide variation	UTR type
MT240479	04-03-2020/Pakistan Gilgit	241C > T	5 UTR
MT123290	05/Feb/2020-China Guangzhou	4A > T	5 UTR
MT007544	25/Jan/2020-Australia Victoria	29749-29759del	3 UTR
NMDC60013002–07	07/JAN/2019-China Wuhan	29869del	3 UTR
NMDC60013002–04	05/Dec/2019-China Wuhan	29856 T > A	3 UTR
NMDC60013002–04	05/Dec/2019-China Wuhan	29854C > T	3 UTR
NMDC60013002–04	05/Dec/2019-China Wuhan	16C > T	5 UTR
MT049951	17/Jan/2019-China Yunnan	75C > A	5 UTR
LC522975	20/JAN/2020-JAPAN	29705G > T	3 UTR
GWHABKG00000001	30/Dec/2019-China Wuhan	124G > A	5 UTR
GWHABKG00000001	30/Dec/2019-China Wuhan	120 T > C	5 UTR
GWHABKG00000001	30/Dec/2019-China Wuhan	119C > G	5 UTR
GWHABKG00000001	30/Dec/2019-China Wuhan	112 T > G	5 UTR
GWHABKG00000001	30/Dec/2019-China Wuhan	111 T > C	5 UTR
GWHABKG00000001	30/Dec/2019-China Wuhan	104 T > A	5 UTR

Table 4.
Non-coding mutation list detected in SARS-CoV-2 genomes [57].

the sequences of proteins obtained worldwide. Some B epitopes (part of an antigen molecule to which an antibody attaches itself), 177-MDLEGKQGNFKNL-189-555-SNKKFLPF-562-656 -VNSYECDIPI-666, 1035- GQSKRVDFC-1043, from the Cons sequence constructed from global protein sequences released between 11 Feb and 06 April, have been found to meet most of the criteria required for real wet application [53]. SARS-CoV is well suited to cultural development and does not seem to be selected in humans. It was also assessed that, in late October 2002, the alleged root of the SARS outbreak was consistent with a previous report of case use from China. The higher structural and antigenic sequence divergence and significant deletions within 3 '– of much of the viral genome indicate that some selection pressures conflict along with the functional structure of these confirmed and suspected ORFs [54]. In three regions the SARS and SARSr of bats-CoVs are largely different: S, ORF8 and ORF3. SARSr-CoVs bats share high sequence with the SARS- COV in the S2 but are highly different in the S1 region. However, bat MERSr-CoVs bats and human and camel MERS-CoVs share similar genomics but are significantly different from their genomic sequences [7]. Comparison of COVID-19, SARS-CoV and MERS-CoV genome sequence showed that COVID-19 has better sequence similarity with SARS-CoV compared to MERS CoV. Nevertheless, the COVID-19 amino acid sequence differed from the other coronavirus in specific areas of 1ab polyprotein and surface glycoprotein or S-protein [31]. Considering the high rate of mutation that characterizes RNA viruses, it is clear that several more mutations will emerge in the viral genome to monitor the spread of SARS-CoV-2 knowing that also their mutations rate are lower than other RNA viruses due to their proofreading activity described above [55, 56] (**Tables 2–4**).

5. Evolution and origin

Most SARS-CoV strains are derived from bats. SARS-CoV bat is a probable progenitor for SARS – CoV that is contagious to humans and civets, and thus it is important to study ACE2 receptor for monitoring origins of SARS-CoV and avoiding and controlling the outbreak. Though palm civets were involved in SARS emergence, most early MERS index cases had contact with dromedary camels. Indeed, the MERS-CoV strains separated from camels were nearly matching to those from humans [7]. The virus shares 96% of its genetic material with a virus detected from a bat found in a cave in Yunnan in China. A persuasive argument that it comes from bats but there is a critical alteration. The coronaviral spike proteins have a unit called a receptor-binding domain that is essential to the successful entry of human cells. Especially powerful is the SARS-CoV-2 binding domain and it varies from the bat virus Yunnan which appears to not affect human. Another Complicating matter, a scaly anteater called the pangolin with a coronavirus which was almost similar to the human version with a receptor-binding domain. However, the majority of the coronavirus was genetically identical just 90%, and some researchers do not believe that pangolin was the intermediary. It is difficult to draw a family tree since both mutations and recombinations are involved [58–60]. An article identifies and uses a machine learning-based alignment-free approach to identify a COVID-19 intrinsic genomic signature for an ultra-fast, scalable, and extremely precise classification of all COVID-19 virus genomes. The technique presented incorporates supervised machine learning with MLDSP for genome analysis, improved by a machine learning component decision tree approach and a Spearman-leading correlation coefficient analysis of tests. These methods are used to examine a broad collection of more than 61.8 million bp, including the 29 COVID-19 virus sequences on 27 January 2020, with over 5,000 unique viral genomic sequences. The findings endorse a bat hypothesis and the COVID-19 virus is classed under Betacoronavirus as the Sarbecovirus. Without any advanced biological expertise, training or genome annotations, our method achieves a 100% precise classification of the COVID-19 virus sequences, and determines the most important relationships between more than 5000 genomes in minutes, from the beginning on, with the sole use of raw DNA sequence details [61]. In a recent research, they have developed a phylogenetic tree, including other members of coronaviridae including Bat coronavirus (BCoV) and extreme acute respiratory 2019 disease, taking advantage of all of the available genomic knowledge. The closest BCoV sequence, with a 96,2% sequence 2019 SARS-CoV2 identity, confirm that all available genomes of the sequence are of zoonotic origin. We have confirmed the high sequence similarity (> 99%) among all available genomes. Given the low 2019 SARS-CoV2 heterogeneity, at least two genomic hyper various hotspots were identified, including one of the Serine/Leucine variations in viral ORF8 Protein encoded, can be detected [62]. (**Figures 6** and 7) In the study a Malayan pangolin-isolated coronavirus showed 100%, 98.6%, 97.8% and 90.7% SARS-CoV-2 amino acid identity in genes E, M, N, and S respectively. Particularly in the S protein of

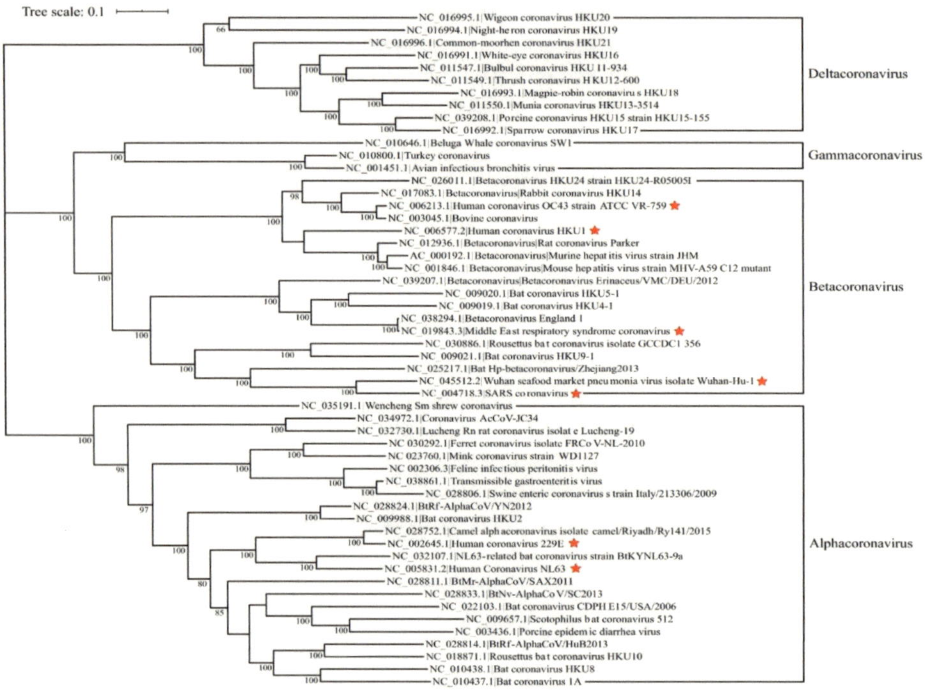

Figure 6.
A coronavirus phylogenetic tree based on full-length genome sequences. Both complete coronavirus genome sequences have been obtained from RefSeq, the NCBI reference sequence database [5].

Pangolin-CoV, the receptor-binding domain is nearly the same as the SARS-CoV-2 with vital one-amino acid alteration. Results of comparative genomic analysis indicate that SARS-CoV-2 may have been the result of a Pangolin-CoV-like virus recombination with a Bat-CoV-RaTG13 virus [63].

6. Immunopathology of SARS-CoV2

Pneumonia, lymphopenia, drained lymphocytes and a cytokine storm are distinguishable symptoms of Extreme Coronavirus Disease 2019 (COVID-19). Major antibody development is detected, but it remains to be determined if this is defensive or pathogenic. Defining the immunopathological changes in COVID-19 patients presents future drug development targets and is critical for clinical management [64]. Asymptomatic condition is found in a large but generally unexplained proportion of the infected people, analogous to many other viral diseases. Usually, a 1-week, self-limiting viral respiratory disease develops in most patients, and ends with the production of neutralizing antiviral T cell and antibody immunity [65]. SARS-CoV-2 has been shown to weaken natural immune responses, resulting in a compromised immune system and an unregulated inflammatory response in extreme and vital COVID-19 patients. These patients display lymphopenia, stimulation and malfunction of lymphocytes, defects of granulocytes and monocytes, elevated levels of cytokines, and higher amounts of immunoglobulin G (IgG) and total antibodies [66] (**Figure 8**). Extreme and fatal COVID-19 is linked with lymphopenia and an elevated amount of blood neutrophils [67]. Lymphocyte counts of 800 cells/μl and a decreased probability of recovery are reported in ICU patients suffering from COVID-19. The mechanism of action and causes of

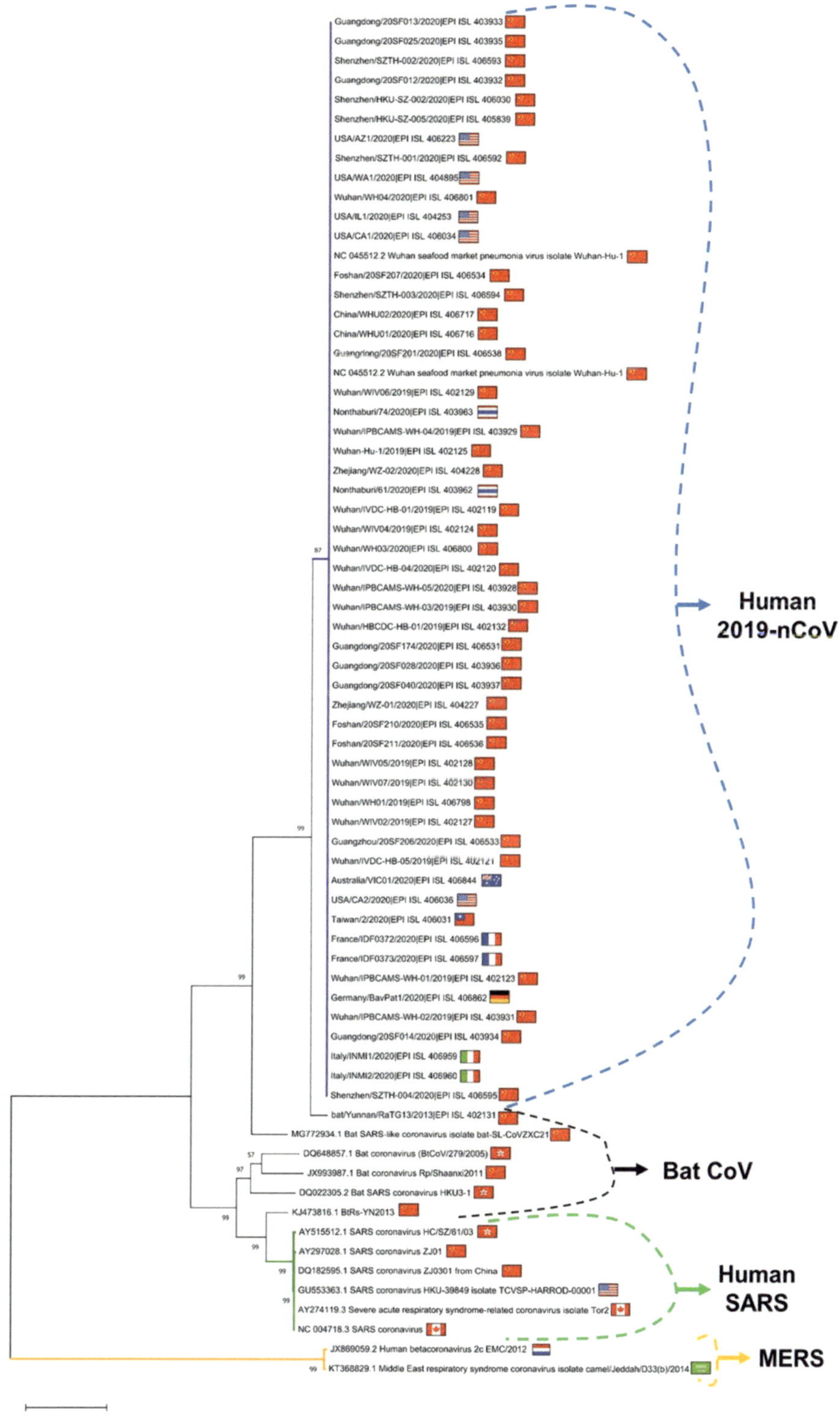

Figure 7.
A phylogenetic tree with all the sequences of SARS- CoV2 available from the 02–Feb-2020 sequence in the blue divisions, plus six Bat coronavirus sequences split in multiple taxa, six human SARS sequences (green) and two MERS sequences (orange); the bootstrap percentage of each branch is recorded [62].

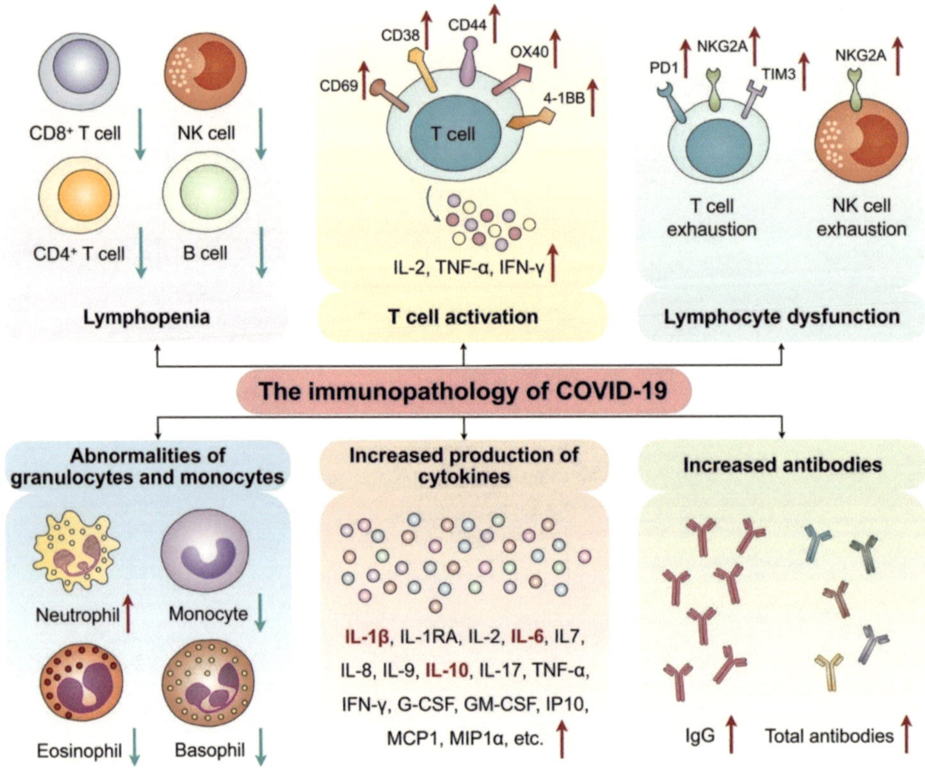

Figure 8.
COVID-19 immunopathology [66].

lymphopenia in patients with COVID-19 are unclear, but SARS-like viral particles and SARS-CoV RNA have been observed in T cells, indicating that the SARS virus may have a detrimental influence on T cells via apoptosis [68]. Accumulating data proves the involvement of T cells in COVID-19 and possibly in the immunological memory that develops after recovering from infection with SARS-CoV-2. Many, but not always, hospitalized patients tend to have both CD8 + and CD4 + T cell responses, and research points to potential T cell responses consistent with extreme disease that are suboptimal, abnormal or otherwise inadequate [5]. In a report, a group of 452 patients with positive test results of COVID-19 in Wuhan, China shows dysregulated immune system. Boosts in NOD-like receptor (NLR) and T lymphopenia, especially a decline in CD4 + T cells, were prominent in COVID-19 patients, and was even more noticeable in extreme cases, but the number of CD8 + cells and B cells did not change significantly. On the basis of these results, it was proposed that COVID-19 may affect lymphocytes, especially T lymphocytes, and that the immune system is disrupted during the infection period [69]. COVID-19 will lead to defects in the routine of peripheral blood parameters. The most noticeable anomalies that are linked to the intensity of the condition and clinical classification are the reduction in lymphocytes and the rise in the NLR ratio. The lower count and delay in eosinophil development can be indicators of weak COVID-19 outcomes. Thus, complex analysis of peripheral blood routine parameters has a significant reference point for COVID-19 progression and prognosis evaluation [70]. Also, In the different stages of COVID-19, multiple cell morphological modifications can be seen. In fact, a strong granulocytic reaction with immaturity, dysmorphism and apoptotic-degenerative morphology was apparent in peripheral blood in the initial stage of symptom aggravation, typically correlating with hospital entry [71].

Cytokine storm plays a crucial role in infected individuals for the pathogenesis of many serious manifestations of the disease. Acute respiratory distress syndrome, thromboembolic disorders such as acute ischaemic strokes caused by myocardial infarction and large vessel occlusion, encephalitis, acute kidney damage, and vasculitis (childhood Kawasaki syndrome and adult renal vasculitis) [72]. Nonetheless, it is uncertain if serious illness is triggered by immune hyperactivity or inability to overcome an inflammatory reaction owing to continuing virus replication or immune dysregulation. However, records of elevated levels of thrombi production and endothelial cell death in patients with COVID-19 suggest disruption to the vascular endothelium and the participation of cytokine elevated activity and immunothrombosis [73]. In response to infection as well as other triggers, cytokine storm is a general term referring to maladaptive cytokine release. The pathogenesis is complicated, but requires the depletion of regulated control at both local and systemic levels of proinflammatory cytokine output. The disease is rapidly progressing, and mortality is elevated. Some data suggests that dysregulated and uncontrolled cytokine release in certain COVID19 patients has been directly correlated with significant deterioration [74].

7. Molecular diagnostics

COVID-19 Test of SARS-CoV-2 is a real-time reverse transcription polymerase chain-reaction (PCR) in upper or lower respiratory samples for the qualitative identification of nucleic acid (such as nasal, nasopharyngeal or oropharyngeal swabs, sputum, lower respiratory tract aspirates, bronchoalveolar lavage, nasopharyngeal wash/aspirate or a nasal aspirate) that is individually obtained from those suspected of COVID-19 by their healthcare provider [75]. The latest COVID-19 outbreak can be detected using qPCR, but insufficient possession of reagents and equipment has hindered the identification of diseases. To assist in making COVID-19 more effective in our diagnostics, a new protocol was suggested for the application of the CRISPR-based SHERLOCK technique for detecting COVID-9. COVID-19 objectives were identified between 20 and 200 aM (10–100 copies per input microlitre) with the use of synthetic COVID- 19 virus RNA fragments. The test can be performed starting with patient-purified RNA as used in qRT-PCR trials and read in less than an hour with a dipstick, without the need for complex instrumentation [58, 59]. GolayMetaMiner, an in-house software, has identified four different regions over 50 nucleotides for the SARS-CoV-2 genome with 96 SARS-CoV-2 and 104 non-SARS-CoV-2 coronaviral genomes. Primers were made to target the longest and previously not targeted nsp2 region and tailored as a reverse transcription-polymerase chain reaction (RT-PCR) test without a probe. The new COVID- 19-nsp2 assay had a detection limit (LOD) of 1.8 TCID5 mL and did not intensify any human coronavirus pathogens and respiratory viruses. The process threshold reproducibility (Cp) values have been adequate and overall imprecision (%CV) values have dropped far below 5%. The latest assay evaluation using 59 clinical samples from 14 reported cases demonstrated a 100% compliance with COVID-19-RdRp/Hel reference assay, which has been previously established. A COVID-19-nsp2, fast sensitive RT-PCR test was developed for SARS-CoV-2 [76].

8. Future perspectives of nucleic acid-based vaccines

Since COVID-19 is new to humanity and the essence of defensive immune responses is incompletely understood, it is unknown which vaccination techniques are going to be most effective. Therefore, designing diverse vaccine platforms and methods in tandem is crucial. Indeed, researchers worldwide have been racing to produce COVID-19 vaccines since the epidemic started, with at least 166 vaccine candidates now in preclinical and clinical production (Draft landscape of COVID-19 candidate vaccines, 2020). A new pandemic vaccine developing framework has been suggested to address the immediate need for a vaccine, compacting the development period from 10 to 15 years to 1 to 2 years [77]. Recombinant plasmid DNA has been investigated as a vaccine model, although lately, mRNA has appeared as a promising platform. Six mRNA-based COVID-19 vaccines and four DNA-based COVID-19 vaccines are currently in clinical trials, with 27 such vaccines (16 mRNA-based and 11 DNA-based) undergoing preclinical production [78]. (Draft landscape of COVID-19 candidate vaccines, 2020). For protein translation and post-translational modifications, antigen-encoding mRNA encapsulated with a carrier such as lipid nanoparticles can be effectively conveyed in vivo into the cytoplasm of host cells, which is a plus over vaccines of the recombinant protein subunit. The mRNA vaccines are non - pathogenic and are synthesized without microbial molecules by in vitro transcription [79]. While no mRNA vaccine has been approved for human use yet, recent reports of influenza, rabies and Zika virus infections in animals support its promise in the covid-19 vaccine development race [80]. Plasmid DNA vaccines share many features, such as safety, ease of development and scalability, with mRNA vaccines, but with the differences of having poor immunogenic and having to be administered in several doses coupled with the addition of an adjuvant. This review provides valuable information that can be redirected to the purpose of working on these nucleic acid-based vaccines which provides a new propitious platform of vaccine production.

9. Conclusion

Coronaviruses have proven themselves to be prevailing and a very high threat to our existence by their unique features and ambiguity that caused catastrophic effects in this pandemic. It is obvious that coronaviruses have high spillover abilities and adaptation to new hosts so that enables more appearances in future. For a vaccine to be made or an anti-viral drug to be produced, it is very mandatory that all is known about the virus family, virus genome and its own central dogma, how it differs from its predecessors, their similarities, mutation rates and screening methods. A treatment that will be effective, long-lasting and prepared for any mutation is needed to be able to fight this virus and eradicate this disease and prevent the emergence of a new pandemic by this family of potential and active killers (The Coronaviruses).

Author details

Maram Adel Abdelghany[1], Sarah Abdullah Gozai Alghamdi[2] and Jehane Ibrahim Eid[3]*

1 Department of Zoology, Faculty of Science, Cairo University, Egypt

2 Bioscience Department, Faculty of Science, University of Jeddah, Jeddah, Saudi Arabia

3 Department of Biotechnology, Faculty of Science, Cairo University, Egypt

*Address all correspondence to: jehaneeid@sci.cu.edu.eg

References

[1] Gallagher, T. M., & Buchmeier, M. J. (2001). Coronavirus spike proteins in viral entry and pathogenesis. Virology, *279*(2), 371-374. https://doi.org/10.1006/viro.2000.0757

[2] de Groot, R. J., Baker, S. C., Baric, R. S., Brown, C. S., Drosten, C., Enjuanes, L., Fouchier, R. A. M., Galiano, M., Gorbalenya, A. E., Memish, Z. A., Perlman, S., Poon, L. L. M., Snijder, E. J., Stephens, G. M., Woo, P. C. Y., Zaki, A. M., Zambon, M., & Ziebuhr, J. (2013). Middle East Respiratory Syndrome Coronavirus (MERS-CoV): Announcement of the Coronavirus Study Group. Journal of Virology, *87*(14), 7790-7792. https://doi.org/10.1128/jvi.01244-13

[3] Woo, P. C. Y., Lau, S. K. P., Huang, Y., & Yuen, K. Y. (2009). Coronavirus diversity, phylogeny and interspecies jumping. In Experimental Biology and Medicine (Vol. 234, Issue 10, pp. 1117-1127). https://doi.org/10.3181/0903-MR-94

[4] Masters, P. S. (2006). The Molecular Biology of Coronaviruses. Advances in Virus Research, *65*(06), 193-292. https://doi.org/10.1016/S0065-3527(06)66005-3

[5] Chen, Z., & John Wherry, E. (2020). T cell responses in patients with COVID-19. Nature Reviews Immunology, *20*(9), 529-536. https://doi.org/10.1038/s41577-020-0402-6

[6] Cyranoski, D. (2020). Profile of a killer: the complex biology powering the coronavirus pandemic. Nature, *581*(7806), 22-26. https://doi.org/10.1038/d41586-020-01315-7

[7] Cui, J., Li, F., & Shi, Z. L. (2019). Origin and evolution of pathogenic coronaviruses. In *Nature Reviews Microbiology* (Vol. 17, Issue 3, pp. 181-192). Nature Publishing Group. https://doi.org/10.1038/s41579-018-0118-9

[8] Sawicki, S. G., Sawicki, D. L., & Siddell, S. G. (2007). A Contemporary View of Coronavirus Transcription. Journal of Virology, *81*(1), 20-29. https://doi.org/10.1128/jvi.01358-06

[9] Torres, J., Surya, W., Li, Y., & Liu, D. X. (2015). Protein-protein interactions of viroporins in coronaviruses and paramyxoviruses: New targets for antivirals? Viruses, *7*(6), 2858-2883. https://doi.org/10.3390/v7062750

[10] Tyrrell, D. A. J., & Bynoe, M. L. (1965). Cultivation of a Novel Type of Common-cold Virus in Organ Cultures. British Medical Journal, *1*(5448), 1467-1470. https://doi.org/10.1136/bmj.1.5448.1467

[11] De Wit, E., Van Doremalen, N., Falzarano, D., & Munster, V. J. (2016). SARS and MERS: Recent insights into emerging coronaviruses. Nature Reviews Microbiology, *14*(8), 523-534. https://doi.org/10.1038/nrmicro.2016.81

[12] WHO COVID-*19 Dashboard*. (n.d.). Retrieved May 1, 2020, from https://covid19.who.int/

[13] Rodriguez-Morales, A. J., Katterine Bonilla-Aldana, D., Balbin-Ramon, G. J., Rabaan, A. A., Sah, R., Paniz-Mondolfi, A., Pagliano, P., & Esposito, S. (2020). History is repeating itself: Probable zoonotic spillover as the cause of the 2019 novel Coronavirus Epidemic. In *Le Infezioni in Medicina, n* (Vol. 1). https://www.researchgate.net/profile/Alfonso_Rodriguez-Morales/publication/338934614_History_is_repeating_itself_a_probable_zoonotic_spillover_as_a_cause_of_an_epidemic_the_case_of_2019_novel_Coronavirus/links/5e345cb492851c7f7f11a0cb/History-is-repeating-itself-a-probable-zoonotic-spillover-as-a-cause-of-an-epidemic-the-case-of-2019-novel-Coronavirus.pdf

[14] Iturriza-Gomara, M., & O'Brien, S. J. (2016). Foodborne viral infections. In Current Opinion in Infectious Diseases. https://doi.org/10.1097/QCO.0000000000000299

[15] Kreuder Johnson, C., Hitchens, P. L., Smiley Evans, T., Goldstein, T., Thomas, K., Clements, A., Joly, D. O., Wolfe, N. D., Daszak, P., Karesh, W. B., & Mazet, J. K. (2015). Spillover and pandemic properties of zoonotic viruses with high host plasticity. Scientific Reports. https://doi.org/10.1038/srep14830

[16] Wang, L. F., & Anderson, D. E. (2019). Viruses in bats and potential spillover to animals and humans. In Current Opinion in Virology. https://doi.org/10.1016/j.coviro.2018.12.007

[17] Goebel, S. J., Hsue, B., Dombrowski, T. F., & Masters, P. S. (2004). Characterization of the RNA Components of a Putative Molecular Switch in the 3' Untranslated Region of the Murine Coronavirus Genome. Journal of Virology, *78*(2), 669-682. https://doi.org/10.1128/jvi.78.2.669-682.2004

[18] Yount, B., Curtis, K. M., & Baric, R. S. (2000). Strategy for Systematic Assembly of Large RNA and DNA Genomes: Transmissible Gastroenteritis Virus Model. In JOURNAL OF VIROLOGY (Vol. 74, Issue 22). http://jvi.asm.org/

[19] Bickerton, E., Keep, S. M., & Britton, P. (2017). Reverse genetics system for the avian coronavirus infectious bronchitis virus. Methods in Molecular Biology, *1602*(24), 83-102. https://doi.org/10.1007/978-1-4939-6964-7_6

[20] Kim, D., Lee, J. Y., Yang, J. S., Kim, J. W., Kim, V. N., & Chang, H. (2020). The Architecture of SARS-CoV-2 Transcriptome. Cell. https://doi.org/10.1016/j.cell.2020.04.011

[21] Phan, T. (2020). Genetic diversity and evolution of SARS-CoV-2. Infection, Genetics and Evolution. https://doi.org/10.1016/j.meegid.2020.104260

[22] Enjuanes, L., Sola, I., Almazan, F., Ortego, J., Izeta, A., Gonzalez, J. M., Alonso, S., Sanchez, J. M., Escors, D., Calvo, E., Riquelme, C., & Sanchez, C. (2001). Coronavirus derived expression systems. Journal of Biotechnology, *88*(3), 183-204. https://doi.org/10.1016/S0168-1656(01)00281-4

[23] Li, F. (2013). Receptor recognition and cross-species infections of SARS coronavirus. In Antiviral Research. https://doi.org/10.1016/j.antiviral.2013.08.014

[24] Li, W., Wong, S.-K., Li, F., Kuhn, J. H., Huang, I.-C., Choe, H., & Farzan, M. (2006). Animal Origins of the Severe Acute Respiratory Syndrome Coronavirus: Insight from ACE2-S-Protein Interactions. JOURNAL OF VIROLOGY, *80*(9), 4211-4219. https://doi.org/10.1128/JVI.80.9.4211-4219.2006

[25] Dimitrov, D. S. (2003). The Secret Life of ACE2 as a Receptor for the SARS Virus. In Cell. https://doi.org/10.1016/S0092-8674(03)00976-0

[26] Fahmi, M., Kubota, Y., & Ito, M. (2020). Nonstructural proteins NS7b and NS8 are likely to be phylogenetically associated with evolution of 2019-nCoV. Infection, Genetics and Evolution. https://doi.org/10.1016/j.meegid.2020.104272

[27] Grifoni, A., Sidney, J., Zhang, Y., Scheuermann, R. H., Peters, B., & Sette, A. (2020). A Sequence Homology and Bioinformatic Approach Can Predict Candidate Targets for Immune Responses to SARS-CoV-2. Cell Host and Microbe. https://doi.org/10.1016/j.chom.2020.03.002

[28] Li, G., Fan, Y., Lai, Y., Han, T., Li, Z., Zhou, P., Pan, P., Wang, W., Hu, D., Liu, X., Zhang, Q., & Wu, J. (2020). Coronavirus infections and immune responses. In *Journal of Medical Virology* (Vol. 92, Issue 4, pp. 424-432). John Wiley and Sons Inc. https://doi.org/10.1002/jmv.25685

[29] Li, W., Moore, M. J., Vasllieva, N., Sui, J., Wong, S. K., Berne, M. A., Somasundaran, M., Sullivan, J. L., Luzuriaga, K., Greeneugh, T. C., Choe, H., & Farzan, M. (2003). Angiotensin-converting enzyme 2 is a functional receptor for the SARS coronavirus. Nature. https://doi.org/10.1038/nature02145

[30] Rota, P. A., Oberste, M. S., Monroe, S. S., Nix, W. A., Campagnoli, R., Icenogle, J. P., Peñaranda, S., Bankamp, B., Maher, K., Chen, M. hsin, Tong, S., Tamin, A., Lowe, L., Frace, M., DeRisi, J. L., Chen, Q., Wang, D., Erdman, D. D., Peret, T. C. T., ... Bellini, W. J. (2003). Characterization of a novel coronavirus associated with severe acute respiratory syndrome. Science. https://doi.org/10.1126/science.1085952

[31] Kannan, S., Shaik Syed Ali, P., Sheeza, A., & Hemalatha, K. (2020). COVID-19 (Novel Coronavirus 2019) - recent trends. European Review for Medical and Pharmacological Sciences. https://doi.org/10.26355/eurrev_202002_20378

[32] Wrapp, D., Wang, N., Corbett, K. S., Goldsmith, J. A., Hsieh, C. L., Abiona, O., Graham, B. S., & McLellan, J. S. (2020). Cryo-EM structure of the 2019-nCoV spike in the prefusion conformation. Science. https://doi.org/10.1126/science.aax0902

[33] Peter Forster, Lucy Forster, Colin Renfrew, Michael Forster. Phylogenetic network analysis of SARS-CoV-2 genomes. Proceedings of the National Academy of Sciences Apr 2020, 117 (17) 9241-9243; DOI: 10.1073/pnas.2004999117

[34] Spagnolo, J. F., & Hogue, B. G. (2000). Host Protein Interactions with the 3′ End of Bovine Coronavirus RNA and the Requirement of the Poly(A) Tail for Coronavirus Defective Genome Replication. Journal of Virology, *74*(11), 5053-5065. https://doi.org/10.1128/jvi.74.11.5053-5065.2000

[35] Ding, Y., He, L., Zhang, Q., Huang, Z., Che, X., Hou, J., Wang, H., Shen, H., Qiu, L., Li, Z., Geng, J., Cai, J., Han, H., Li, X., Kang, W., Weng, D., Liang, P., & Jiang, S. (2004). Organ distribution of severe acute respiratory syndrome (SARS) associated coronavirus (SARS-CoV) in SARS patients: Implications for pathogenesis virus transmission pathways. Journal of Pathology, *203*(2), 622-630. https://doi.org/10.1002/path.1560

[36] Jiang, S., Lu, L., Liu, Q., Xu, W., & Du, L. (2012). Receptor-binding domains of spike proteins of emerging or re-emerging viruses as targets for development of antiviral vaccines. In Emerging Microbes and Infections (Vol. 1). https://doi.org/10.1038/emi.2012.1

[37] Lai, M. (1990). Coronavirus: Organization, Replication And Expression Of Genome. Annual Review of Microbiology, *44*(1), 303-333. https://doi.org/10.1146/annurev.micro.44.1.303

[38] Lai, M. M. C., & Cavanagh, D. (1997). *The Molecular Biology of Coronaviruses* (Vol. 48, pp. 1-100). Academic Press. https://doi.org/10.1016/S0065-3527(08)60286-9

[39] Almazan, F., DeDiego, M. L., Galan, C., Escors, D., Alvarez, E., Ortego, J., Sola, I., Zuniga, S., Alonso, S., Moreno, J. L., Nogales, A., Capiscol, C., & Enjuanes, L. (2006). Construction of a Severe Acute Respiratory Syndrome Coronavirus Infectious cDNA Clone and a Replicon To Study Coronavirus RNA Synthesis. Journal of Virology, *80*(21), 10900-10906. https://doi.org/10.1128/jvi.00385-06

[40] Zuniga, S., Sola, I., Alonso, S., & Enjuanes, L. (2004). Sequence Motifs Involved in the Regulation of Discontinuous Coronavirus Subgenomic RNA Synthesis. Journal of Virology, *78*(2), 980-994. https://doi.org/10.1128/jvi.78.2.980-994.2004

[41] Perlman, S., Ries, D., Bolger, E., Lung-Ji, C., & Stoltzfus, C. M. (1986). MHV nucleocapsid synthesis in the presence of cycloheximide and accumulation of negative strand MHV RNA. Virus Research. https://doi.org/10.1016/0168-1702(86)90074-2

[42] Sawicki, S. G., & Sawicki, D. L. (1986). Coronavirus minus-strand RNA synthesis and effect of cycloheximide on coronavirus RNA synthesis. Journal of Virology. https://doi.org/10.1128/jvi.57.1.328-334.1986

[43] Sawicki, S. G., & Sawicki, D. L. (1990). Coronavirus transcription: subgenomic mouse hepatitis virus replicative intermediates function in RNA synthesis. Journal of Virology. https://doi.org/10.1128/jvi.64.3.1050-1056.1990

[44] Boursnell, M. E. G., Binns, M. M., & Brown, T. D. K. (1985). Sequencing of coronavirus IBV genomic RNA: Three open reading frames in the 5′ "unique" region of mRNA D. Journal of General Virology. https://doi.org/10.1099/0022-1317-66-10-2253

[45] Fischer, F., Stegen, C. F., Masters, P. S., & Samsonoff, W. A. (1998). Analysis of Constructed E Gene Mutants of Mouse Hepatitis Virus Confirms a Pivotal Role for E Protein in Coronavirus Assembly. Journal of Virology, *72*(10), 7885-7894. https://doi.org/10.1128/jvi.72.10.7885-7894.1998

[46] Skinner, M. A., Ebner, D., & Siddell, S. G. (1985). Coronavirus MHV-JHM mRNA 5 has a sequence arrangement which potentially allows translation of a second, downstream open reading frame. Journal of General Virology. https://doi.org/10.1099/0022-1317-66-3-581

[47] Peng, D., Koetzner, C. A., McMahon, T., Zhu, Y., & Masters, P. S. (1995). Construction of murine coronavirus mutants containing interspecies chimeric nucleocapsid proteins. Journal of Virology, *69*(9), 5475-5484. https://doi.org/10.1128/jvi.69.9.5475-5484.1995

[48] Denison, M. R., Graham, R. L., Donaldson, E. F., Eckerle, L. D., & Baric, R. S. (2011). Coronaviruses: An RNA proofreading machine regulates replication fidelity and diversity. In RNA Biology. https://doi.org/10.4161/rna.8.2.15013

[49] Hu, J., Wang, J., Xu, J., Li, W., Han, Y., Li, Y., Ji, J., Ye, J., Xu, Z., Zhang, Z., Wei, W., Li, S., Wang, J., Wang, J., Yu, J., & Yang, H. (2003). Evolution and variation of the SARS-CoV genome. Genomics, Proteomics & Bioinformatics / Beijing Genomics Institute, *1*(3), 216-225. https://doi.org/10.1016/S1672-0229(03)01027-1

[50] Lau, S. K. P., Woo, P. C. Y., Li, K. S. M., Huang, Y., Wang, M., Lam, C. S. F., Xu, H., Guo, R., Chan, K. hung, Zheng, B. jian, & Yuen, K. yung. (2007). Complete genome sequence of bat coronavirus HKU2 from Chinese horseshoe bats revealed a much smaller spike gene with a different evolutionary lineage from the rest of the genome. Virology, *367*(2), 428-439. https://doi.org/10.1016/j.virol.2007.06.009

[51] Stavrinides, J., & Guttman, D. S. (2004). Mosaic Evolution of the Severe Acute Respiratory Syndrome Coronavirus. Journal of Virology, *78*(1), 76-82. https://doi.org/10.1128/jvi.78.1.76-82.2004

[52] Yeh, S. H., Wang, H. Y., Tsai, C. Y., Kao, C. L., Yang, J. Y., Liu, H. W., Su, I. J., Tsai, S. F., Chen, D. S., Chen, P. J., Lee, Y. T., Teng, C. M., Yang, P. C., Ho,

H. N., Chang, M. F., Wang, J. T., Chang, S. C., Kao, C. L., Wang, W. K., ... Hsueh, P. R. (2004). Characterization of severe acute respiratory syndrome coronavirus genomes in Taiwan: Molecular epidemiology and genome evolution. Proceedings of the National Academy of Sciences of the United States of America, *101*(8), 2542-2547. https://doi.org/10.1073/pnas.0307904100

[53] Al-khafaji, Z. M., Mahmood, A. B., & Mahmood, M. B. (n.d.). *COVID-19 Outbreak Part I : Evolution of the Virus and Designing B Epitopes*. 1-23.

[54] Vega, V. B., Ruan, Y., Liu, J., Lee, W. H., Wei, C. L., Se-Thoe, S. Y., Tang, K. F., Zhang, T., Kolatkar, P. R., Ooi, E. E., Ling, A. E., Stanton, L. W., Long, P. M., & Liu, E. T. (2004). Mutational dynamics of the SARS coronavirus in cell culture and human populations isolated in 2003. BMC Infectious Diseases, *4*, 1-9. https://doi.org/10.1186/1471-2334-4-32

[55] Grubaugh, N. D., Ladner, J. T., Lemey, P., Pybus, O. G., Rambaut, A., Holmes, E. C., & Andersen, K. G. (2019). Tracking virus outbreaks in the twenty-first century. In *Nature Microbiology* (Vol. 4, Issue 1, pp. 10-19). Nature Publishing Group. https://doi.org/10.1038/s41564-018-0296-2

[56] Minskaia, E., Hertzig, T., Gorbalenya, A. E., Campanacci, V., Cambillau, C., Canard, B., & Ziebuhr, J. (2006). Discovery of an RNA virus 3′→5′ exoribonuclease that is critically involved in coronavirus RNA synthesis. Proceedings of the National Academy of Sciences of the United States of America, *103*(13), 5108-5113. https://doi.org/10.1073/pnas.0508200103

[57] Khailany, R. A., Safdar, M., & Ozaslan, M. (2020). Genomic characterization of a novel SARS-CoV-2. Gene Reports, *19*(April), 100682. https://doi.org/10.1016/j.genrep.2020.100682

[58] Zhang, F., Abudayyeh, O. O., Gootenberg, J. S., Sciences, C., & Mathers, L. (2020a). A protocol for detection of COVID-19 using CRISPR diagnostics. *Bioarchive*.

[59] Zhang, T., Wu, Q., & Zhang, Z. (2020b). Probable Pangolin Origin of SARS-CoV-2 Associated with the COVID-19 Outbreak. Current Biology. https://doi.org/10.1016/j.cub.2020.03.022

[60] Zhou, P., Yang, X. Lou, Wang, X. G., Hu, B., Zhang, L., Zhang, W., Si, H. R., Zhu, Y., Li, B., Huang, C. L., Chen, H. D., Chen, J., Luo, Y., Guo, H., Jiang, R. Di, Liu, M. Q., Chen, Y., Shen, X. R., Wang, X., ... Shi, Z. L. (2020). A pneumonia outbreak associated with a new coronavirus of probable bat origin. Nature. https://doi.org/10.1038/s41586-020-2012-7

[61] Randhawa, G. S., Soltysiak, M. P. M., El Roz, H., de Souza, C. P. E., Hill, K. A., & Kari, L. (2020). Machine learning using intrinsic genomic signatures for rapid classification of novel pathogens: COVID-19 case study. PLoS ONE. https://doi.org/10.1371/journal.pone.0232391

[62] Ceraolo, C., & Giorgi, F. M. (2020). Genomic variance of the 2019-nCoV coronavirus. Journal of Medical Virology, *92*(5), 522-528. https://doi.org/10.1002/jmv.25700

[63] Xiao, K., Zhai, J., Feng, Y., Zhou, N., Zhang, X., Zou, J.-J., Li, N., Guo, Y., Li, X., Shen, X., Zhang, Z., Shu, F., Huang, W., Li, Y., Zhang, Z., Chen, R.-A., Wu, Y.-J., Peng, S.-M., Huang, M., ... Shen, Y. (2020). Isolation of SARS-CoV-2-related coronavirus from Malayan pangolins. *Nature*, 1-7. https://doi.org/10.1038/s41586-020-2313-x

[64] Cao, X. (2020). COVID-19: immunopathology and its implications for therapy. Nature Reviews Immunology, *20*(5), 269-270. https://doi.org/10.1038/s41577-020-0308-3

[65] Azkur, A. K., Akdis, M., Azkur, D., Sokolowska, M., van de Veen, W., Brüggen, M. C., O'Mahony, L., Gao, Y., Nadeau, K., & Akdis, C. A. (2020). Immune response to SARS-CoV-2 and mechanisms of immunopathological changes in COVID-19. *Allergy: European* Journal of Allergy and Clinical Immunology, *75*(7), 1564-1581. https://doi.org/10.1111/all.14364

[66] Yang, L., Liu, S., Liu, J., Zhang, Z., Wan, X., Huang, B., Chen, Y., & Zhang, Y. (2020). COVID-19: immunopathogenesis and Immunotherapeutics. Signal Transduction and Targeted Therapy, 5(1), 1-8. https://doi.org/10.1038/s41392-020-00243-2

[67] McGee, M. C., August, A., & Huang, W. (2020). BTK/ITK dual inhibitors: Modulating immunopathology and lymphopenia for COVID-19 therapy. *Journal of Leukocyte Biology*, JLB.5COVR0620-306R. https://doi.org/10.1002/JLB.5COVR0620-306R

[68] Mortaz, E., Tabarsi, P., Varahram, M., Folkerts, G., & Adcock, I. M. (2020). The Immune Response and Immunopathology of COVID-19. In *Frontiers in Immunology* (Vol. 11). Frontiers Media S.A. https://doi.org/10.3389/fimmu.2020.02037

[69] Qin, C., Zhou, L., Hu, Z., Zhang, S., Yang, S., Tao, Y., Xie, C., Ma, K., Shang, K., Wang, W., & Tian, D. S. (2020). Dysregulation of Immune Response in Patients With Coronavirus 2019 (COVID-19) in Wuhan, China. *Clinical Infectious Diseases : An Official Publication of the Infectious Diseases Society of America*. https://doi.org/10.1093/cid/ciaa248

[70] Sun, S., Cai, X., Wang, H., He, G., Lin, Y., Lu, B., Chen, C., Pan, Y., & Hu, X. (2020). Abnormalities of peripheral blood system in patients with COVID-19 in Wenzhou, China. *Clinica Chimica Acta*. https://doi.org/10.1016/j.cca.2020.04.024

[71] Zini, G., Bellesi, S., Ramundo, F., & d'Onofrio, G. (2020). Morphological anomalies of circulating blood cells in COVID-19. American Journal of Hematology. https://doi.org/10.1002/ajh.25824

[72] Bhaskar, S., Sinha, A., Banach, M., Mittoo, S., Weissert, R., Kass, J. S., Rajagopal, S., Pai, A. R., & Kutty, S. (2020). Cytokine Storm in COVID-19—Immunopathological Mechanisms, Clinical Considerations, and Therapeutic Approaches: The REPROGRAM Consortium Position Paper. Frontiers in Immunology. https://doi.org/10.3389/fimmu.2020.01648

[73] Mangalmurti, N., & Hunter, C. A. (2020). Cytokine Storms: Understanding COVID-19. In *Immunity* (Vol. 53, Issue 1, pp. 19-25). Cell Press. https://doi.org/10.1016/j.immuni.2020.06.017

[74] Ye, Q., Wang, B., & Mao, J. (2020). The pathogenesis and treatment of the 'Cytokine Storm" in COVID-19.' In Journal of Infection. https://doi.org/10.1016/j.jinf.2020.03.037

[75] Laboratory Corporation of America. (2020). Accelerated Emergency Use Authorization (EUA) Summary COVID-19 RT-PCR Test. *LabCorp COVID-19 RT-PCR Test EUA Summary*. https://doi.org/10.1007/978-1-4939-0758-8

[76] Yip, C. C. Y., Ho, C. C., Chan, J. F. W., To, K. K. W., Chan, H. S. Y., Wong, S. C. Y., Leung, K. H., Fung, A. Y. F., Ng, A. C. K., Zou, Z., Tam, A. R., Chung, T. W. H., Chan, K. H., Hung, I. F. N., Cheng, V. C. C., Tsang, O. T. Y., Tsui, S. K. W., & Yuen, K. Y. (2020). Development of a novel, genome subtraction-derived, sars-cov-2-specific covid-19-nsp2 real-time rt-pcr assay and its evaluation using clinical specimens. International Journal of Molecular

Sciences. https://doi.org/10.3390/ijms21072574

[77] Lurie, N., Saville, M., Hatchett, R., & Halton, J. (2020). Developing Covid-19 Vaccines at Pandemic Speed. New England Journal of Medicine, 382(21), 1969-1973. https://doi.org/10.1056/nejmp2005630

[78] https://www.who.int/publications/m/item/draft-landscape-of-covid-19-candidate-vaccines

[79] Pardi, N.; Hogan, M.J.; Porter, F.W.; Weissman, D. mRNA vaccines—A new era in vaccinology. Nat. Rev. Drug Discov. 2018, 17, 261-279

[80] Jeyanathan, M., Afkhami, S., Smaill, F. *et al.* Immunological considerations for COVID-19 vaccine strategies. *Nat Rev Immunol* 20, 615-632 (2020). https://doi.org/10.1038/s41577-020-00434-6

Chapter 7

Molecular Dynamics of Mosquito-*Plasmodium vivax* Interaction: A Smart Strategy of Parasitism

Charu Chauhan, Sanjay Tevatiya, Seena Kumari, Punita Sharma, Jyoti Rani and Rajnikant Dixit

Abstract

Parallel to *Plasmodium falciparum*, *P. vivax* is a fast emerging challenge to control malaria in South-East Asia regions. Owing to unique biological differences such as the preference for invading reticulocytes, early maturation of sexual stages during the infection, the formation of hypnozoites, unavailability of *in-vitro* culture, the molecular relation of *P. vivax* development inside the mosquito host is poorly known. In this chapter, we briefly provide a basic overview of Mosquito-*Plasmodium* interaction and update current knowledge of tissue-specific viz. midgut, hemocyte, and salivary glands- molecular dynamics of *Plasmodium vivax* interaction during its developmental transformation inside the mosquito host, in specific.

Keywords: malaria, mosquito, *Anopheles*, *Plasmodium vivax*, host–parasite interaction

1. Introduction

The apicomplexan parasite *Plasmodium*, which is accountable for malaria, has a complex life cycle that includes both vertebrate hosts and invertebrate mosquitoes. Adaptation to blood-feeding in mosquitoes has made it inadvertently a carrier of various diseases. A blood meal is indispensable for adult female mosquitoes to nourish its egg, and maintain the gonotrophic cycle. But during blood feeding, ingestion of *Plasmodium* gametocyte from an infected person's blood results in the onset of 18-20 days long sporogonic cycle that culminates in the production of infectious sporozoites in the mosquito host [1]. These infectious sporozoites are then delivered into the human body through salivary discharge, which initiates the intricate stages of the asexual process causing malaria. In humans, malaria is caused by five *Plasmodium* species i.e., *Plasmodium falciparum, P. vivax, P. malariae, P. ovale,* and *P. knowlesi* [2].

P. falciparum and *P. vivax*, vectored by the adult female *Anopheline* mosquitoes, are two principal parasites of human malaria [3]. Of the five *Plasmodium* species that cause human malaria, *Plasmodium vivax* is the most geographically widespread [4]. The parasite could survive quiescent for extended periods when circumstances are not conducive to its ongoing transmission [5]. According to the current report by WHO, in the year 2019 around 75% of malaria cases were caused by *P. vivax* in the WHO Region of the Americas. An approximated 52% of the global burden of *P. vivax* emerged from the WHO South-East Asia Region, among which 47% were contributed from India [6].

P. vivax is considered as a less fatal parasite, but the recent emergence of more *P. vivax* infected cases in *P. falciparum* endemic areas, and increased mortality, morbidity rates are drawing our attention to this least studied parasite. It is more difficult to monitor and eradicate the *P. vivax* than *P. falciparum*, because of limited information, and associated biological complexities of its development in the mosquito as well as the human host [7, 8].

P. vivax normally circulates at low peripheral parasite densities, but still, they are transmissible by the mosquito vectors, and hence presents major challenges for the diagnosis of infected peoples. *P. vivax* has adapted to live with varying *Anopheles* vectors in different ecological conditions. Unlike other *Plasmodium* species, *P. vivax* has the potential to form dormant hypnozoites inside the host liver, and these liver-stage parasites are accountable for malaria relapses for weeks or months after initial infection [5]. Lastly, the lack of long-term *in-vitro* culture further restricts our understanding of the biological consequences of *P. vivax* development and transmission [9]. Nevertheless, for the last two decades, the integration and utilization of high-throughput molecular technologies such as genomics, RNA-Seq/transcriptomics, proteomics, have been valuable to decode and trace the genetic variation and diversity in the *P. vivax* population collected from different geographical origins [10, 11]. Efforts are continuing to uncover molecular and functional correlation of tissue/stage-specific *P. vivax* biology in the vertebrate host, identify genetic signatures to develop new diagnosis tools, anti-*P. vivax* drugs, or vaccine development. However, the biological complexity of the *P. vivax* development cycle in the mosquito vector-host is too limited, and therefore in this article, we highlight the current progress made so far in the understanding of the Mosquito-*P. vivax* interaction biology.

2. A general overview of the sporogonic cycle in mosquito host

The transmission of the parasite from human host to mosquito transpire when a female mosquito acquires gametocyte containing blood meal from the infected vertebrate host. When the parasite enters the midgut lumen it faces the dynamically changing environment, where both male and female gametocytes get differentiated into male and female gametes [12–15]. Ingested gametocytes also encounter proteolytic enzymes released by midgut epithelium in the midgut environment to digest the blood meal, which may have an agonistic or antagonistic effect on parasite growth. Fertilization of male and female gametes results in zygote formation, which rapidly transforms into motile ookinetes [16]. After exiting from the blood bolus, ookinete traverses the midgut epithelium either through intracellular or intercellular route and then rests beneath the epithelial cell at basal lamina. Later ookinetes transform into replicative oocyst stage which undergoes an umpteen round of nuclear division to produce thousands of sporozoites within a time period of one to two weeks. Once in the hemolymph circulation, the free circulatory sporozoites (*fcSPZ*) target to invade salivary glands, but most of them are rapidly cleared off by hemocytes, the immune blood cells of the mosquitoes [17]. Thus tracking of molecular, biochemical, and cellular events during *Plasmodium* developmental transition from one stage to another stage, is of particular interest. Several laboratory studies on mosquito-parasite interaction involving *P. berghei* or *P. falciparum*, demonstrate that the developmental kinetics of the *Plasmodium* population is significantly altered, though the mechanism is not fully understood [18–20]. The last two decades of research highlights the crucial role of the tissue-specific mosquito immune system to control the parasite load, though the physiological relevance is yet to be investigated [21–24].

3. *Plasmodium* population dynamics and their immune regulation in the mosquito host

During *Plasmodium* development inside the mosquito host, the parasite population undergoes various bottlenecks. Previous investigations demonstrated that if a female mosquito takes ~1000 gametocytes through its infected blood meal, ~100 can be transformed into ookinetes, and among them, only 1–5 can successfully form oocysts. Furthermore, these survived oocysts will form millions of sporozoites, but only 19-20% can successfully invade the salivary glands for further transmission [25]. In refractory strains, not a single ookinete could transform into oocysts [26]. In general, a substantial loss of parasite population occurs at each developmental stage of the parasite, and this major parasite loss can be attributed to both human as well as mosquito components, which are harmful to *Plasmodium*.

The human component includes cytokines, complement protein, and reactive nitrogen species that are ingested along with the gametocytes during blood meal intake, and detrimental to the parasite within the midgut lumen of vector [26]. During the parasite transition through midgut epithelium, the mosquito mounts early immune response by increasing midgut nitration and activation of the signaling pathway. The nitration process modifies the ookinetes surface, and mark them to be recognized by the mosquito complement system when they emerge toward the basal side of the midgut [27]. Signaling pathways provide varying responses to various species of *Plasmodium*, such as the IMD pathway acts more efficiently against *P. falciparum* than *P. berghei*, and the Toll pathway is more responsive against *P. berghei*, and *P. gallinaceum* [28]. The proliferation of microbiota following blood meal also exacerbated the mosquito immune response, which in turn is detrimental to parasite development. Plasmodium parasite faces population bottlenecks throughout their development (in vertebrate as well as invertebrate host) but the mosquito midgut serves as the major site of extermination, where the number of parasites is minimal during the oocyst stage which makes it the most susceptible stage to identify molecular targets to disrupt the transmission [26]. Parallel to gut-immune interaction, several factors have been identified from mosquito hemocyte and salivary glands that interact with *Plasmodium sporozoites*; a bulk of literature is available on the mosquito innate immune system against *P. berghei* and *P. falciparum*, and therefore readers may refer to many excellent reviews [29–31]. Here we update the reports on the Mosquito-*P. vivax* interactions, and highlight their relevance for future implications.

4. Mosquito-*P. vivax* interaction

Undoubtedly, advanced omics technologies, especially genome sequencing and transcriptome analysis, has now become a basic method in living organisms for the assessment of genome-scale gene identification. The expression of large scale identified genes is currently being explored to decode the molecular complexity of *P. vivax* development in the vertebrate host. Earlier, a high-density tiling microarray-based study showed the gene expression variation of *P. vivax* from human and mosquito stages such as sporozoites, gametes, zygotes, ookinetes, and *in-vivo* asexual blood stages. Their comparison to *P. falciparum* and *P. yoelii* further reveals conserved and species-specific patterns highlighting the metabolic state of parasites growing within humans and identifies many orthologs of *P. falciparum* transcripts that are needed for exoerythrocytic development, which may also likely help in hypnozoite formation in the *P. vivax* [32].

4.1 *Plasmodium vivax* strategy to adapt in the mosquito *Anopheles stephensi*

The successful development of *P. vivax* within the midgut of a susceptible strain of *Anopheles stephensi* can be divided into two phases: pre-invasion (within midgut lumen) and post-invasion strategy i.e. development of oocyst stage which depends upon the nutrient availability within the host. During the pre-midgut invasion phase, *P. vivax* imparts an intricate mechanism to evade the mosquito immune response. It indirectly attenuates the mosquito immune response by dramatically suppressing the bacterial population, and whereas in the post- midgut invasion phase i.e. during the development of oocyst it modulates the expression of genes that are directly or indirectly involved with nutrition physiology to fulfill their nutritional requirement. We have limited information about the phases beyond oocysts maturation and their strategies to evade the mosquito immune system and promote their transmission.

4.1.1 Pre- invasion strategy of P. vivax

The midgut of Anopheles mosquitoes is housed by a complex and diverse community of bacteria, protozoa, fungi, etc. collectively referred to as the microbiota, and this microbiota is believed to shape the vector competency of mosquito. The gut bacteria of Anopheles mosquitoes adversely affect the *Plasmodium* infection [33, 34]. These tripartite interactions have been studied between the mosquito, its microbiota, and the *Plasmodium* parasites, but the precise relationship between the three remains unknown.

Numerous research reports have revealed that microbiota of specific bacterial species, particularly gram-negative bacteria, in many *Anopheles* species have an inhibitory effect on various *Plasmodium* species. The elimination of midgut bacteria through antibiotic treatment enhances oocyst load and parasite prevalence in different species of Anopheles. There are two mechanisms by which microbiota interfere with the *Plasmodium* development in the midgut lumen:-(i) indirectly by triggering the immune response of the mosquito (Imd Pathway) that guides the synthesis of AMP and other immune effectors that interferes with the development of parasites, and (ii) directly by certain bacterial species producing the metabolites that interfere with *Plasmodium* development and survival [33]. Recently, we have demonstrated that *P. vivax* plays a unique strategy to steer clear off the mosquito immune response during its pre-invasive phase, by dramatic suppression of the gut-bacterial population [35] (**Figure 1**). This study hypothesizes that the parasites outcompete the midgut microbiota presumably by scavenging the iron from the blood meal which is necessary for bacterial growth [35].

4.1.2 Post- invasion strategy of P. vivax (development of oocyst)

During the *Plasmodium* transit through midgut epithelium within the susceptible strain of Anopheles, some of the ookinetes successfully manage to escape the mosquito immune response [36], and reach the basal lamina of midgut to further differentiate into oocyst, and rests there nearly for two weeks. The sessile oocyst

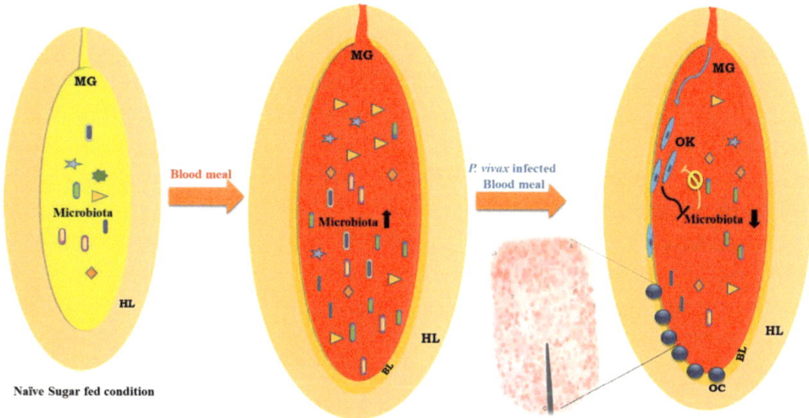

Figure 1.
Alteration of midgut microbiota proliferation by P. vivax. *Blood meal induces midgut microbiota proliferation within 24 hours. But during* P. vivax *infection, somehow this parasite restricts this microbiota proliferation after blood-meal, to avoid nutritional competition and immune defense exerted by the microbiota. This smart strategy of restriction helps the parasite to survive and proliferate better.*

stage is metabolically active, and follows an umpteen rounds of the nuclear division to transform into sporozoites. A single oocyst is capable of producing thousands of haploid sporozoites [37, 38]. Limited research has been undertaken on the underlying mechanism of *P. vivax* oocyst development (transition from a small oocyst of 7-8 μm to a large oocyst of 35-40 μm) in the mosquitoes. A few recent RNA-Seq analyses of *P. vivax* infected mosquitoes have been valuable to understand the ookinete and oocyst stage of *P. vivax* which reveals the alteration of several transcripts in the gut after 18 hours and 7 days post-infection in mosquito *Anopheles dirus* [39]. Notably, the authors identified several genes such as *Anoctamin 6* (ANO6; ADIR005670) and *Fibroblast Growth Factor* (FGF; ADIR008464), which may likely have immune regulation of *P. vivax* growth in the gut of the mosquito.

The parasite scavenges the nutrients from the host, and thus one of the main deciding factors of the infection outcome is likely dependent on the availability of nutritional resources of the host [40]. Our ongoing tissue-specific RNA-Seq analysis of *An. stephensi* infected with *P. vivax* oocyst identifies several unique sets of transcripts/genes, which have not yet find associated with any other *Plasmodium* infection. This study revealed the expression of genes involved in maintaining glucose homeostasis (*Trehalase*), nutrient transport (*Sterol Carrier* protein), energy, and nutrient homeostasis (*Folliculin*) during *P. vivax* infection [24]. We noticed that *P. vivax* infection modulates the *Trehalase* and *Sterol Carrier protein* expression in the midgut and salivary gland (SCP) for its own development and maturation. *Trehalase*, a glucosidase enzyme, catalyze the hydrolysis of disaccharide trehalose sugar into glucose units. Glucose is the main source of energy for the extensive proliferation of malarial parasites during both the blood and liver stages of malaria infection [41–44]. *Plasmodium* obtains the host glucose via hexose transporter. However, the role of sugar metabolism on *Plasmodium* infection in the mosquito vector remains poorly known. A multifold enriched expression of *Trehalase* transcript during early to late-stage oocysts in the gut as well as salivary glands, in addition to retrieval of *Plasmodium* hexose transcript in the midgut during oocyst stage, suggests that *Trehalase* may significantly contribute to hydrolyze the trehalose to provide glucose for the rapid proliferation of parasites, and also affect the reproductive capacity of adult female mosquito *An. stephensi* [45].

Similar to sugar requirement, *Plasmodium* also relies heavily on the host's cholesterol for its growth when maturing from small oocysts to large oocysts in the gut. Since *Plasmodium* is incapable to synthesize *de-novo* cholesterol [46], and *P. vivax* infection induces a multifold expression of SCP after seven days of infection in the gut, likely indicates its role in cholesterol transport. Currently, there is no functional correlation exists between SCP and *Plasmodium* infection, however, with the current observation of SCP enrichment in the midgut as well as salivary gland, we propose that besides a possible role of supplying cholesterol to developing oocyst, it is possible that *As-SCP* may impart an anti-*Plasmodium* immune response, as increased lipid droplets have been shown in the midgut of *Ae. aegypti* during bacterial and viral infection [47]. *Folliculin* (FLCN) is a tumor suppressor protein associated with Birt-Hogg-Dube(BHD) syndrome [48, 49]. It is involved in many biological processes including vesicular trafficking, energy, and nutrient homeostasis, and monitors E-cadherin protein level [50, 51]. Late induction of *FLCN* in response to *P. vivax* infection (unpublished) suggests that it might also play an important role in maintaining the integrity of midgut epithelial cells during oocyst bursting or acquisition of nutrients by developing oocyst, though further studies needed to support this hypothesis.

4.2 *P. vivax* infection and immune strategy of the *Anopheles stephensi*

As described earlier, the parasite population undergoes several bottlenecks throughout their development inside the mosquito host. These bottlenecks are achieved because of the mosquito immune system [26]. Once the *Plasmodium* parasite transforms in the ookinete, midgut nitration modifies the parasite surface, which is then recognized by the hemocyte encoded pattern recognition receptors (like *TEP1*) circulating in the hemolymph [52]. Studies in the mosquito *An. gambiae* suggest that the complex of *LRIM1/APL1C* and *TEP1* bind to the parasite surface and activate the complement system, and in turn, the circulating hemocytes kill the parasite through cell lysis, phagocytosis, melanization, etc [53–55]. This whole phase is completed within 24 hours after infective blood meal uptake and is known as the

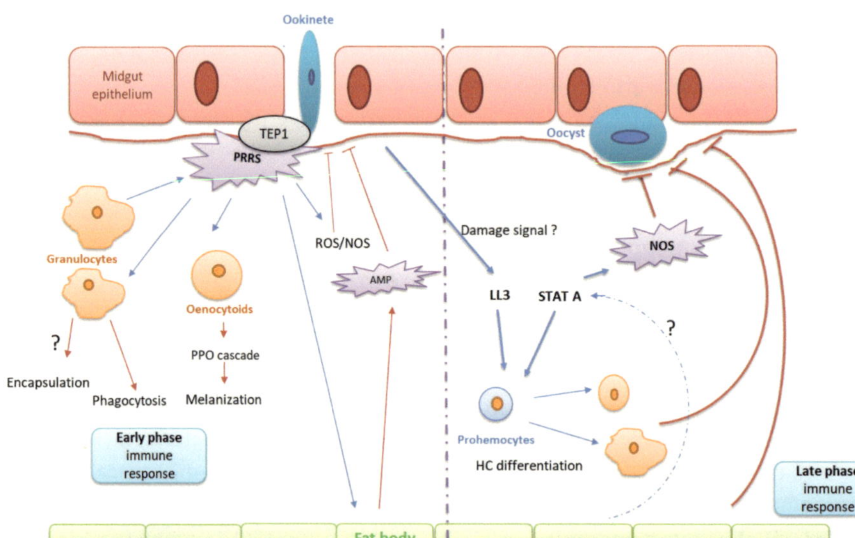

Figure 2.
Systematic representation of events occurring during early and late phase immunity in malaria parasite-infected mosquito: Once the ookinete invade the midgut epithelium, PRRs (pattern recognition receptors) like TEP1 *recognize the pathogen and activate the complement system, which further triggers the hemocytes for phagocytosis, melanization, etc.*

"early phase" immune response (**Figure 2**). Once the ookinetes reach the midgut epithelium, they get transformed into oocysts, and the immune system working against these transformed parasites is known as "late phase" immune response [56, 57]. Although very little is known about this phase but recent literature suggests that *LL3* mediated hemocyte differentiation, and *STAT* pathway activation, together helps in the restriction of the oocysts development [58]. Post oocysts maturation, millions of sporozoites evade the midgut lamina and circulate in the hemolymph, in order to reach and invade salivary glands for their successful transmission. Current literature suggests that among thousands of sporozoites only 19% can successfully invade the salivary gland, the rest are eliminated by the hemocyte mediated mosquito immune system [25]. But we have very limited information about this direct cell (hemocytes)-cell (free circulating sporozoites) interaction and elimination mechanism [29].

Altogether this information is restricted to the model organisms, and due to problems in culturing of *P. vivax* and extraction of hemocytes the exact species-specific interaction biology of this neglected parasite is still unknown [29]. As hemocytes play a crucial role in immune regulation, decoding the direct or indirect immune interactions between hemocytes and *P. vivax* parasite, will help us to figure out the parasite population control strategies of the mosquito hosts.

4.2.1 Hemocytes: the cellular immune army of the mosquito host

Mosquitoes have an open circulatory system, and hemocytes are the tiny blood cells circulating across the body reaching every mosquito tissue. These are the major immune elicitors working against a diverse range of pathogens [29]. Hemocytes are the core of the mosquito immune system which can induce both cellular as well as humoral immune responses [30, 59, 60]. Mosquito hemocytes population can be discriminated on the basis of their anatomical location (circulatory and sessile), DNA content (euploid and polyploid), morphology, and functions (granulocytes, oenocytoids, and prohemocytes) [61–63]. Granulocytes are the phagocytic cells, which engulf the invaded parasite and kill them by lysozyme activity [64, 65]. Oenocytoids are the producers of the *Pro-phenoloxidases*, the rate-limiting enzyme of the melanization pathway [66]. Melanization is the systematic enzymatic process, which ultimately produces the melanin protein. When a foreign invader infects the mosquito, hemocytes cover the parasite in the melanin envelop, which will cut-off the parasite from the outside environment, nutrition, and also induces oxidative stress which results in the killing of the parasite. Prohemocytes are considered as the progenitor cells, which produce granulocytes and oenocytoids, although the actual function is not known yet about these tiny cells [64, 67]. Previous literature illustrated various hemocyte encoded molecules, like *TEP1*, *FBN30*, *LRR3*, etc. are vital for the early and late phase immune responses [55, 68–71]. Researchers have also successfully tracked the involvement of phagocytosis and melanization events for the removal of parasites [17]. But we do not have much information about the direct cell–cell interaction of the hemocytes and *P. vivax* free-circulating sporozoites (*fcSPZ*).

Recently we conducted a transcriptome based study, to understand that how hemocytes control the *P. vivax* free circulatory sporozoites (*fcSPZ*) population before salivary invasion [24]. Here we found that hemocyte encoded transcripts undergo a major shift during *P. vivax* infection. A detailed comparison of the *P. vivax* infected and uninfected hemocyte transcriptomes revealed that transcripts of organelle organization and riboprotein complex biogenesis have exclusively emerged during *P. vivax fcSPZ* infection. Altogether these findings suggested that the hemocyte population undergoes dynamic changes i.e., differentiate and increase the population in response to the *fcSPZ*. Through the immune database comparison, we found that AMPs like *Defensin and Gambicin* were exclusively induced when *fcSPZ* were circulating in the hemolymph. These findings were further validated by the real-time based experiments and depicted that *Defensin3* and *Gambicin* may likely play a crucial role during *P. vivax* late-phase immune function against *fcSPZ* infection. Hence, conclusively current findings illustrate that hemocytes rapidly proliferate and impart humoral immune responses against the parasite to limit the *fcSPZ* population before salivary gland invasion (**Figure 3**).

Apart from global transcriptomic changes undergone by the hemocyte population to manage *P. vivax* infection, we also found the species-specific molecular differences among the hemocyte encoded immune transcripts. *FBN9* which was previously considered as the potent anti-*Plasmodium* molecule and showed multi-fold upregulation during *P. berghei*/ *P. falciparum* infection [71, 72] was found to be downregulated during *P. vivax* infection. Novel molecules like *FREP12* and *FREP50* were predicted to be involved in the clearance of *P. vivax* sporozoites. Furthermore, storage proteins like *ApolipophorinIII*, *Hexamerin* were also found to be highly induced during *P. vivax* oocysts development, which further supported the previous evidence of host nutrient scavenging by the maturing oocysts [73–75].

4.2.2 Mosquito salivary glands: Gatekeeper of entry and exit for the parasite

The salivary glands are the crucial organ for the development and transmission of the *Plasmodium* to a vertebrate host. Salivary glands are paired epithelial organs that are located in the thorax, and consist of three lobes namely, two lateral and one median, where each lateral lobe is comprised of proximal and distal lobes [76, 77]. The proximal portion of the female glands produces enzymes involved in sugar metabolism, where distal lobes are shown to play roles in blood meal acquisition, *Plasmodium* invasion, and transmission. Although, studies suggest that only 10–20% hemolymph circulating sporozoites, manage to invade the salivary glands, however, the mechanism of this drastic reduction of 80–90% sporozoites is poorly

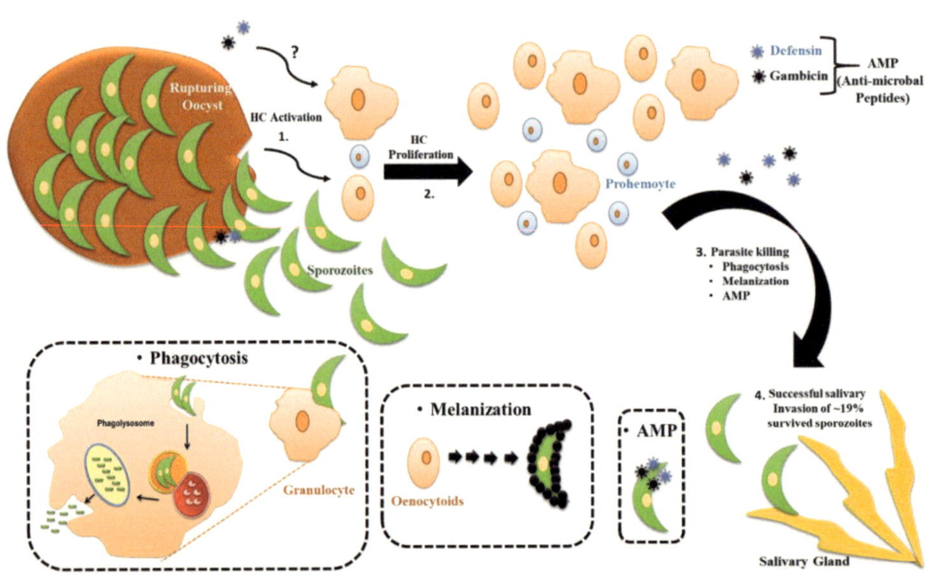

Figure 3.
*Direct interaction of free-circulatory sporozoites and hemocytes. Post oocyst maturation, sporozoites circulate freely in the hemolymph, in order to reach salivary glands for further transmission. But oocyst rupture triggers the mosquito immune system and activates the hemocyte proliferation, which leads to the sporozoite clearance by phagocytosis, melanization, and AMP (*Defensin *and* Gambicin*) production. Although the role of AMPs in hemocyte activation is still unclear.*

known [25, 78]. Accumulating evidence highlights that sporozoite invasion into the glands is mediated by salivary specific receptor-ligand interactions [18, 79].

The sporozoites must leave (egress) the oocyst after maturation to invade the salivary gland and to be transmitted to the next vertebrate host. The egress of sporozoite is mediated by a protease named *Cysteine protease* (ECP1) which ruptures the oocyst [80, 81]. The sporozoites are released into the hemolymph and carried to the salivary glands by the circulation of hemolymph in an anterior direction from the abdomen to head, and facilitate the sporozoite invasion to the salivary gland [82]. The salivary gland epithelium forms a physical barrier that pathogens must cross, and *Plasmodium* parasites are evolved with unique proteins that drive invasion by first binding to the salivary gland specific surface receptors [83]. The salivary invasion process completion occurs in two stages, where first, sporozoite binds to invade the salivary gland basal lamina; and second, then interacts with the plasma membrane of the epithelial cells favoring sporozoite internalization. During the invasion, sporozoites attach and invade the distal and medial lobe of the salivary glands, and this attachment and invasion are highly specific to the nature of *Plasmodium* species [84, 85].

Empirical evidence showing that the salivary glands serve as an active immune organ is largely lacking, except some studies highlighting that a *Serine Protease Inhibitor* (SRPN6) produced in the salivary epithelium limits gland invasion by *Plasmodium sporozoites*, and thus *SRPN6* serves as an important salivary invasion immune-marker [86]. Several putative salivary encoded factors such as *Saglin*, *CSP* binding proteins which effectively binds with sporozoites surface antigens such as *TRAP*, *CSP* are well known salivary receptors for sporozoites invasion [87–91]. However, several other salivary factors such as *Plasmodium Responsive Salivary1* (PRS1), *ESP*, *Peptide-O-xylosyl Transferase 1* (OXT1) have also been identified to play a crucial role in parasite invasion of both midgut and salivary glands [92–95]. Once inside the salivary glands, the parasite undergoes transcriptional reprogramming before its transmission to the next mammalian host.

Transmission of many viral and protozoan parasites to a vertebrate host requires their salivary injection with the mosquito saliva during blood-feeding, and thus the migration of sporozoites needs duct for the continuation of the life cycle. Mosquito saliva has a pleiotropic property such as anti-hemostatic, vasodilator, or anti-inflammatory properties and immune modulators, and basic function to facilitate blood-feeding [96–98]. However, saliva proteins can also have an indirect impact on pathogen development and transmission. For example, a recent study in mosquito *An. gambiae* shows that mosquito saliva proteins such as *AgTRIO* and *mosGILT* serve as an important mediator of the transmission of *P. falciparum*, and inhibition of this protein can reduce the parasite burden in the human host [99, 100]. Although, a major study on salivary-sporozoites interaction is restricted to *P. berghei* and/or *P. falciparum*, however, very limited information is available on the salivary-*P. vivax* interaction.

A comparative RNA-Seq analysis of uninfected and *P. vivax*-infected mosquito salivary glands suggests that salivary transcripts undergo substantial changes during *P. vivax* infection. The maturation of sporozoite seems to coincide with the change in gene expression essential for invasion and transmission. Findings of several classes of immune proteins such as *Heme-peroxidase*, *FADD*, *Gambicin*, *GNBP*, and multiple family proteins of *Serine proteases*, and *SCRC* in the *P. vivax* sporozoites invaded salivary glands highlighted their anti-*Plasmodium* immune role of salivary glands. The transcriptome of the infected salivary glands also revealed that *P. vivax* infection decreased the expression of apyrase significantly which suggests that *P. vivax* interferes with salivary secretion before probing and feeding to ensure their delivery into the next human host. These findings offer valuable new insights into the biology of malaria parasites. Manipulating tissue-specific immuno-physiology of the mosquitoes may halt the *Plasmodium vivax* development and hence the transmission (**Figure 4**).

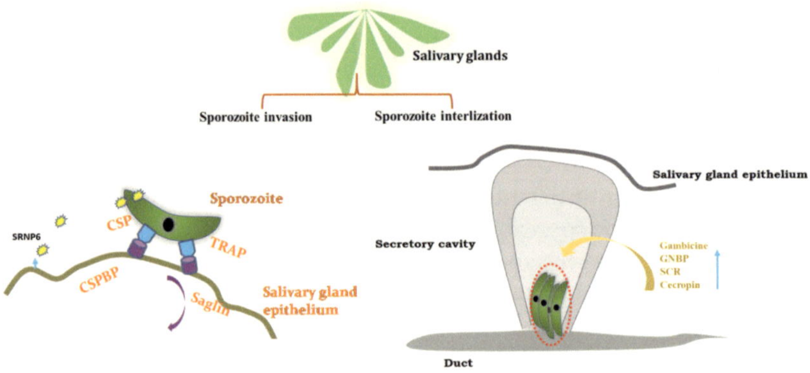

Figure 4.
*Proposed hypothesis of salivary gland-*Plasmodium *interaction and transmission: Free circulating sporozoite in the hemolymph recognize and attach to the basal lamina of the salivary gland receptor-ligand interaction;(1) initial attachment of the sporozoite mediated by interactions of carbohydrate residues on the basal lamina with a parasite CS, SGS1, MABEL. CSP binding protein and* Saglin *bind with CSP and TRAP respectively are an important component of salivary gland invasion. (2) Sporozoite internalization: After invasion sporozoite passes into the secretory cavity and sporozoites begin to assembly there as a large bundle form. Within the salivary duct component of the mosquito immune responses* Gambicin, Cecropin, GNBP, *and* SCR *family members presumably act upon sporozoite and limit the number.*

5. Conclusion

Plasmodium and mosquito host both are involved in the dynamic molecular relationship, where parasite tries to dodge the host immune system and utilize its nutrients for their successful proliferation/ transmission. On the contrary, the mosquito host immune system tries to restrict the parasite development and eliminate the remnants. During this ultimate battle, some host species defeat the parasite through its active immune system and become resistant but in others, the parasite smartly manipulates the host system and defends itself for successful transmission.

P. vivax is one of the neglected parasites which successfully manipulated the host system for its efficient transmission. *P. vivax* suppresses the microbiota proliferation to avoid nutritional competition as well as early immune responses. Different nutrient transport proteins like *Trehalase*, *Sterol Carrier*, *ApolipophorinIII*, etc. were modulated by the parasite for fulfilling its nutritional requirements. But still, mosquito hosts also developed species-specific immune effector molecules like *FREP50*, *FREP12*, *LRIM17*, etc. to block the parasite development. Likewise finding salivary-specific factors such as *Heme-peroxidase*, *SP24D* that are crucial to sporozoite invasion and survival, may further help to halt the progression of *Plasmodium* development and malaria transmission.

In summary, future functional exploration of the novel *P. vivax* specific host factors, will help in the development of transmission-blocking vaccines and the generation of new intervention techniques or modify current ones.

Acknowledgements

CC, ST, SK, RKD conceptualize the idea and drafted MS. CC, ST, SK, PS, JR and RKD reviewed, edited, corrected and revised the MS. We would like to thank ICMR-NIMR and funding agencies, CSIR/UGC/DST for infrastructural and financial support to conduct the research. CC, ST, and SK are recipients of DST (DST/INSPIRE/03/2014/003463), UGC (22/12/2013(II)EU-V), and CSIR (09/905(0015)/2015-EMR-1), fellowships, respectively.

Conflict of interest

No competing interests were disclosed.

Author details

Charu Chauhan, Sanjay Tevatiya, Seena Kumari, Punita Sharma, Jyoti Rani and Rajnikant Dixit*
Laboratory of Host-Parasite Interaction Studies, ICMR-National Institute of Malaria Research, Dwarka, New Delhi, India

*Address all correspondence to: rkd1976.rajnikant@gmail.com; dixitrk@mrcindia.org

References

[1] Antinori, S., Galimberti, L., Milazzo, L. & Corbellino, M. Biology of human malaria plasmodia including *Plasmodium knowlesi*. *Mediterranean Journal of Hematology and Infectious Diseases* (2012). doi:10.4084/MJHID.2012.013

[2] Singh, B. & Daneshvar, C. Human infections and detection of *Plasmodium knowlesi*. *Clinical Microbiology Reviews* (2013). doi:10.1128/CMR.00079-12

[3] White, M. T. *et al. Plasmodium vivax* and *Plasmodium falciparum* infection dynamics: Re-infections, recrudescences and relapses. *Malar. J.* (2018). doi:10.1186/s12936-018-2318-1

[4] Howes, R. E. *et al.* Global epidemiology of *Plasmodium vivax*. *American Journal of Tropical Medicine and Hygiene* (2016). doi:10.4269/ajtmh.16-0141

[5] White, N. J. Determinants of relapse periodicity in *Plasmodium vivax* malaria. *Malaria Journal* (2011). doi:10.1186/1475-2875-10-297

[6] Global Malaria Programme: WHO Global. *World malaria report 2019. WHO Regional Office for Africa* (2019).

[7] Mueller, I. *et al.* Key gaps in the knowledge of *Plasmodium vivax*, a neglected human malaria parasite. *The Lancet Infectious Diseases* (2009). doi:10.1016/S1473-3099(09)70177-X

[8] Mendis, K., Sina, B. J., Marchesini, P. & Carter, R. The neglected burden of *Plasmodium vivax* malaria. in *American Journal of Tropical Medicine and Hygiene* (2001). doi:10.4269/ajtmh.2001.64.97

[9] Bermúdez, M., Moreno-Pérez, D. A., Arévalo-Pinzón, G., Curtidor, H. & Patarroyo, M. A. *Plasmodium vivax* in vitro continuous culture: The spoke in the wheel. *Malaria Journal* (2018). doi:10.1186/s12936-018-2456-5

[10] Garrido-Cardenas, J. A., González-Cerón, L., Manzano-Agugliaro, F. & Mesa-Valle, C. *Plasmodium* genomics: an approach for learning about and ending human malaria. *Parasitol. Res.* (2019). doi:10.1007/s00436-018-6127-9

[11] Bourgard, C., Albrecht, L., Kayano, A. C. A. V., Sunnerhagen, P. & Costa, F. T. M. *Plasmodium vivax* biology: Insights provided by genomics, transcriptomics and proteomics. *Frontiers in Cellular and Infection Microbiology* (2018). doi:10.3389/fcimb.2018.00034

[12] Aly, A. S. I., Vaughan, A. M. & Kappe, S. H. I. Malaria Parasite Development in the Mosquito and Infection of the Mammalian Host. *Annu. Rev. Microbiol.* (2009). doi:10.1146/annurev.micro.091208.073403

[13] Tuteja, R. Malaria - An overview. *FEBS Journal* (2007). doi:10.1111/j.1742-4658.2007.05997.x

[14] Wiser, M. *Plasmodium* Life Cycle. *Tulane Univ.* (5) 1-4. (2009).

[15] Delves, M. J. *et al.* Male and female *Plasmodium falciparum* mature gametocytes show different responses to antimalarial drugs. *Antimicrob. Agents Chemother.* (2013). doi:10.1128/AAC.00325-13

[16] Bennink, S., Kiesow, M. J. & Pradel, G. The development of malaria parasites in the mosquito midgut. *Cellular Microbiology* (2016). doi:10.1111/cmi.12604

[17] Hillyer, J. F., Schmidt, S. L. & Christensen, B. M. Rapid phagocytosis and melanization of bacteria and *Plasmodium sporozoites* by hemocytes of the mosquito *Aedes aegypti*. *J.*

Parasitol. (2003). doi:10.1645/0022-3395(2003)089[0062:rpamob]2.0.co;2

[18] Mueller, A. K., Kohlhepp, F., Hammerschmidt, C. & Michel, K. Invasion of mosquito salivary glands by malaria parasites: Prerequisites and defense strategies. *International Journal for Parasitology* (2010). doi:10.1016/j.ijpara.2010.05.005

[19] Volohonsky, G. *et al.* Transgenic Expression of the Anti-parasitic Factor TEP1 in the Malaria Mosquito *Anopheles gambiae*. *PLoS Pathog.* (2017). doi:10.1371/journal.ppat.1006113

[20] Garcia, J. E., Puentes, A. & Patarroyo, M. E. Developmental biology of sporozoite-host interactions in *Plasmodium falciparum* malaria: Implications for vaccine design. *Clinical Microbiology Reviews* (2006). doi:10.1128/CMR.00063-05

[21] Smith, R. C., Barillas-Mury, C. & Jacobs-Lorena, M. Hemocyte differentiation mediates the mosquito late-phase immune response against *Plasmodium* in *Anopheles gambiae*. *Proc. Natl. Acad. Sci.* **112**, E3412–E3420 (2015).

[22] Dixit, R. *et al.* Salivary gland transcriptome analysis during Plasmodium infection in malaria vector *Anopheles stephensi*. *Int. J. Infect. Dis.* (2009). doi:10.1016/j.ijid.2008.07.027

[23] Qayum, A. A. & Telang, A. A protocol for collecting and staining hemocytes from the yellow fever mosquito *Aedes aegypti*. *J. Vis. Exp.* 2-5 (2011). doi:10.3791/2772

[24] Kumari S, Chauhan C, Tevatiya S, et al (2021) Current Research in Immunology Genetics changes of Plasmodium vivax tempers host tissue-specific responses in Anopheles stephensi. Curr Res Immunol 2:12-22. https://doi.org/10.1016/j.crimmu.2021.02.002

[25] Hillyer, J. F., Barreau, C. & Vernick, K. D. Efficiency of salivary gland invasion by malaria sporozoites is controlled by rapid sporozoite destruction in the mosquito haemocoel. *Int. J. Parasitol.* **37**, 673-681 (2007).

[26] Smith, R. C., Vega-Rodríguez, J. & Jacobs-Lorena, M. The *Plasmodium* bottleneck: Malaria parasite losses in the mosquito vector. *Mem. Inst. Oswaldo Cruz* **109**, 644-661 (2014).

[27] Oliveira, G. D. A., Lieberman, J. & Barillas-Mury, C. Epithelial nitration by a peroxidase/NOX5 system mediates mosquito antiplasmodial immunity. *Science (80).* (2012). doi:10.1126/science.1209678

[28] Garver, L. S., Dong, Y. & Dimopoulos, G. Caspar controls resistance to *Plasmodium falciparum* in diverse anopheline species. *PLoS Pathog.* (2009). doi:10.1371/journal.ppat.1000335

[29] Hillyer, J. F. & Strand, M. R. Mosquito hemocyte-mediated immune responses. *Curr. Opin. Insect Sci.* **3**, 14-21 (2014).

[30] Strand, M. R. The insect cellular immune response. *Insect Science* (2008). doi:10.1111/j.1744-7917.2008.00183.x

[31] Peterson, T. M. L., Gow, A. J. & Luckhart, S. Nitric oxide metabolites induced in *Anopheles stephensi* control malaria parasite infection. *Free Radic. Biol. Med.* (2007). doi:10.1016/j.freeradbiomed.2006.10.037

[32] Westenberger, S. J. *et al.* A systems-based analysis of *Plasmodium vivax* lifecycle transcription from human to mosquito. *PLoS Negl. Trop. Dis.* (2010). doi:10.1371/journal.pntd.0000653

[33] Romoli, O. & Gendrin, M. The tripartite interactions between the mosquito, its microbiota and

Plasmodium. *Parasites and Vectors* **11**, 1-8 (2018).

[34] Mukherjee, D., Chora, Â. F. & Mota, M. M. Microbiota, a Third Player in the Host–*Plasmodium* Affair. *Trends in Parasitology* (2020). doi:10.1016/j.pt.2019.11.001

[35] Sharma, P. *et al.* Altered Gut Microbiota and Immunity Defines *Plasmodium vivax* Survival in *Anopheles stephensi*. *Front. Immunol.* (2020). doi:10.3389/fimmu.2020.00609

[36] Molina-Cruz, A. *et al. Plasmodium falciparum* evades immunity of anopheline mosquitoes by interacting with a Pfs47 midgut receptor. *Proc. Natl. Acad. Sci. U. S. A.* (2020). doi:10.1073/pnas.1917042117

[37] Volohonsky, G. *et al.* Kinetics of *Plasmodium* midgut invasion in *Anopheles* mosquitoes. *PLoS Pathog.* **16**, 1-19 (2020).

[38] Smith, R. C. & Barillas-Mury, C. *Plasmodium* Oocysts: Overlooked Targets of Mosquito Immunity. *Trends Parasitol.* **32**, 979-990 (2016).

[39] Boonkaew, T. *et al.* Transcriptome analysis of *Anopheles dirus* and *Plasmodium vivax* at ookinete and oocyst stages. *Acta Trop.* (2020). doi:10.1016/j.actatropica.2020.105502

[40] Zuzarte-Luís, V. & Mota, M. M. Parasite Sensing of Host Nutrients and Environmental Cues. *Cell Host and Microbe* (2018). doi:10.1016/j.chom.2018.05.018

[41] Meireles, P. *et al.* GLUT1-mediated glucose uptake plays a crucial role during *Plasmodium* hepatic infection. *Cell. Microbiol.* (2017). doi:10.1111/cmi.12646

[42] Olszewski, K. L. & Llinás, M. Central carbon metabolism of *Plasmodium* parasites. *Molecular and Biochemical Parasitology* (2011). doi:10.1016/j.molbiopara.2010.09.001

[43] Stanway, R. R. *et al.* Genome-Scale Identification of Essential Metabolic Processes for Targeting the *Plasmodium* Liver Stage. *Cell* (2019). doi:10.1016/j.cell.2019.10.030

[44] Wang, A. *et al.* Glucose metabolism mediates disease tolerance in cerebral malaria. *Proc. Natl. Acad. Sci. U. S. A.* (2018). doi:10.1073/pnas.1806376115

[45] Tevatiya, S. *et al.* Molecular and Functional Characterization of Trehalase in the Mosquito *Anopheles stephensi*. *Front. Physiol.* (2020). doi:10.3389/fphys.2020.575718

[46] Labaied, M. *et al. Plasmodium* salvages cholesterol internalized by LDL and synthesized de novo in the liver. *Cell. Microbiol.* (2011). doi:10.1111/j.1462-5822.2010.01555.x

[47] Barletta, A. B. F. *et al.* Emerging role of lipid droplets in *Aedes aegypti* immune response against bacteria and Dengue virus. *Sci. Rep.* (2016). doi:10.1038/srep19928

[48] Zong, D., Li, J., Liu, X., Guo, T. & Ouyang, R. Identification of a Novel Pathogenic Folliculin Variant in a Chinese Family With Birt–Hogg–Dubé Syndrome (Hornstein-Knickenberg Syndrome). *Front. Genet.* (2020). doi:10.3389/fgene.2020.565566

[49] Nickerson, M. L. *et al.* Mutations in a novel gene lead to kidney tumors, lung wall defects, and benign tumors of the hair follicle in patients with the Birt-Hogg-Dubé syndrome. *Cancer Cell* (2002). doi:10.1016/S1535-6108(02)00104-6

[50] Schmidt, L. S. & Linehan, W. M. FLCN: The causative gene for Birt-Hogg-Dubé syndrome. *Gene* (2018). doi:10.1016/j.gene.2017.09.044

[51] Yang, T. *et al.* Folliculin Controls the Intracellular Survival and Trans-Epithelial Passage of *Neisseria gonorrhoeae*. *Front. Cell. Infect. Microbiol.* (2020). doi:10.3389/fcimb.2020.00422

[52] Jaramillo-Gutierrez, G. *et al.* Mosquito immune responses and compatibility between *Plasmodium* parasites and anopheline mosquitoes. *BMC Microbiol.* (2009). doi:10.1186/1471-2180-9-154

[53] Baxter, R. H. G. *et al.* A heterodimeric complex of the LRR proteins LRIM1 and APL1C regulates complement-like immunity in *Anopheles gambiae*. *Proc. Natl. Acad. Sci. U. S. A.* (2010). doi:10.1073/pnas.1010575107

[54] Ruiz, V. M. R. *et al.* Stimulation of a protease targeting the LRIM1/APL1C complex reveals specificity in complement-like pathway activation in *Anopheles gambiae*. *PLoS One* **14**, 1-14 (2019).

[55] Fraiture, M. *et al.* Two Mosquito LRR Proteins Function as Complement Control Factors in the TEP1-Mediated Killing of *Plasmodium*. *Cell Host Microbe* (2009). doi:10.1016/j.chom.2009.01.005

[56] Kwon, H., Arends, B. R. & Smith, R. C. Late-phase immune responses limiting oocyst survival are independent of TEP1 function yet display strain specific differences in *Anopheles gambiae*. *Parasites and Vectors* (2017). doi:10.1186/s13071-017-2308-0

[57] Subbarao, S. K., Nanda, N., Rahi, M. & Raghavendra, K. Biology and bionomics of malaria vectors in India: Existing information and what more needs to be known for strategizing elimination of malaria. *Malaria Journal* (2019). doi:10.1186/s12936-019-3011-8

[58] Gupta, L. *et al.* The STAT Pathway Mediates Late-Phase Immunity against *Plasmodium* in the Mosquito *Anopheles gambiae*. *Cell Host Microbe* **5**, 498-507 (2009).

[59] Hillyer, J. F. Insect immunology and hematopoiesis. *Dev. Comp. Immunol.* **58**, 102-118 (2016).

[60] De, T. Das *et al.* Interorgan molecular communication strategies of 'Local' and 'Systemic' innate immune responses in mosquito *Anopheles stephensi*. *Front. Immunol.* (2018). doi:10.3389/fimmu.2018.00148

[61] Castillo, J. C., Robertson, A. E. & Strand, M. R. Characterization of hemocytes from the mosquitoes *Anopheles gambiae* and *Aedes aegypti*. *Insect Biochem. Mol. Biol.* **36**, 891-903 (2006).

[62] Bryant, W. B. & Michel, K. *Anopheles gambiae* hemocytes exhibit transient states of activation. *Dev. Comp. Immunol.* (2016). doi:10.1016/j.dci.2015.10.020

[63] Honti, V., Csordás, G., Kurucz, É., Márkus, R. & Andó, I. The cell-mediated immunity of *Drosophila melanogaster*: Hemocyte lineages, immune compartments, microanatomy and regulation. *Dev. Comp. Immunol.* (2014). doi:10.1016/j.dci.2013.06.005

[64] Smith, R. C. *et al.* Molecular profiling of phagocytic immune cells in *Anopheles gambiae* reveals integral roles for hemocytes in mosquito innate immunity. *Mol. Cell. Proteomics* **15**, 3373-3387 (2016).

[65] Kwon, H. & Smith, R. C. Chemical depletion of phagocytic immune cells reveals dual roles of mosquito hemocytes in *Anopheles gambiae* anti-*Plasmodium* 2 immunity 3. (2018). doi:10.1101/422543

[66] Hillyer, J. F., Schmidt, S. L. & Christensen, B. M. Hemocyte-mediated phagocytosis and melanization in the

mosquito *Armigeres subalbatus* following immune challenge by bacteria. *Cell Tissue Res.* **313**, 117-127 (2003).

[67] King, J. G. & Hillyer, J. F. Spatial and temporal in vivo analysis of circulating and sessile immune cells in mosquitoes: Hemocyte mitosis following infection. *BMC Biol.* **11**, (2013).

[68] Blandin, S. *et al.* Complement-like protein TEP1 is a determinant of vectorial capacity in the malaria vector *Anopheles gambiae. Cell* **116**, 661-670 (2004).

[69] Yan, Y. & Hillyer, J. F. Complement-like proteins TEP1, TEP3 and TEP4 are positive regulators of periostial hemocyte aggregation in the mosquito *Anopheles gambiae. Insect Biochem. Mol. Biol.* (2019). doi:10.1016/j.ibmb.2019.01.007

[70] Niu, G. *et al.* FBN30 in wild *Anopheles gambiae* functions as a pathogen recognition molecule against clinically circulating *Plasmodium falciparum* in malaria endemic areas in Kenya. *Sci. Rep.* 7, (2017).

[71] Simões, M. L. *et al.* The *Anopheles* FBN9 immune factor mediates *Plasmodium* species-specific defense through transgenic fat body expression. *Dev. Comp. Immunol.* **67**, 257-265 (2017).

[72] Oliveira, S. B. *et al.* Identification of a fibrinogen-related protein (FBN9) gene in neotropical anopheline mosquitoes. *Malar. J.* **10**, 1-8 (2011).

[73] Whitten, M. M. A., Tew, I. F., Lee, B. L. & Ratcliffe, N. A. A Novel Role for an Insect Apolipoprotein (Apolipophorin III) in β-1,3-Glucan Pattern Recognition and Cellular Encapsulation Reactions. *J. Immunol.* (2004). doi:10.4049/jimmunol.172.4.2177

[74] Gupta, L. *et al.* Apolipophorin-III mediates antiplasmodial epithelial responses in *Anopheles gambiae* (G3) mosquitoes. *PLoS One* (2010). doi:10.1371/journal.pone.0015410

[75] Chauhan C, Das De T, Kumari S, Rani J, Sharma P, Tevatiya S, Pandey KC, Pande V, Dixit R Hemocyte-specific FREP 13 abrogates the exogenous bactorial population in the hemolymph and promotes midgut endosymbionts in Anopheles stephensi. Immunol Cell Biol. 2020 oct;98 (9):757-769. doi: 10.1111/imcb.12374

[76] Wells, M. B. & Andrew, D. J. Salivary gland cellular architecture in the Asian malaria vector mosquito *Anopheles stephensi*. *Parasites and Vectors* (2015). doi:10.1186/s13071-015-1229-z

[77] Dhar, R. & Kumar, N. Role of mosquito salivary glands. *Current Science* **85(9)**:1308- 1313 (2003).

[78] Akaki, M. & Dvorak, J. A. A chemotactic response facilitates mosquito salivary gland infection by malaria sporozoites. *J. Exp. Biol.* (2005). doi:10.1242/jeb.01756

[79] Sreenivasamurthy, S. K. *et al.* A compendium of molecules involved in vector-pathogen interactions pertaining to malaria. *Malar. J.* (2013). doi:10.1186/1475-2875-12-216

[80] Aly, A. S. I. & Matuschewski, K. A malarial cysteine protease is necessary for *Plasmodium* sporozoite egress from oocysts. *J. Exp. Med.* (2005). doi:10.1084/jem.20050545

[81] Boysen, K. E. & Matuschewski, K. Inhibitor of cysteine proteases is critical for motility and infectivity of *Plasmodium sporozoites*. *MBio* (2013). doi:10.1128/mBio.00874-13

[82] Rodriguez, M. H. & Hernández-Hernández, F. D. L. C. Insect-malaria parasites interactions: The salivary gland. in *Insect Biochemistry and*

Molecular Biology (2004). doi:10.1016/j.ibmb.2004.03.014

[83] Brennan, J. D. G., Kent, M., Dhar, R., Fujioka, H. & Kumar, N. *Anopheles gambiae* salivary gland proteins as putative targets for blocking the transmission of malaria parasites. *Proc. Natl. Acad. Sci. U. S. A.* (2000). doi:10.1073/pnas.250472597

[84] Sterling, C. R., Aikawa, M. & Vanderberg, J. P. The passage of *Plasmodium berghei* sporozoites through the salivary glands of *Anopheles stephensi*: An electron microscope study. *J. Parasitol.* (1973). doi:10.2307/3278847

[85] Rosenberg, R. Inability of *Plasmodium knowlesi* sporozoites to invade *Anopheles freeborni* salivary glands. *Am. J. Trop. Med. Hyg.* (1985). doi:10.4269/ajtmh.1985.34.687

[86] Pinto, S. B., Kafatos, F. C. & Michel, K. The parasite invasion marker SRPN6 reduces sporozoite numbers in salivary glands of *Anopheles gambiae*. *Cell. Microbiol.* (2008). doi:10.1111/j.1462-5822.2007.01091.x

[87] Ghosh, A. K. *et al.* Malaria parasite invasion of the mosquito salivary gland requires interaction between the *Plasmodium* TRAP and the *Anopheles* saglin proteins. *PLoS Pathog.* (2009). doi:10.1371/journal.ppat.1000265

[88] Okulate, M. A. *et al.* Identification and molecular characterization of a novel protein Saglin as a target of monoclonal antibodies affecting salivary gland infectivity of *Plasmodium sporozoites*. *Insect Mol. Biol.* (2007). doi:10.1111/j.1365-2583.2007.00765.x

[89] Sultan, A. A. *et al.* TRAP is necessary for gliding motility and infectivity of *Plasmodium sporozoites*. *Cell* (1997). doi:10.1016/S0092-8674(00)80511-5

[90] Sidjanski, S. P., Vanderberg, J. P. & Sinnis, P. *Anopheles stephensi* salivary glands bear receptors for region I of the circumsporozoite protein of *Plasmodium falciparum. Mol. Biochem. Parasitol.* (1997). doi:10.1016/S0166-6851(97)00124-2

[91] Myung, J. M., Marshall, P. & Sinnis, P. The *Plasmodium* circumsporozoite protein is involved in mosquito salivary gland invasion by sporozoites. *Mol. Biochem. Parasitol.* (2004). doi:10.1016/j.molbiopara.2003.09.002

[92] Rodrigues, J. *et al.* An epithelial serine protease, AgESP, is required for plasmodium invasion in the mosquito *Anopheles gambiae*. *PLoS One* (2012). doi:10.1371/journal.pone.0035210

[93] Chertemps, T. *et al. Anopheles gambiae* PRS1 modulates *Plasmodium* development at both midgut and salivary gland steps. *PLoS One* (2010). doi:10.1371/journal.pone.0011538

[94] Armistead, J. S., Wilson, I. B. H., Van Kuppevelt, T. H. & Dinglasan, R. R. A role for heparan sulfate proteoglycans in *Plasmodium falciparum* sporozoite invasion of anopheline mosquito salivary glands. *Biochem. J.* (2011). doi:10.1042/BJ20110694

[95] Ramakrishnan, C. *et al.* Salivary gland-specific *P. berghei* reporter lines enable rapid evaluation of tissue-specific sporozoite loads in mosquitoes. *PLoS One* (2012). doi:10.1371/journal.pone.0036376

[96] Ribeiro, J. M. Role of saliva in blood-feeding by arthropods. *Annual review of entomology* (1987). doi:10.1146/annurev.en.32.010187.002335

[97] Ribeiro, J. M., Rossignol, P. A. & Spielman, A. Role of mosquito saliva in blood vessel location. *J. Exp. Biol.* 108: 1-7; (1984).

[98] Fontaine, A. *et al.* Implication of haematophagous arthropod salivary

proteins in host-vector interactions. *Parasites and Vectors* (2011). doi:10.1186/1756-3305-4-187

[99] Schleicher, T. R. *et al.* A mosquito salivary gland protein partially inhibits *Plasmodium* sporozoite cell traversal and transmission. *Nat. Commun.* (2018). doi:10.1038/s41467-018-05374-3

[100] Dragovic, S. M. *et al.* Immunization with AgTRIO, a Protein in *Anopheles* Saliva, Contributes to Protection against *Plasmodium* Infection in Mice. *Cell Host Microbe* (2018). doi:10.1016/j.chom.2018.03.008

Chapter 8

Role of Dendritic Cells in Pathogen Infections: A Current Perspective

José Luis Muñoz-Carrillo, Juan Francisco Contreras-Cordero, Oscar Gutiérrez-Coronado, Paola Trinidad Villalobos-Gutiérrez, Luis Guillermo Ramos-Gracia and Jazmín Monserrat Vargas-Barboza

Abstract

Dendritic cells (DC) represent an important link between innate and adaptive immunity, which play an important role during the immune response against pathogens. There are several populations and subpopulations of DC, but mainly two subpopulations are characterized: the classic DC specialized in the processing and presentation of the antigen; and the plasmacytoid DC that have a high phagocytic activity and capacity for the production of cytokines. This chapter aims to present the current aspects related to the most relevant characteristics and functions of DC, as well as their role in host defense against infections by viruses, parasites, bacteria, and fungi.

Keywords: dendritic cells, pathogen infections, innate immune response, inflammation

1. Introduction

DCs represent an important link between innate and adaptive immunity. DCs are heterogeneous population of antigen-presenting cells that are crucial to initiate and polarize the immune response. Although, all DCs are capable of capturing, processing, and presenting antigens to T cells, DCs subtypes differ in origin, location, migration patterns, and specialized immunological roles [1]. All the DCs are continuously renewed by hematopoietic stem cell progenitor cell located in bone marrow, except of Langerhans cells (LCs) that develop from embryonic macrophages in the yolk sac and fetal liver, that are recruited in the epidermis during embryonic life. The process is not clearly, but hematopoietic stem cell is differentiated into granulocyte-macrophage progenitors (GMP) and multilymphoid progenitors (MLP), that have the potential to differentiate into macrophage-dendritic precursor (MPD) or common dendritic cell progenitor (CDP) like progenitor. These progenitors are subsequently differentiated into common monocyte progenitor (cMoPs), plasmacytoid dendritic cells (pDCs) and human equivalent of pre-DC, those are the most important to differentiate all subsets of DCs. cMoPs develop into blood monocytes, which differentiate into monocyte-derived DCs (MoDCs) in inflamed tissues, and

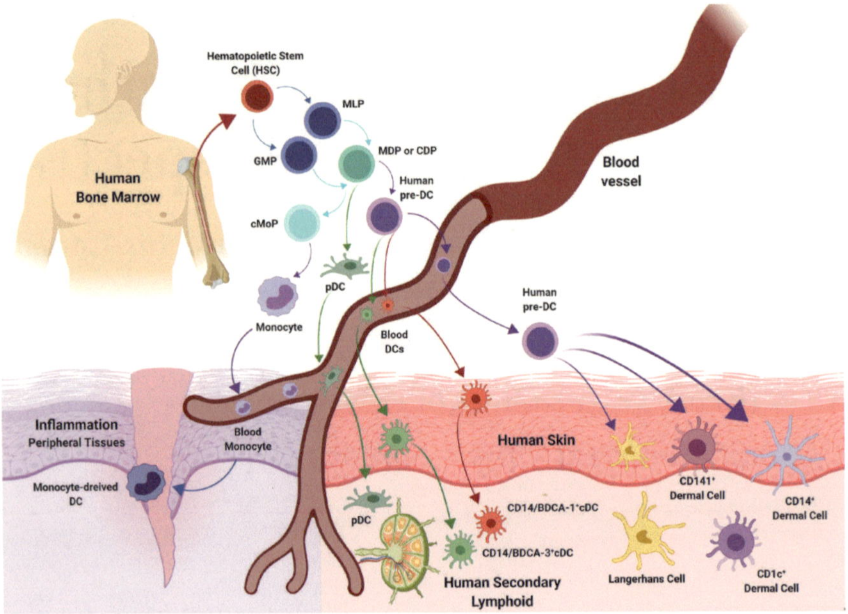

Figure 1.
Dendritic cell lineage development. The hematopoietic stem cell located in bone morrow is the progenitor of all DCs. Here the differentiation in multi-lymphoid progenitor and granulocyte-macrophage can become the human equivalent of macrophage-dendritic precursor (MPD) or dendritic cell progenitor (CDP). From this cell arise three important progenitor cells (cMoPs), pDCs and pre-DC, these last cells migrate to bloodstream or target tissue/organ to maturate and differentiate to become one of the different subsets of DCs. Explanation in the text. Figure modified by the authors from reference [3] and authorized to be published by bio-Techne (figure created by Muñoz-Carrillo et al., *with BioRender.com).*

fully mature pDCs along with unmatured pre-DCs migrate through the blood tissue. Immature human pre-DCs differentiate either in the bloodstream or in tissues following migration, developing thus in different DCs subsets (**Figure 1**) [2–4].

2. Dendritic cell subsets

The DCs are present in lymph organs and non-lymphoid organs, also in blood stream, afferent lymph, and mucous membranes. There are different ways to classify DCs, by its linage, as mentioned above there are cMoPs and pDCs. The cMocPs express typical myeloid antigens as CD11c, but lack of CD14 or CD16 and may be split in CD1c + and CD141+ fractions. While pDCs have expression of CD123, CD303 and CD304, with high or low expression of CD123, CD303 or CD304; the cluster of differentiation is determined in the differentiation of their precursor. These cells cMoPs and pDCs are classified into blood DCs [5, 6].

Inflammatory DCs derived from classical CD14+ blood monocytes, using granulocyte-macrophage colony-stimulating factor (GM-CSF) and interleukin (IL)-4. Monocytes are highly plastic, and they differentiate into DCs or different forms of macrophages (M1/M2). Human inflammatory exudates contain distinct inflammatory DC-like and macrophage-like cells and transcriptional profiling suggests a common monocyte origin. Key features of these cells are the expression of CD1c, CD1a, CD206, FcεR1, Sirpα but lack of CD16 and CD209. Non-classical monocytes and antigen 6-Sulpho LacNac DCs are a heterogeneous population and CD16+ monocytes possess distinct characteristics including higher major histocompatibility complex (MHC) class II and co-stimulatory antigen expression, classify as a type of blood DCs [5].

The functional-anatomical classification of DCs is widely vast, the classification of DCs are dependent of anatomical location or function, for example, DCs in heart are known as interstitial cells, in ganglia are known as interdigitating cells, but when DCs are in the afferent lymph are called veiled cells. Also, the function of these are different but sequential [5, 6]. Intestinal DCs are found in small intestine, lamina propia and gut associated lymphoid tissue. This DCs express CD103 and Sirpα in three different ways, such as CD103 + Sirpα- DCs, The CD103 + Sirpα+ DCs and CD103- Sirpα+ DCs. Most of these cells are located deeper into lamina propia, and express CD45, human leukocyte antigen-DR isotype (HLA-DR), CD14, CD64 and high levels of CX3C chemokine receptor 1(CX3CR1), and since these cells do not migrate to the lymph nodes, they have been depicted as intestinal macrophages. In the mesenteric lymph node DCs are a mixture of cells found in the peripheral blood. Such as peripheral blood, where soluble food bioactives may also be directly available for internalization by DCs in the draining lymph nodes *via* the conduit system [7].

LCs and microglia are two specialized self-renewing DCs, found them in stratified squamous epithelium and parenchyma of the brain, respectively. The LCs differentiate into migratory DCs, whereas microglia are considered as a type of specialized macrophage. There are DCs found in tissues and lymph nodes with marker CD14+, a subset of CD11c+, found in interstitial DCs; but they are more monocyte-like or macrophage-like, that may arise from classical monocytes [5].

2.1 Morphology

Immatures and matures DCs have different morphologic, immatures DCs monocyte-derived are spherical, irregular shape, with little cytoplasmatic projections, also abundant phase-dense granules (birbeck's granules or bodies) and irregular nucleus with small nucleoli. Once the DCs maturates shows stellate process, giving veiled appearance, with more extended dendrites projecting in many directions from the body cell [6, 8].

2.2 Maturation

The DCs have 3 stages precursor, immature and mature stage, but some authors do not count the precursor phase [6, 9]. Precursor phase course with any of the principal precursor as cMoPs, pDCs or Human equivalent of pre-DCs. It migrates from bone morrow to specific tissue or area, process leaded by chemokine chemoreceptors such as C-C chemokine receptor type 1 (CCR1), CCR5 and CCR6 and by adhesion molecules CD26P ligand. When the cell arrives to the corresponding tissue or place, it becomes immature DC. The immature DC express CCR1 and CCR3, where its ligand is in endothelium and inflammatory cells, promoting its migration to different organs and inflammatory tissues. This immature DC is capable of capture antigens by different receptors like Fc receptor, integrins, type C lectin and scavenger receptors such as lectin-type oxidized LDL receptor 1 (LOX-1) and CD91. Immature DC is characterized for various amounts of chemokines, so it can be extravasated to inflammatory tissue [6].

Once the DC has captured the antigen, this one is degraded to peptides that will get bind to MHC class I or class II. The endogenous antigens are processed by MHC class I, while exogenous antigens are processed by MHC class II. The lipidic antigens are presented by different molecules CD1(a-d) to T cell receptor (TCR) or natural killer T (NKT) cell. The differentiation process of immature DC to mature DC needs different signals to complete the process. To the immature DC gets mature, needs to stimulate T lymphocyte. This is possible when the antigen is presented to T lymphocyte by MHC class I or class II to TCR receptor and interaction of costimulatory molecules (CD28, CD40, CD54, CD58, CD80, CD83 and CD86) to activate T lymphocyte. Other molecules like adhesion (CD58, CD54) danger signal (CD40 ligand, tumor necrosis factor (TNF)-α, IL-1, IL-6, Interferon (INF)-α and Toll-like receptors (TLRs) agonist) [6, 8]. When the DC becomes mature, decreases the chemokine receptor expression of CCR1 and CCR5, whereas CCR7 increases. The CCR7 ligand is in ganglia walls and ganglionic paracortical zone. There, the mature DC secretes chemokines such as thymus and activation-regulated chemokine (TARC), macrophage-derived chemokine (MDC) or interferon gamma-induced protein 10 (IP-10), which recruits different types of T lymphocytes, monocytes, regulated on activation, normal T cell expressed and secreted (RANTES), macrophage inflammatory protein (MIP)-1α and MIP-β, to the local microenvironment [6].

2.3 Functions

DCs cells have many functions, but these can be globed within three functions. The first one is the main function as antigen presentation and activation of T lymphocytes as inducing adaptative immunity, with important release of cytokines for example IL-12 to differentiate T lymphocytes in T helper cell or cytotoxic lymphocytes. DCs have a wide range of properties including potent stimulation of native CD4+ T cells, cross-presentation to $CD8^+$ T cells and production of pro-inflammatory cytokines IL-1, IL-6, TNF-α, IL-12 and IL-23 [5, 9, 10]. The second function to induce tolerance. There are 2 types of tolerance central and peripheral. Central tolerance develops in thymus where a tolerance upon our own antigens occurs, and the reactive T lymphocytes to those antigens are destroyed, this also happen in bone morrow for B lymphocytes. The peripheral tolerance occurs when costimulatory molecules, last step of antigen presentation is not complete, there is a failure in activation of T lymphocyte, so the T lymphocyte become tolerogenic [6, 9, 10]. The third function to maintain immune memory in tandem with B cells. As mentioned before, there are population of DCs in ganglia, in the germinal center are found the follicular DCs which seems to be a reservoir of antigen and antibody complexes, that last an exceptionally long time up to months or years. This allows a constant stimulation of B cells to maintain memory [9].

There are others important functions of DCs, as their role in innate immunity, the DCs have pattern recognition receptor (PRR) and pathogen-associated molecular pattern (PAMPs) [10]. These receptor patterns activate TLR pathways, type C lectins and release pro-inflammatory cytokines to activate innate immunity system [8]. Also, DCs have been related to B lymphocytes proliferation and induction of T lymphocytes to suppress the immune response by missing of costimulatory molecules without IL-12, inducing T lymphocytes to secrete IL-10 and transforming growth factor (TGF)-β [6, 9].

3. Role of dendritic cells in viral infection

Since the discovery of DCs [11], the knowledge of the innate and adaptive immune response has been increasing significantly. At present, DCs are considered a key cell in immune response activation with multiple functions including the virus recognition, processing of viral antigen and as antigen-presenting cells to cells of specific immune response, serving as a bridge between innate and adaptive response [12]. DCs are bone marrow-derived cells and they can be found in different parts of the organism including mucous membrane, the skin, and lymphoid tissue [13]. Depending on surface markers, DCs can be classified as immature or mature myeloid DCs and plasmacytoid DCs [14, 15].

Immature DCs are inactive cell with high capacity to capture viral antigen. They are present in virtually all tissue with high probability to capture invading viruses. Immature DCs lack the capacity of antigen presentation. On the other hand, mature MDC is generated by an immature DC that was activated when recognized and processed viral antigen. Mature DCs function as antigen presenting cells (APCs). They express MHC-II molecules and different co-stimulators surface molecules that give them the antigen presentation capacity. Mature DCs also produce different cytokines to initiate antiviral immune response [16].

Likewise, plasmacytoid DCs also sense viral pathogen. They are called plasmacytoid DCs by its high resemblance to plasma cells. Although pDC has the ability of antigen-presenting, this function is low compared with MDC. However, pDCs contribute to both inflammatory process and antiviral state. They are specialized DCs that produce proinflammatory cytokines and high levels of IFN type I [17]. Both MDC and pDCs are present in lymphatic nodes where they are capable to present viral antigen to naïve T cell [18, 19].

3.1 From immature to mature cDCs in viral infection

Immature DCs are considered the sentinels of the immune response. These cells are distributed in practically all the body where they have the capacity of interact with the invading virus. They carry out the function against viral infection by different mechanisms. They can be infected by viruses or they can respond to molecules produced and secreted by other virus infected cells. When they are infected, DCs can respond in various ways, firstly, DCs have different receptors distributed on cell surface, cytoplasm, and specialized endosomes. TLRs and C-type lectins receptors (CLRs) are present in cell surface and some TLRs in endosomes, while retinoic acid-inducible gene (RIG), melanoma differentiation-associated protein 5 (MDA5) and nucleotide-binding oligomerization domain 2 (NOD2) are only present in cytosol [20–22]. TLRs have N-terminal ectodomains (ECDs) which recognize molecules of viruses. This ECDs are constructed by a tandem motif of leucine-rich repeats (LRRs) and forms a horseshoe structures [23]. Binding of TLRs with their ligand depends on these structures [24]. However, diverse receptors respond to an extensive repertoire of viral PAMPs. These viral PAMPs can be glycoproteins present on the viral external surface, viral genome, or replication intermediates formed during viral replication [25].

Depending on the activated receptor, DCs can produce proinflammatory cytokines or IFN. During maturation process DCs interact with the antigen and upregulate MHC-II to present antigen to naïve $CD4^{+}$T cells. In addition, DCs produce diverse surface molecules such as CCR7 which is necessary in trafficking into lymphatic nodes and CD40, CD80, and CD86 which are co-stimulatory surface factors that enable them to activate T naïve cell to initiate the adaptive immune responses [26, 27].

3.2 Differential PRR activation on dendritic cells

DCs is the main cell used to establish an effective immune response. At present, four subsets with different functions have been identify in human. Each subset of DC has different markers and a functional distinction that enable them to participate in different states to orchestrate an antiviral immune response. Each type of DC expresses different receptors that can be membrane-associated molecules or free in the cytoplasm. Activation of these receptors ends in different cytokine-proinflammatory production and interferon. Depending on cytokine produced, naïve CD4⁺T cells is differentiated into T helper effector cell [14].

Myeloid DCs, called classical or conventional DCs (cDCs) detect viral proteins through expression of membrane surface receptors such TLR-4 and DC-specific intercellular adhesion molecule 3 (ICAM3)-grabbing non-integrin (DC-SIGN) (see **Figure 2**) [28]. DC-SIGN support the initial immune response between T cells and DCs, but when DC-SIGN have contact with viral glycoproteins results in activation of signal transduction pathways than cause modulation of immune responses [29]. The signaling pathway triggered by DC-SIGN recruits Ras and the subsequent phosphorylation of the kinase RAF1 which is mediated by p21-activated kinases (PAKs) and Src Kinases. The activation of RAF1 induces phosphorylation of nuclear factor (NF)-κB increasing the transcriptional activation from IL-18, IL-10 and IL-12 promoter [29, 30].

The association of viral proteins through concave surface of TLR4-ECD induces two different pathways [31]. Myeloid differentiation primary response 88 (MyD88)-Dependent Pathway initiates with the recruitment of MyD88 adapter and subsequent activation of tumor necrosis factor receptor (TNFR)-associated factor 6 (TRAF6). Then TRAF6 activates the NF-κB essential modulator (NEMO), which is the regulatory subunit IKK complex and activates NF-κB causing its translocation to the nucleus, where induces gene expression such as IL-6 and IL-12 [21]. MyD88-Independent pathway recruits TIR-domain-containing adapter-inducing interferon-β (TRIF) [32]. TRIF activates TRAF3 and finally induce interferon regulatory transcription factor (IRF-3) activation and the subsequent IFN-β expression [21].

In addition to membrane surface receptors cDCs also have endosomal TLRs such as TLR-3 and TLR-7/TLR-8 which sense dsRNA and ssRNA respectively. Each receptor has a specific signaling pathways [14]. TLR-3 sense viral dsRNA through its largely uniform and flat horseshoe structure of TLR-ECD [33]. TLR3 has the same MyD88-Independent pathway with the activation of TRAF3 and subsequent IFN-β expression [32]. Viral ssRNA are sense by TLR-7 and TLR-8, these receptors activate MyD88 pathway with the recruitment of TRAF6 and TRAF3. Finally, activation of IRF-3 and IRF-7 induces IFN-β and IFN-α expression respectively (see **Figure 2A**) [21, 34].

In addition to DC-SIGN and TLRs, the viral genome can be exposed in the cytoplasm during the replicative processes or during direct penetration into the cell. NOD2 and RNA helicases such melanoma differentiation-associated protein 5 (MDA5) and RIG-1 detect dsRNA in the cytoplasm [35]. Interferon promoter

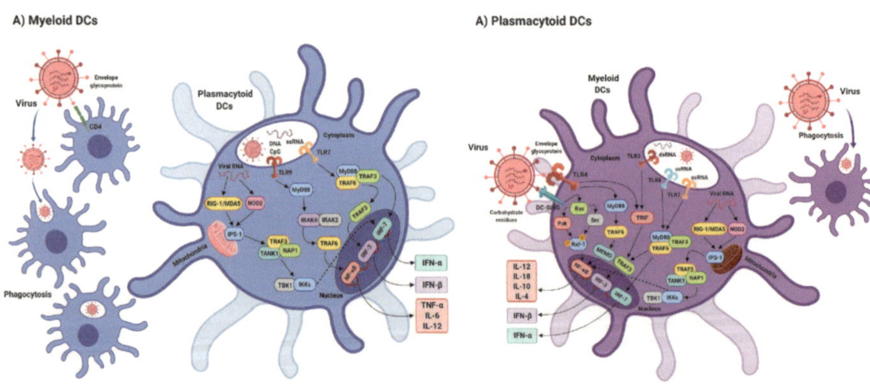

Figure 2.
Signaling pathway and cytokines production of DCs during viral infection. (A) Myeloid DCs and (B) Plasmacytoid DCs. Description in the text (figure created by Muñoz-Carrillo et al., *with BioRender.com).*

stimulator-1 (IPS-1) interacts with MDA5, RIG-1 and NOD2 *via* caspase activation and recruitment (CARD) domain. IPS-1 localizes in mitochondria and interacts with TRAF3. TRAF family member associated NF-κB activator (TANK) is recruited from TAF3 and interacts with TANK Binding Kinase 1 (TBK1) and the kinase IKKε [36–38]. Finally, TBK1 and IKKε interact *via* their C termini with NFκB activating kinase (NAK)-associated protein 1 (NAP1) [39]. This signaling pathway activates NFκB, IRF-3 and IRF-7 to express IL-12, IFN-β and IFN-α [38, 39].

On the other hand, pDCs not express DC-SIGN but express CD4 that can sense glycoproteins of viruses as human immunodeficiency virus (HIV). The viruses can enter through direct fusion with the cell membrane or through receptor-mediated endocytosis and activates different signaling pathways (see **Figure 2B**) [40, 41]. The endosomal receptors TLR-7 and TLR-9 are selectively express in pDCs and sense RNA or DNA respectively. This engage activates downstream signaling pathway [42]. TLR-9 and TLR-7 activates IRF-3 and IRF-7 like in cDCs signaling with final IFN-β and IFN-α expression respectively [43]. TLR-9 signaling pathways include the recruitment of Interleukin-1 receptor-associated kinase 4 (IRAK4) through its death domain. Activated IRAK4 interacts with IRAK2. This complex associates with TRAF6 to final activation and nucleus translocation of NF-κB and leads TNF-α and IL-6 production [17, 44, 45]. pDCs can also be infected by direct penetration of virus and the viral genome can be sense by RIG-1, MDA5 and NOD2. The signaling in the pDCs is through IPS-1 pathways as the same way that on cDCs [20, 22]. This pathway activates NFκB, IRF-3 and IRF-7 to express IL-12, IFN-β and IFN-α respectively [38, 39].

Other subsets of DCs are the LCs and Interstitial DCs (IDCs), these kinds of DCs are commonly the first DCs that have contact with some virus [46]. LCs are localized in mucosal stratified squamous epithelium and skin epidermis. LCs express different CLR: CD207 or Langerin. Moreover, LC has a low expression of TLR4 and expression of TLR-3, –7 and – 8 [14, 47]. LCs activated finally express IL-8, IL-6, TNF-α [48]. On the other hand, the IDCs are localized in the epidermis and express similar receptors that cDCs like DC-SIGN and TLR-3, -4, -7 and -8 and have similar signaling pathways [14].

Activation of the antiviral response generated by immune system depends largely on the activation of dendritic cells. Each subtype of this family of antigen-presenting cells have an important role by processing viral antigens that trigger different signaling pathways through their distinct receptors. The consequence of this signaling pathway results in the expression of various cytokines involved in the activation of immune cells. For this reason, a better knowledge about how different immune cells subtypes can induce distinct pathways is required for a better vision of whole antiviral response.

4. Role of dendritic cells in parasitic infection

In parasitic infections is difficult to generalize about the mechanisms of anti-parasitic immunity because there is a great variety of different parasites that have different morphology and reside in different locations of tissues and hosts during their life cycles [49]. For this reason, we will talk about the role of dendritic cells in protozoa and helminths infection, two of the main parasites of medical importance for human health.

DCs have the capacity to recognize different molecules in the surface of parasites and are efficient phagocytes; thus, several intracellular parasites reside inside DCs. Once DCs phagocytose intracellular parasites, they can exert their microbicidal capacities, although it has been shown that they are not as efficient in the destruction of microorganisms as other phagocytes such as macrophages and neutrophils. Once internalized, DCs process antigens for presentation to T cells [50].

4.1 Parasitic protozoan infections

Protozoan parasites are pathogens that have developed additional and sophisticated strategies to escape the immune attack of the host. This is because their life cycles generally involve several stages of specific antigenicity, which facilitates their survival and propagation within different cells, tissues, and hosts [51]. Frequently, the host is unable to eliminate protozoan infections, which often results in chronic disease or irreparable infections, in which the host continues to act as a reservoir of parasites, a cause of great concern due to their prevalence, morbidity and mortality [52, 53]. This host resistance to protozoa infections depends mainly on the development of a T helper type 1 (Th1) response and on the production of IL-12 by APCs [54]. Therefore, the classical reaction of the host to infections by protozoan parasites is the maturation of different subsets of DC, and in some cases, the activity of these cells leads to a response that is effective in controlling the infection [55].

Among the most important protozoan parasites are those that living in human blood and tissues, which can cause fatal diseases. The immune response against protozoan infections involves a strong innate immune response followed by predominantly a Th1 response. The innate immune system is comprised of several cell types, including DCs. Recognition of pathogens by these cell types leads to phagocytosis in some cases, and the production of pro-inflammatory cytokines, which assist in shaping the subsequent adaptive immune response (see **Figure 3**) [56].

During the parasitic protozoan infections different PRRs present in DCs are involved in the recognition PAMPs of parasites. In trypanosomiasis, the

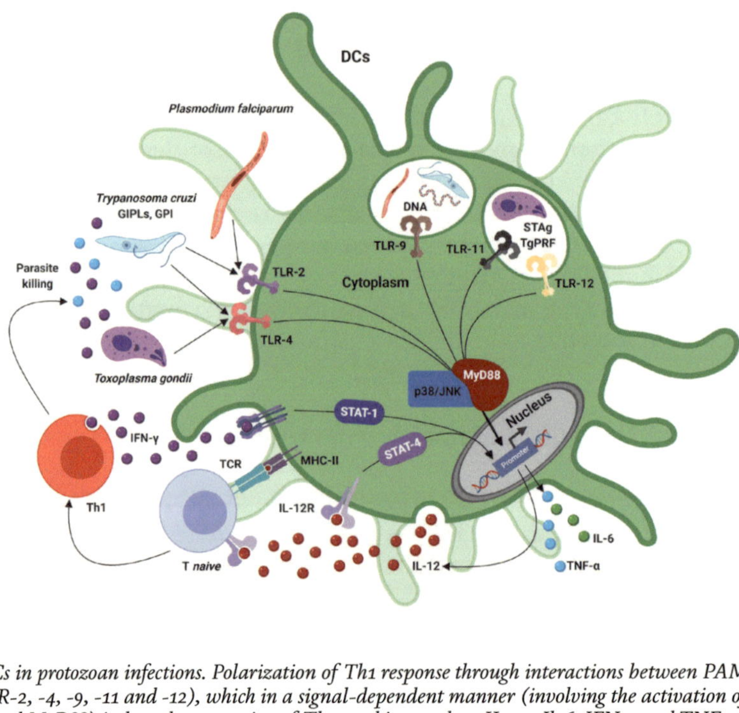

Figure 3.
Role of DCs in protozoan infections. Polarization of Th1 response through interactions between PAMPs and PRRs (TLR-2, -4, -9, -11 and -12), which in a signal-dependent manner (involving the activation of MAPKs p38/JNK and MyD88) induce the expression of Th1 cytokines such as IL-12, Il-6, IFN-γ and TNF-α. the PRRs from protozoa induce the presentation of antigens, the co-stimulation, and the expression of the cytokine IL-12, IFN-γ production by DCs during Ag presentation, by signaling pathway STAT-4. Description in the text. (figure created by Muñoz-Carrillo et al., *with BioRender.com).*

glycoinositolophospholipids (GIPLs) and glucosylphosphatidylinositol (GPI) anchors from Trypanosoma cruzi are recognized by TLR-4 and TLR-2, respectively, inducing the inflammatory cytokines production [57, 58]. Likewise, the DNA of *T. cruzi* stimulates the production of cytokines in a manner dependent on TLR-9 and synergizes with the GPI anchor of TLR-2 in the induction of cytokines [59], such as IL-12 by activation of the p38 pathway [60].

Toxoplasma gondii is a parasite that can infect any nucleated host cell, but it has a preference for cells of the immune system, including DCs [61]. Currently, the participation of TLRs in the recognition of *T. gondii* is not very clear. On the one hand, studies have shown that the soluble parasite extract (STAg) of *T. gondii* induces the production of IL-12 through the binding of *Toxoplasma profilin* (TgPRF) with TLR11 in DCs, signaling pathway MyD88 [62–65]. In fact, it has been shown that TgPRF is not required for the intracellular growth of *T. gondii*, but it is indispensable for host cell invasion and active egress from cells [65], and it is critical for the IL-12 production, especially in plasmacytoid DCs [66]. On the other hand, studies show that the absence of either TLR-2 or TLR-4 in DCs does not modify the production of IL-12 in response to STAg [62]. Other authors have reported the involvement of the TLR4-dependent signaling pathway in *T. gondii* independent of the MyD88 pathway [67]. However, reports have shown that mice deficient for TLR-2, TLR-4 or TLR-11 survive *T. gondii* infection, suggesting that *T. gondii* recognition may be associated with an additional signaling pathway MyD88-TLR-dependent. This additional signaling pathway could be by binding of TgPRF with TLR-12, since it has been observed that TLR-12-deficient mice succumb rapidly to *T. gondii* infection [62, 63, 66, 68]. On the other hand, *T. gondii* is capable to activate the JAK/STAT signaling pathway to facilitate survival within the host, blocking IFN-γ-mediated-STAT1-dependent proinflammatory gene expression in APCs. This is through sustained STAT-1 phosphorylation and nuclear translocation in bone marrow-derived DCs (BMDCs). However, in combination with IFN-γ, T. gondii simultaneously blocks IFN-γ-induced STAT 1 transcriptional activity avoiding the DCs activation by IFN-γ [69].

Plasmodium falciparum is capable to activate DCs through TLR-2 [58, 70, 71] and TLR-9, inducing the production of proinflammatory cytokines [72]. Depending on the DCs population that are activated during Plasmodium infection, it will be the type of cellular immune response that the host will mount against the infection. On the one hand, it has been observed that DCs subpopulations such as $CD8^+CD11b^-DC$ (located in the peripheral lymph nodes), mature ($CD40^+$) spleen DC and ($CD8\alpha^+CD11b^-$ and $CD8\alpha^-CD11b^+$) DCs [73, 74], are associated to the protective effect of $CD8^+$ T-cells, which produce INF-γ and induce parasite death, reducing the parasite burden in hepatocytes [75–78]. On the other hand, during the acute phase of infection $CD8\alpha$-$CD11b^+DC$ activates $CD4^+$ T-cells, inducing the production of IL-12, IL-6, IFN-γ and TNF-α [79–83].

4.2 Parasitic helminth infections

Helminth parasites, like protozoan parasites, have significant differences in their biological life cycles, which are reflected in the differences in clinical outcomes seen among helminth parasites. Pathological consequences of most helminth infections have been associated with both with the parasite intensity (or burden) and the relative acuteness or chronicity of the infection, because the helminth parasites modulate/regulate the host response to themselves (parasite-specific immunoregulation) [84].

The immune response against helminths is characterized by the induction of an early immune response of type Th1, with subsequent predominance of a Th2 type immune response, resulting in a mixture of both Th1/Th2 responses [85, 86], which are dependent on the immune responses mediated by CD4⁺ T cells [87]. These CD4⁺ T cells can function as APCs and play a key role in establishment the cytokine environment, thus directing their differentiation either by suppressing or favoring the inflammatory response at the intestinal level, which is crucial for the expulsion and elimination of the parasite (see **Figure 4**) [88].

This implies that the helminths have developed strategies, such as the evasion or suppression of the host immune response, which prevent their expulsion and allow their long-term survival. It is believed that the modulating effects of the immune system arise from the ability of the helminth to regulate the host immune response, developing mechanisms for the modulation of DCs as key players in the initiation and polarization of adaptive immune responses [89–91].

During the intestinal infection by helminths, the polarization of the cellular immune response to a Th1 type immune response depends on the type of signal derived from DCs. For example, *Trichinella spiralis* larvae group (TSL-1) antigens induce the DCs maturation [92], leading to the expression of MHC II [93, 94], promoting the development of a Th1 type cellular immune response [95]. Several studies, both *in vitro* and *in vivo*, have shown that during the early stage of intestinal infection by *T. spiralis* there is a significant increase of Th1 cytokines such as IL-12 [96, 97], INF-γ [95–98], IL-1β [97–99] and TNF-α [96, 97, 100]. It is possible that this Th1 response is mediated through the TLR-4 activation in DCs by TSL-1, through the signaling pathway TLR4/MyD88/NF-κB [101, 102]. Another example is double-stranded RNA from schistosome eggs has been implicated in the activation of DCs *via* TLR-3, resulting in a Th1-polarized response [103, 104].

Intestinal DCs are classified according to their unique or combined expression of CD11b and CD103, as well as the dependence on either interferon regulatory factor 4 or 8 (IRF4 or IRF8) for their development and/or survival. The intestinal DCs are

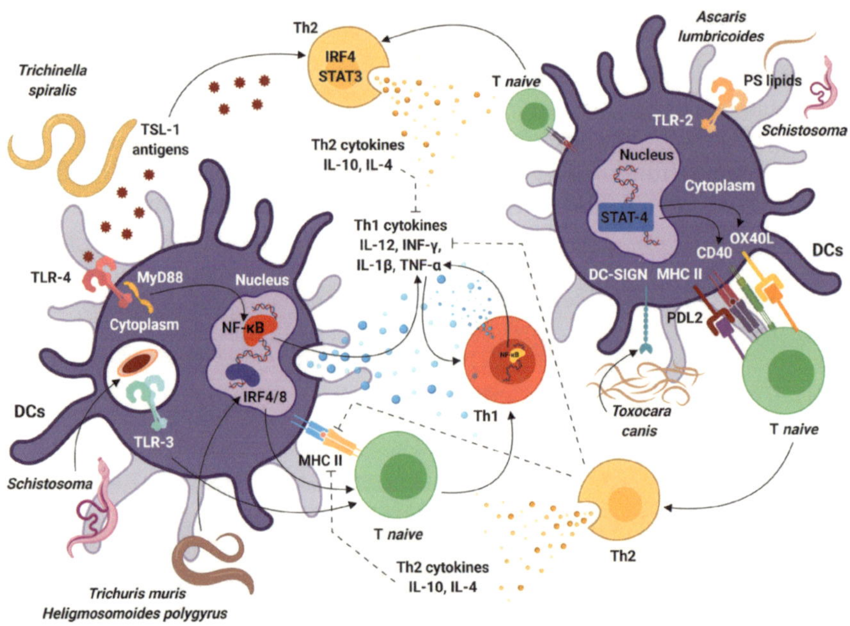

Figure 4.
Role of DCs in helminth infections. The immune response against helminths is characterized by the induction of an early immune response of type Th1, with subsequent predominance of a Th2 type immune response, resulting in a mixture of both Th1/Th2 responses. The polarization of the cellular immune response to a Th1/Th2 type immune response depends on the type of signal derived from DCs. Description in the text. (figure created by Muñoz-Carrillo et al.*, with BioRender.com).*

capable of process antigens, migrating to mesenteric lymph nodes upon activation, and priming naive T cells. However, IRF8-dependent CD103$^+$ DCs are important for the generation of type 1 responses of both helper and cytotoxic T cells, thus promoting *Trichuris muris* and *Heligmosomoides polygyrus* chronicity. In contrast, IRF4-dependent CD11b$^+$ DCs in the induction of Th2 immunity, notably during infection with *Nippostrongylus brasiliensis*, *T. muris*, and the parasitic trematode *Schistosoma mansoni* [105].

On the other hand, the PRRs from helminths can also activate the DCs for the induction of the Th2 response by interacting with the TLR and CLR. This interaction may promote Th2 responses by suppressing antigen presentation, co-stimulation and/or expression of Th1-promoting cytokines by directly interfering with these pathways. DCs that drive Th2 responses typically exhibit specialized markers, such as CD301b, PDL2, and CD11b, and several receptors for the Th2-related cytokines IL-4R, IL-13R, IL-25R, TSLP-R, and IL-33R. Additionally, the extracellular signal-regulated kinase (ERK) and signal transducer and activator of transcription 4 (STAT4) pathway upregulates the costimulatory molecules, CD40, OX40L, and Jagged. Activation of the major transcription factors interferon regulatory factor 4 (IRF4) and KLF4 inhibits IL-12 production and increased IL-10 secretion. These factors typically act individually or in concert to orchestrate Th2 responses in helminth infections [106–108].

In *T. spiralis* infection, the initial exposure of TSL-1 antigens of *T. spiralis* activated CD4$^+$ T cells, as well as DCs, leading to the secretion of large amounts of IL-10. IL-10 suppress cell markers, the proliferation and antigen presentation by DCs and inhibition of IL-12 secretion. In addition, TSL-1 increased the both IL-4 and IL-10 production derived from Th2 cells with a decrease in INF-γ production, polarizing the immune response to a strong Th2 cellular immune response, protective and responsible for the *T. spiralis* expulsion [109]. In addition, it has been shown that phosphatidylserine (PS) lipids derived from schistosomes and ascaris worms, which carry TLR2-activating molecules, promote Th2 responses through DCs [110]. Further, it was found that antigens of *Toxocara canis* were recognized by DC-SIGN expressed on DCs [111], and the induction of a Th2 response *in vivo* by antigens of the parasitic nematode *Brugia malayi*, as well as the free-living nematode *Caenorhabditis elegans*, was found to be dependent on intact glycans [112]. These findings together suggest that certain helminth glycans can serve as PAMPs that instruct DCs through CLR to boost polarized Th2 responses [113].

5. Role of dendritic cells in bacterial infection

Activated DCs are involved in the response to infections, which induces an increase in MHC expression, adhesion, and costimulatory molecules. The recognition of intracellular pathogens derived from mycobacterial cell wall components (lipids/glycolipids) such as phosphatidyl-myo-inositol mannoside, lipo-mannan, lipoarabinomannan, mycolic acids, lipopeptides, and phosphoinositol-containing lipids is given through the TLR-2, TLR-4, TLR-9, TLR-8 and the TLR1/TLR6 that heterodimerize with the TLR-2 [114, 115]. The signaling pathway that occurs in almost all TLRs is through MYD88, while for TLR4 the signaling pathway can be through MYD88 and TRIF [116, 117]. The activation of these receptors induces the activation of mitogen-activated protein kinase (MAPK) and NF-κB producing proinflammatory cytokines in DCs (see **Figure 5**). Other antigens derived from *Mycobacterium tuberculosis* such as lipoamide dehydrogenase C (Rv0462) induce the maturation and activation of DCs, increasing the expression of costimulatory molecules, MHC II and proinflammatory cytokines such as TNF-α, IL-1β, IL-6,

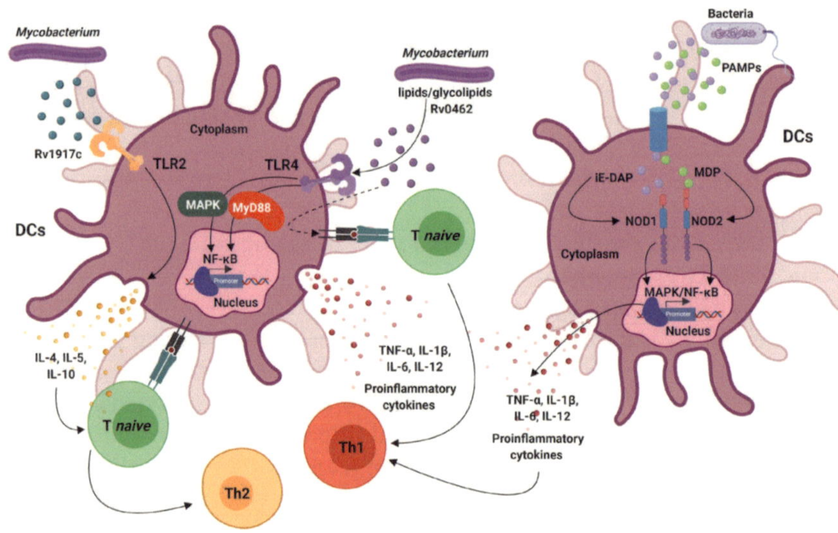

Figure 5.
Role of DCs in bacterial infections. The TLRs are involved in the recognition of mycobacterial antigens. The activation of TLR-4 and TLR-2 by these antigens leads to an intracellular signaling pathway, leading to a Th1 and Th2 response, respectively. NOD-like receptors (NOD 1 and NOD 2) recognize bacterial peptidoglycans (DAP and MDP), the downstream signaling activates NF-κB and MAPK generating a Th1 response. Description in the text (figure created by Muñoz-Carrillo et al., *with BioRender.com).*

and IL-12, which leads to a Th1 immune response [118, 119]. Another protein that induces the maturation of DCs is RV2220 is a glutamine synthetase (GS) type I enzyme derived from M. tuberculosis, which induces the upregulation of MHC I and MHC II as well as CD80 and CD86, which leads to a Th1 response or Th2 or to regulatory T cell, through the secretion of cytokines such as, TNF-α, IL-6, IL-1β, IL-12 or IL-10, activating the MAPK and NF-κB pathway [120]. Different proteins that derive from *M. tuberculosis* trigger different responses, as cell wall-associated/secretory Rv1917c antigen acts as a ligand of TLR-2, which induces the maturation of DCs secreting IL-10 and inducing the production of IL-4, IL-5 and IL-10 in CD4+ T cell which leads to a Th2 response (see **Figure 5**) [121].

On the other hand, DCs infection with other bacteria of the type *Listeria monocytoges*, *Shigella flexneri*, *Salmonella typhimurium* and *Francisella tularensis*, can activate inflammasome receptors [122]. The inflammasome is a multiprotein complex that contains one or more Nod-like receptors (NLRs) and regulates caspase-1 activity [123, 124], this complex is formed by at least three elements: (1) an inflammatory caspase (caspase-1, caspase- 11); (2) an adapter molecule such as apoptosis-associated speck-like protein containing a CARD, caspase recruitment domain (ASC); and (3) a sensor protein such as NLR Family Pyrin Domain Containing 1 (NLRP1), NLRP3, NLRP12, NAIP1, NAIP2, NAIP5, or absent in melanoma 2 (AIM2) [125]. The NLRP1 inflammasome is activated by anthrax lethal toxin, a toxin produced by *Bacillus anthracis* [126]. The toxin is composed of a protective antigen and lethal factor, the protective antigen generates pores in the membrane of the host while the lethal factor enters the cell and short NLRP1b and leads to inflammasome activation [127]. The NLRP3 inflammasome is activated by ligands derived from pathogens such as microbial cell wall components, nucleic acids, and pore-forming toxins [128]. Activation NLRP3 inflammasome require two signals: the priming which occurs when cells are activated by a PRR and activates the NF-kB, that induce the production of NLPR3, pro-IL-1β and pro-IL-18 and cytokines proinflammatory drugs such as IL-6, IL-8 and TNF-α. Subsequently the second signal carrying the assembly for inflammasome activation of caspase-1 occurs, which gives rise to the production of

IL-1β and IL-18 responsible for maintaining the inflammatory response [129]. The NLRC4 inflammasome is activated by the bacterial flagellar protein flagellin, as well as the Salmonella type III secretion system, this inflammasome does not interact directly with its activator, the NAIPs proteins do (NLR family), which recognize the ligands and induce activation of the NLRC4 inflammasome [130, 131]. The double chains of microbial DNA present in the cytosol are recognized by the AIM2 inflammasome, this receptor contributes to host defense when pathogens do not have ligands that stimulate PRRs such as flagellin and LPS, such as *Brucella spp* and *Francisella spp*. This receptor binds to DNA and oligomerizes with ASC to then form the caspase-1 activating inflammasome, which leads to the secretion of cytokines such as IL-1β and IL-18 [132]. The cytokines that are produced through the inflammasome not only contribute to the defense of the host against infections, they also induce a Th17 response, this differentiation is driven by IL-1β, and is regulated by the factors NF-κB, activator protein 1 (AP-1) or the signaling way of the MAPK [133]. After the binding of IL-1β to IL-1R, signaling occurs through MYD88 until activating NF-κB, which induces the production of proinflammatory cytokines leading to a Th17 phenotype, in this differentiation IL-1β synergizes with IL-6 which upregulates the master transcription factor of Th17 cells, such as STAT3, IRF4 and RAR-related orphan receptor gamma (RORγt) [134]. The Th17 response is a typical response that occurs against extracellular bacteria such as *Klebsiella pneumoniae*, *Bordetella pertussis*, or *Streptococcus pneumoniae* and is characterized by a vigorous response of neutrophils which is coordinated by the Th17 cells, an alteration in IL-17 signaling increases the susceptibility to infection of these bacteria [135]. Although the defense of the host against extracellular bacteria is considered mainly associated with the

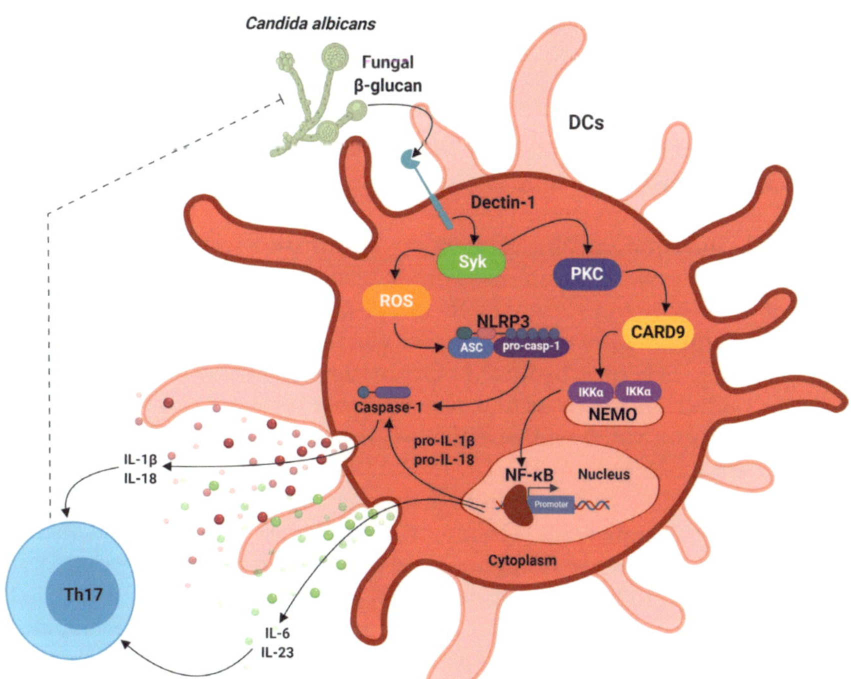

Figure 6.
Role of DCs in fungal infection. Antigens derived from fungi such as b-glucan which are recognized by Dectin-1, this leads to a downstream signaling pathway activating NF-kB producing IL-6 and IL-23 leading to a Th17 phenotype. The union of Dectin-1 whit b-glucan also leads to the activation of ROS, which can NLRP3 inflammasome assembly activating caspase-1 which cuts the pro-IL-1 and pro-IL-18 generating its active forms, which together with IL-23 activates the Th17 phenotype. Description in the text (figure created by Muñoz-Carrillo et al., *with BioRender.com).*

Th17, some authors indicate that effective protection requires the synergism of Th1 and Th17 cells, as it is for *Bordetella perussis* that induces the production of IFN-γ in the phase maximum infection and decreases its expression as time passes reaching basal levels at 14 days post-infection, however the Th17 response is persistent and production of IL-17 remains high even when the infection has been eliminated [136].

Other receptors involved in the response to pathogens are NOD1 and NOD2 receptors make up the family of NOD-like receptors containing a CARD domain (NLRC) [137]. These receptors are highly expressed in DCs and act as intracellular PRRs that recognize bacterial peptidoglycans [138–140]. NOD1 mainly recognizes γ-D-Glu-meso-diaminopimelic acid (DAP) while NOD2 recognizes muramyl dipeptide (MDP) [141]. Once the activation of these receptors occur, the downstream signaling activates NFκB through the union of its CARD domain to the protein kinase RIP2, which in turn recruits IRAK2, TRAF6, TAK1 binding protein (TAB1) and transforming growth factor-β-activated kinase 1 (TAK1) to activate the IKK complex, these events result in the degradation of IκBα inhibitor which leads to the translocation of NFκB to the nucleus and induce the expression of proinflammatory mediators [142]. In addition to the NFκB pathway, the stimulation of NOD1 and NOD2 leads to the activation of MAP kinases p38, ERK, and JNK pathway *via* RIP2. This event facilitates the formation of a multiprotein complex called "Nodosome" that leads to the production of inflammatory and antimicrobial agents mediated by NFκB and MAPK (see **Figure 6**) [143].

6. Role of dendritic cells in fungal infections

Infections caused by opportunistic fungal pathogens include *Aspergillus fumigatus*, *Cryptococcus neoformans* and thermal dimorphic fungi (*Histoplasma capsulatum*, *Blastomyces dermatitidis*, *Paracoccidioides brasiliensis*, *Coccidioides immitis*, *Penicillium marneffei* and *Sporothrix schenckii*) and *Candida albicans*, the latter being a normal inhabitant of the human intestine, however as a pathogen has been associated with various serious diseases ranging from severe mucocutaneous allergy to bloodstream infections [144, 145].

DCs are the only ones capable of decoding information related to fungi [146]. The activation of the various immunity mechanisms is carried out efficiently by the DC that decode the signals sent by the fungi and translate them into an immune response of T helper (Th) *in vitro* and *in vivo* where the DC recognize each fungal morphotype of specific form by means of different recognition receptors which triggers the production and co-stimulation of cytokines [144]. For the immunological processes to be activated against different classes of fungi, the differentiation of the naive CD4 + T cells towards the Th1 or Th17 subtype is essential, which occurs by interaction with dendritic cells through different cytokines, these subsets of cells Th1 and Th17 play an important role in protection against various fungal diseases [147]. To be contained and resistant to fungal infections it is necessary that DC are activated since they produce cytokines of the IL-1 family, such as IL-1β and IL-18 and which activate other innate immune cells, or they modulate the development of the acquired immune response. IL-1β plays an important role in the inflammatory immune response and polarization of Th17 cells, whereas IL-18 participates in the differentiation of Th1 cells, but may also be responsible for the expansion of Th2 cells in the absence of IL-18 [148] IL-12 and IFN-γ promote Th1 differentiation, while TGF-β, IL-6, IL-1, IL-21 and IL-23 promote the differentiation and maintenance of Th17. The release of these cytokines by DCs is in turn regulated by innate receptors activated in response to fungal infection [149]. In order for the effective response of the host to the fungi to occur, the Th17 cells are indispensable [147].

Inflammatory DCs generate the responses of Th17 and Th2 antifungal cells *in vivo* by means of signaling pathways in which the TLR adapter MYD88 participates, while tolerogenic DCs promote regulatory differentiation programs of Th1 and Treg cells through processes in which the signaling adapter TRIF participates. In addition, STAT3, which alters the balance between the canonical and non-canonical activation of NF-κB and, therefore, the expression of the enzyme indoleamine 2,3-dioxygenase (IDO), has a key role to DCs plasticity and functional specialization. The multiple, functionally distinct receptor signaling pathways in DCs affect the balance between $CD4^+$ effector T cells and Treg cells and, therefore, are likely to be harnessed by fungi to allow them to establish commensalism or infection [146]. In contrast some studies have shown that suppressive silencing of cytokine signaling 1 (SOCS1) can induce maturation of DCs and initiate the immune response find *C. albicans in vitro*. In which DC silenced by SOCS1 extend mouse survival and significantly decrease the colonization of fungi in the kidneys and the differentiation of $CD4^+$ T cells producing IL-4 (Th2) or $CD4^+$ T cells producing IL-17 (Th17 cells) are not affected under the same treatment, suggesting that DC silenced by SOCS1 significantly affect the $CD4^+$ producer of IFN-γ cells (Th1). However, in the later stages of infection, when differentiation of Th1, Th2 and Th17 cells decreases in mice treated with DCs silenced with SOCS1, all serum cytokines (IFN-γ, IL-4 and IL-17) also reduced [150].

It has also been reported that NLRP3 linked with ASC and caspase 1, is triggering inflammation activated by pathogenic fungi such as *C. albicans*, *Aspergillus fumigatus* and *Cryptococcus neoformans*. Inflammasome NLRP3 responds to various stimuli, such as crystalline and particulate matter, extracellular ATP, pore-forming toxins, reactive oxygen species (ROS) (see **Figure 6**), endosome destabilization and cathepsin release, changes in intracellular calcium levels and K^+ efflux [148].

Many types of cells, including macrophages and DCs, produce IL-1β induces the differentiation of Th17 cells, which are necessary for effective defense of the host against *C. albicans* when producing IL-17 through the stimulation of PRRs like Dectin-1 and Dectin 2, and both types of cells are indispensable for host defense against *C. albicans*. Dectin 1 is activated through the binding of the b-glucan of the fungal cells, and triggers intracellular signaling recruitment of Syk, activation of NF-kB *via* CARD9, the phosphorylation of IκB is mediated by the IκB kinase (IKK) complex, this complex consists of NF-κB essential modulator (NEMO, or IKKγ), IKKα, and IKKβ, to release the IκBα from NF-κB (see **Figure 6**). In the early stages of candidiasis, DCs are also essential in the antifungal response, since they are responsible for detecting fungal PAMP through their PRR, secreting cytokines and chemokines into the environment, retaining fungal particles by phagocytosis and presenting antigens to T cells to induce an adaptive immune response [147, 151].

7. Dendritic cells and its potential benefits to combat different diseases

DCs are considered key cells as the first line of defense against viruses and to induce adaptive defense. In the innate immune response, they can exert virus phagocytosis and produce cytokines to activate NK cells to eliminate virus infected cells. In adaptive immune response, DCs induce differentiation of Th1-cells that in turn induce activation of antigen specific cytotoxic cells, macrophages, and antibody production to participate in viral clearance.

For the elimination of bacteria, a specific immune response is required, for intracellular bacteria a Th1 response is required as well as cytotoxic T lymphocytes, the latter to produce IFN-γ and can kill the cells that have been infected, in this response the Il −12 is important and its production by DCs requires stimuli derived from pathogens as well as from $CD4^+$ T-cells; on the other hand, for extracellular bacteria a Th17 response is required, in this response DCs play an important role in producing pro-inflammatory cytokines so that a Th17 response can be given, thus these cells coordinate the recruitment of neutrophils that phagocytize extracellular bacteria and thus eliminate the bacterial infection.

DCs participate in the immune response against different opportunistic fungi, the latter are capable of producing different diseases including vulvovaginal candidiasis, oral candidiasis or disseminated candidiasis (*Candida albicans*), invasive pulmonary asperilosis (*Aspergillus fumigatus*), pneumonia (*Pneumocystis carinii*), cryptococcosis (*Cryptococcosis neoformans*). DCs recognize specific structures of fungi such as carbohydrates, proteins, and nucleic acids. This recognition through the PPR activates signaling pathways that lead the DCs to a state of maturation and secretion of cytokines which play an important role in host defense against fungal infections, generating a response either of the Th1 type or Th17.

During parasitic infections, DCs play an important role, since, through them, the body can mount a specific immune response, mainly mediated by T lymphocytes. The DCs recognize the antigens of the parasites, and in the first instance, they induce a Th1-type immune response, characterized mainly by the production of pro-inflammatory cytokines and mediators. Nevertheless, parasites have the ability to polarize, through the activation of DCs, towards a Th2-type immune response, characterized mainly by the production of anti-inflammatory cytokines, eosinophilia and mastocytosis. However, due to the great diversity of parasites that exist, as well as their phenotypic variability, which involves different stages of antigenicity, conditioned by the life cycle of the parasite itself, these microorganisms have the ability to develop strategies that allow them to evade the immune system and facilitate their survival and spread in the host. Despite the different immune responses that the host assembles in contact with the different diseases caused by these microorganisms, DCs are very important, since they represent the junction point between the innate and adaptive immune responses, allowing the host to differentiate the type of microorganism by which it has been invaded and thus be able to mount a specific immune response.

8. Conclusions

Dendritic cells are a key cell type in the recognition of intracellular and extracellular pathogens through the different receptors that they express. The maturation of the DCs is an important event since through this mechanism these cells acquire the ability to express MHC as well as costimulatory molecules, thus conditioning the presentation of the antigen, producing cytokines and mounting immune in order to kill the invading pathogen. The response can be mediated by the PRRs as they will recognize different structures of the invading microorganism and execute a defensive response with the purpose of eliminating the invading microorganism through the production of antimicrobial cytokines and intermediaries, as well as activating transcription factors to produce cytokines that have an important role in the polarization of the T helper cell during priming by DCs.

Acknowledgements

Thanks to the authors who collaborated in the writing of this chapter: Dr. en C. José Luis Muñoz-Carrillo; Dr. en C. Oscar Gutiérrez-Coronado; Dr. en C. Juan Francisco Contreras-Cordero; Dra. en C. Paola Trinidad Villalobos-Gutiérrez; Dr. Luis Guillermo Ramos-Gracia, and Dra. Jazmín Monserrat Vargas-Barboza; as well as the Universities involved: Cuauhtémoc University Aguascalientes, University of Guadalajara, and Autonomous University of Nuevo Leon for financial support for chapter publication.

Conflict of interest

We have no conflict of interest related to this work.

Author details

José Luis Muñoz-Carrillo[1*], Juan Francisco Contreras-Cordero[2], Oscar Gutiérrez-Coronado[3], Paola Trinidad Villalobos-Gutiérrez[3], Luis Guillermo Ramos-Gracia[1] and Jazmín Monserrat Vargas-Barboza[1]

1 Laboratory of Basic Sciences, Faculty of Odontology, School of Biomedical Sciences of the Cuauhtémoc University Aguascalientes, Aguascalientes, Mexico

2 Laboratory of Immunology and Virology, Faculty of Biological Sciences of the Autonomous University of Nuevo Leon, San Nicolás de Los Garza, Nuevo León, Mexico

3 Laboratory of Immunology, Department of Earth and Life Sciences, University Center of Lagos de Moreno of the University of Guadalajara, Lagos de Moreno, Jalisco, Mexico

4 Faculty of Medicine, School of Biomedical Sciences of the Cuauhtémoc University Aguascalientes, Aguascalientes, Mexico

*Address all correspondence to: mcbjlmc@gmail.com; investigacionodontologia@ucuauhtemoc.edu.mx

References

[1] Muñoz-Carrillo JL, Castro-García FP, Gutiérrez-Coronado O, Moreno-García MA and Contreras-Cordero JF. Physiology and pathology of innate immune response against pathogens. In: Rezaei N, editor. Physiology and Pathology of Immunology. London: InTechOpen; 99-134, 2017 p. DOI: 10.5772/intechopen.70556

[2] R&D Systems a bio-techne brand. Dendritic cell Lineage Development Pathways, Bio-Techne [Internet]. 2017. Available from: https://www.rndsystems.com/pathways/dendritic-cell-lineage-development-pathways.

[3] R&D Systems a bio-techne brand. DC: Development Lineage Pathway Human, Bio-Techne. [Internet]. 2017. Available from: https://www.rndsystems.com/pathways/dendritic-cell-lineage-development-pathways.

[4] Tamoutounour S, Guilliams M, Sanchis FM, Liu H, Terhorst D, Malosse C, Pollet E, Ardouin L, Luche H, Sanchez C, Dalod M, Malissen B, Henri S. Immunity. 2013;39:925-938. DOI: 10.1016/j.immuni.2013.10.004.

[5] Collin M, McGovern N, Haniffa M. Human dendritic cell subsets. Immunology. 2013;140: 22-30. DOI: 10.1111/imm.12117.

[6] Vázquez MB, Sureda M, Rebollo J. Células dendríticas I: aspectos básicos de su biología y funciones. Inmunología. 2012;31:21-30. DOI: 10.1016/j.inmuno.2011.10.001

[7] M. Plantinga, C. de Haar, S. Nierkens. Dendritic Cells. In: K. Verhoeckx, P. Cotter, I. López-Expósito, C. Kleiveland, T. Lea, A. Mackie, T. Requena, D. Swiatecka, H. Wichers, editors. The impact of food bioactives on health: *in vitro* and *ex vivo* models. Cham. Springer: 2015. 181-196 p. DOI: https://doi.org/10.1007/978-3-319-16104-4

[8] Y. F. Tan, C. F. Leong, S. K. Cheong. Observation of dendritic cell morphology under light, phase-contrast or confocal laser scanning microscopy. Malays J Pathol. 2010;32:97-102.

[9] Romero-Palomo F, Sanchez-Cordon PJ, Risalde MA, Pedrera M, Molina V, Ruiz-Villamor E, Gomez-Villamandos JC. Funciones y clasificación de las células dendríticas. Anales. 2011;24:167-191.

[10] Muñoz-Carrillo JL, Castro-García FP, Chávez-Rubalcaba F, Martínez-Rodríguez JL, Hernández-Ruiz ME. Immune System Disorders: Hypersensitivity and Autoimmunity. In: Seyyed SA, editor. Immunoregulatory Aspects of Immunotherapy. London: InTechOpen; 2018. p. 1-30. DOI: 10.5772/intechopen.75794.

[11] Steinman RM, Cohn ZA. Identification of a novel cell type in peripheral lymphoid organs of mice. I. Morphology, quantitation, tissue distribution. J Exp Med. 1973;137:1142-1146. DOI: 10.1084/jem.137.5.1142

[12] Banchereau J, Steinman RM. Dendritic cells and the control of immunity. Nature. 1998;392:245-252. DOI: 10.1038/32588

[13] Wu L, Liu YJ. Development of dendritic-cell lineages. Immunity. 2007;26:741-750. DOI: 10.1016/j.immuni.2007.06.006

[14] Lambotin M, Raghuraman S, Stoll-Keller F, Baumert TF, Barth H. A look behind closed doors: interaction of persistent viruses with dendritic cells. Nat Rev Microbiol. 2010;8:350-360. DOI: 10.1038/nrmicro2332

[15] Liu YJ. Dendritic cell subsets and lineages, and their functions in innate and adaptive immunity. Cell.

2001;106:259-262. DOI: 10.1016/s0092-8674(01)00456-1

[16] Ho LJ, Wang JJ, Shaio MF, Kao CL, Chang DM, Han SW, Lai JH. Infection of human dendritic cells by dengue virus causes cell maturation and cytokine production. J Immunol. 2001;166:1499-1506. DOI: 10.4049/jimmunol.166.3.1499

[17] Bao M, Liu YJ. Regulation of TLR7/9 signaling in plasmacytoid dendritic cells. Protein Cell. 2013;4:40-52. DOI: 10.1007/s13238-012-2104-8

[18] Swiecki M, Colonna M. Unraveling the functions of plasmacytoid dendritic cells during viral infections, autoimmunity, and tolerance. Immunol Rev. 2010;234:142-162. DOI: 10.1111/j.0105-2896.2009.00881.x

[19] Cella M, Sallusto F, Lanzavecchia A. Origin, maturation and antigen presenting function of dendritic cells. Curr Opin Immunol. 1997;9:10-16. DOI: 10.1016/s0952-7915(97)80153-7

[20] Szabo A, Rajnavolgyi E. Collaboration of Toll-like and RIG-I-like receptors in human dendritic cells: tRIGgering antiviral innate immune responses. Am J Clin Exp Immunol. 2013;2:195-207. PMID: 24179728

[21] Hemmi H, Akira S. TLR signalling and the function of dendritic cells. Chem Immunol Allergy. 2005;86:120-135. DOI: 10.1159/000086657

[22] Baños-Lara MDR, Ghosh A, Guerrero-Plata A. Critical role of MDA5 in the interferon response induced by human metapneumovirus infection in dendritic cells and *in vivo*. J Virol. 2013;87(2):1242-1251. DOI: 10.1128/JVI.01213-12

[23] Bell JK, Mullen GE, Leifer CA, Mazzoni A, Davies DR, Segal DM. Leucine-rich repeats and pathogen recognition in Toll-like receptors. Trends in immunology. 2003;24:528-533. DOI: 10.1016/S1471-4906(03)00242-4

[24] Botos I, Segal DM, Davies DR. The structural biology of Toll-like receptors. Structure. 2011;19:447-459. DOI: 10.1016/j.str.2011.02.004

[25] Takeuchi O, Akira S. Pattern recognition receptors and inflammation. Cell. 2010;140:805-820. DOI: 10.1016/j.cell.2010.01.022

[26] Van Montfoort N, van der Aa E, Woltman AM. Understanding MHC class I presentation of viral antigens by human dendritic cells as a basis for rational design of therapeutic vaccines. Front Immunol. 2014;5:182. DOI: 10.3389/fimmu.2014.00182

[27] López-Albaitero A, Mailliard R, Hackman T, Andrade Filho PA, Wang X, Gooding W, Ferrone S, Kalinski P, Ferris RL. Maturation pathways of dendritic cells determine TAP1 and TAP2 levels and cross-presenting function. J Immunother. 2009;32:465-473. DOI: 10.1097/CJI.0b013e3181a1c24e

[28] Geijtenbeek TB, Torensma R, van Vliet SJ, van Duijnhoven GC, Adema GJ, van Kooyk Y, Figdor CG. Identification of DC-SIGN, a novel dendritic cell-specific ICAM-3 receptor that supports primary immune responses. Cell. 2000;100:575-585. DOI: 10.1016/s0092-8674(00)80693-5

[29] Gringhuis SI, den Dunnen J, Litjens M, van Het Hof B, van Kooyk Y, Geijtenbeek TB. C-type lectin DC-SIGN modulates Toll-like receptor signaling *via* Raf-1 kinase-dependent acetylation of transcription factor NF-kappaB. Immunity. 2007;26:605-616. DOI: 10.1016/j.immuni.2007.03.012

[30] Puig-Kröger A, Serrano-Gómez D, Caparrós E, Domínguez-Soto A, Relloso M, Colmenares M, Martínez-Muñoz L, Longo N,

Sánchez-Sánchez N, Rincon M, Rivas L, Sánchez-Mateos P, Fernández-Ruiz E, Corbí AL. Regulated expression of the pathogen receptor dendritic cell-specific intercellular adhesion molecule 3 (ICAM-3)-grabbing nonintegrin in THP-1 human leukemic cells, monocytes, and macrophages. J Biol Chem. 2004;279:25680-25688. DOI: 10.1074/jbc.M311516200

[31] Kim HM, Park BS, Kim JI, Kim SE, Lee J, Oh SC, Enkhbayar P, Matsushima N, Lee H, Yoo OJ, Lee JO. Crystal structure of the TLR4-MD-2 complex with bound endotoxin antagonist Eritoran. Cell. 2007;130:906-917. DOI: 10.1016/j.cell.2007.08.002

[32] Yamamoto M, Sato S, Mori K, Hoshino K, Takeuchi O, Takeda K, Akira S. Cutting edge: a novel Toll/IL-1 receptor domain-containing adapter that preferentially activates the IFN-beta promoter in the Toll-like receptor signaling. J Immunol. 2002;169:6668-6672. DOI: 10.4049/jimmunol.169.12.6668

[33] Liu L, Botos I, Wang Y, Leonard JN, Shiloach J, Segal DM, Davies DR. Structural basis of toll-like receptor 3 signaling with double-stranded RNA. Science. 2008;320:379-381. DOI: 10.1126/science.1155406

[34] Brown J, Wang H, Hajishengallis GN, Martin M. TLR-signaling networks: an integration of adaptor molecules, kinases, and cross-talk. J Dent Res. 2011;90:417-427. DOI: 10.1177/0022034510381264

[35] Kato H, Takeuchi O, Sato S, Yoneyama M, Yamamoto M, Matsui K, Uematsu S, Jung A, Kawai T, Ishii KJ, Yamaguchi O, Otsu K, Tsujimura T, Koh CS, Reis e Sousa C, Matsuura Y, Fujita T, Akira S. Differential roles of MDA5 and RIG-I helicases in the recognition of RNA viruses. Nature. 2006;441:101-105. DOI: 10.1038/nature04734

[36] Hemmi H, Takeuchi O, Sato S, Yamamoto M, Kaisho T, Sanjo H, Kawai T, Hoshino K, Takeda K, Akira S. The roles of two IkappaB kinase-related kinases in lipopolysaccharide and double stranded RNA signaling and viral infection. J Exp Med. 2004;199:1641-1650. DOI: 10.1084/jem.20040520

[37] Kawai T, Takahashi K, Sato S, Coban C, Kumar H, Kato H, Ishii KJ, Takeuchi O, Akira S. IPS-1, an adaptor triggering RIG-I- and Mda5-mediated type I interferon induction. Nat Immunol. 2005;6:981-988. DOI: 10.1038/ni1243

[38] Saha SK, Pietras EM, He JQ, Kang JR, Liu SY, Oganesyan G, Shahangian A, Zarnegar B, Shiba TL, Wang Y, Cheng G. Regulation of antiviral responses by a direct and specific interaction between TRAF3 and Cardif. EMBO J. 2006;25:3257-3263. DOI: 10.1038/sj.emboj.7601220

[39] Clark K, Takeuchi O, Akira S, Cohen P. The TRAF-associated protein TANK facilitates cross-talk within the IkappaB kinase family during Toll-like receptor signaling. Proc Natl Acad Sci U S A. 2011;108:17093-17098. DOI: 10.1073/pnas.1114194108

[40] O'Brien M, Manches O, Bhardwaj N. Plasmacytoid dendritic cells in HIV infection. Adv Exp Med Biol. 2013;762:71-107. DOI: 10.1007/978-1-4614-4433-6_3

[41] Beignon AS, McKenna K, Skoberne M, Manches O, DaSilva I, Kavanagh DG, Larsson M, Gorelick RJ, Lifson JD, Bhardwaj N. Endocytosis of HIV-1 activates plasmacytoid dendritic cells *via* Toll-like receptor-viral RNA interactions. J Clin Invest. 2005;115:3265-75. DOI: 10.1172/JCI26032

[42] Kadowaki N, Ho S, Antonenko S, Malefyt RW, Kastelein RA, Bazan F, Liu YJ. Subsets of human dendritic

cell precursors express different toll-like receptors and respond to different microbial antigens. J Exp Med. 2001;194:863-869. DOI: 10.1084/jem.194.6.863

[43] Asselin-Paturel C, Trinchieri G. Production of type I interferons: plasmacytoid dendritic cells and beyond. J Exp Med. 2005;202:461-465. DOI: 10.1084/jem.20051395

[44] Li S, Strelow A, Fontana EJ, Wesche H. IRAK-4: a novel member of the IRAK family with the properties of an IRAK-kinase. Proc Natl Acad Sci U S A. 2002;99:5567-5572. DOI: 10.1073/pnas.082100399

[45] Lin SC, Lo YC, Wu H. Helical assembly in the MyD88-IRAK4-IRAK2 complex in TLR/IL-1R signalling. Nature. 2010;465:885-890. DOI: 10.1038/nature09121

[46] Peressin M, Holl V, Schmidt S, Decoville T, Mirisky D, Lederle A, Delaporte M, Xu K, Aubertin AM, Moog C. HIV-1 replication in Langerhans and interstitial dendritic cells is inhibited by neutralizing and Fc-mediated inhibitory antibodies. J Virol. 2011;85:1077-1085. DOI: 10.1128/JVI.01619-10

[47] Cunningham AL, Carbone F, Geijtenbeek TB. Langerhans cells and viral immunity. Eur J Immunol. 2008;38:2377-2385. DOI: 10.1002/eji.200838521

[48] Flacher V, Bouschbacher M, Verronèse E, Massacrier C, Sisirak V, Berthier-Vergnes O, de Saint-Vis B, Caux C, Dezutter-Dambuyant C, Lebecque S, Valladeau J. Human Langerhans cells express a specific TLR profile and differentially respond to viruses and Gram-positive bacteria. J Immunol. 2006;177:7959-7967. DOI: 10.4049/jimmunol.177.11.7959

[49] Murray PR, Rosenthal KS, Pfaller MA. Medical Microbiology E-Book. 9th ed. Philadelphia, PA: Elsevier Health Sciences; 2015. 89 p.

[50] Gutiérrez-Kobeh L, Rodríguez-González J, Argueta-Donohué J, Vázquez-López R, Wilkins-Rodríguez AA. Role of Dendritic Cells in Parasitic Infections. In: Chapoval SP, editor. Dendritic Cells. London: InTechOpen. 2018. p. 47-77. DOI: 10.5772/intechopen.79491

[51] Muñoz-Carrillo JL, Contreras-Cordero JF, Gutiérrez-Coronado O, Villalobos-Gutiérrez P T, Ramos-Gracia LG, Hernández-Reyes VE. Cytokine profiling plays a crucial role in activating immune system to clear infectious pathogens. In: Tyagi RK, Bisen PS, editors. Immune Response Activation and Immunomodulation. London: InTechOpen. 2018. p. 1-30. DOI: 10.5772/intechopen.80843

[52] Schnittger L, Florin-Christensen M. Introduction into parasitic protozoa. In: Florin-Christensen M, Schnittger L, editors. Parasitic Protozoa of Farm Animals and Pets. Switzerland: Springer International Publishing. 2018. p. 1-1. DOI: 10.1007/978-3-319-70132-5_1

[53] Hotez PJ. Human Parasitology and Parasitic Diseases: Heading Towards 2050. Adv Parasitol. 2018;100:29-38. DOI: 10.1016/bs.apar.2018.03.002

[54] Sibley LD. Invasion and intracellular survival by protozoan parasites. Immunol Rev. 2011;240:72-91. DOI: 10.1111/j.1600-065X.2010.00990.x

[55] Motran CC, Ambrosio LF, Volpini X, Celias DP, Cervi L. Dendritic cells and parasites: from recognition and activation to immune response instruction. Semin Immunopathol. 2017;39:199-213. DOI: 10.1007/s00281-016-0588-7

[56] Ghosh D, Stumhofer JS. Do you see what I see: Recognition of protozoan

parasites by Toll-like receptors. Curr Immunol Rev. 2013;9:129-140. DOI: 10.2174/1573395509666131203225929

[57] Oliveira AC, Peixoto JR, de Arruda LB, Campos MA, Gazzinelli RT, Golenbock DT, Akira S, Previato JO, Mendonça-Previato L, Nobrega A, Bellio M. Expression of functional TLR4 confers proinflammatory responsiveness to *Trypanosoma cruzi* glycoinositolphospholipids and higher resistance to infection with *T. cruzi*. J Immunol. 2004;173:5688-5696. DOI: 10.4049/jimmunol.173.9.5688

[58] Campos MA, Almeida IC, Takeuchi O, Akira S, Valente EP, Procópio DO, Travassos LR, Smith JA, Golenbock DT, Gazzinelli RT. Activation of Toll-like receptor-2 by glycosylphosphatidylinositol anchors from a protozoan parasite. J Immunol. 2001;167:416-423. DOI: 10.4049/jimmunol.167.1.416

[59] Bafica A, Santiago HC, Goldszmid R, Ropert C, Gazzinelli RT, Sher A. Cutting edge: TLR9 and TLR2 signaling together account for MyD88-dependent control of parasitemia in *Trypanosoma cruzi* infection. J Immunol. 2006;177(6):3515-3519. DOI: 10.4049/jimmunol.177.6.3515

[60] Terrazas CA, Huitron E, Vazquez A, Juarez I, Camacho GM, Calleja EA, Rodriguez-Sosa M. MIF synergizes with *Trypanosoma cruzi* antigens to promote efficient dendritic cell maturation and IL-12 production *via* p38 MAPK. Int J Biol Sci. 2011;7:1298-1310. DOI: 10.7150/ijbs.7.1298

[61] Sanecka A, Frickel EM. Use and abuse of dendritic cells by Toxoplasma gondii. Virulence. 2012;3:678-689. DOI: 10.4161/viru.22833

[62] Scanga CA, Aliberti J, Jankovic D, Tilloy F, Bennouna S, Denkers EY, Medzhitov R, Sher A. Cutting edge: MyD88 is required for resistance to *Toxoplasma gondii* infection and regulates parasite-induced IL-12 production by dendritic cells. J Immunol. 2002;168:5997-6001. DOI: 10.4049/jimmunol.168.12.5997

[63] Yarovinsky F, Zhang D, Andersen JF, Bannenberg GL, Serhan CN, Hayden MS, Hieny S, Sutterwala FS, Flavell RA, Ghosh S, Sher A. TLR11 activation of dendritic cells by a protozoan profilin-like protein. Science. 2005;308:1626-9. DOI: 10.1126/science.1109893

[64] Aliberti J, Jankovic D, Sher A. Turning it on and off: regulation of dendritic cell function in *Toxoplasma gondii* infection. Immunol Rev. 2004;201:26-34. DOI: 10.1111/j.0105-2896.2004.00179.x

[65] Plattner F, Yarovinsky F, Romero S, Didry D, Carlier MF, Sher A, Soldati-Favre D. *Toxoplasma profilin* is essential for host cell invasion and TLR11-dependent induction of an interleukin-12 response. Cell Host Microbe. 2008;3:77-87. DOI: 10.1016/j.chom.2008.01.001

[66] Koblansky AA, Jankovic D, Oh H, Hieny S, Sungnak W, Mathur R, Hayden MS, Akira S, Sher A, Ghosh S. Recognition of profilin by Toll-like receptor 12 is critical for host resistance to *Toxoplasma gondii*. Immunity. 2013;38:119-130. DOI: 10.1016/j.immuni.2012.09.016

[67] Aosai F, Rodriguez Pena MS, Mun HS, Fang H, Mitsunaga T, Norose K, Kang HK, Bae YS, Yano A. *Toxoplasma gondii*-derived heat shock protein 70 stimulates maturation of murine bone marrow-derived dendritic cells *via* Toll-like receptor 4. Cell Stress Chaperones. 2006;11:13-22. DOI: 10.1379/csc-138r.1

[68] Debierre-Grockiego F, Campos MA, Azzouz N, Schmidt J, Bieker U, Resende MG, Mansur DS, Weingart R, Schmidt RR, Golenbock DT,

Gazzinelli RT, Schwarz RT. Activation of TLR2 and TLR4 by glycosylphosphatidylinositols derived from *Toxoplasma gondii*. J Immunol. 2007;179:1129-1137. DOI: 10.4049/jimmunol.179.2.1129

[69] Schneider AG, Abi Abdallah DS, Butcher BA, Denkers EY. *Toxoplasma gondii* triggers phosphorylation and nuclear translocation of dendritic cell STAT1 while simultaneously blocking IFNγ-induced STAT1 transcriptional activity. PLoS One. 2013;8:e60215. DOI: 10.1371/journal.pone.0060215

[70] Naik RS, Branch OH, Woods AS, Vijaykumar M, Perkins DJ, Nahlen BL, Lal AA, Cotter RJ, Costello CE, Ockenhouse CF, Davidson EA, Gowda DC. Glycosylphosphatidylinositol anchors of *Plasmodium falciparum*: molecular characterization and naturally elicited antibody response that may provide immunity to malaria pathogenesis. J Exp Med. 2000;192:1563-1576. DOI: 10.1084/jem.192.11.1563

[71] Kumar S, Gowda NM, Wu X, Gowda RN, Gowda DC. CD36 modulates proinflammatory cytokine responses to *Plasmodium falciparum* glycosylphosphatidylinositols and merozoites by dendritic cells. Parasite Immunol. 2012;34:372-382. DOI: 10.1111/j.1365-3024.2012.01367.x

[72] Wu X, Gowda NM, Kumar S, Gowda DC. Protein-DNA complex is the exclusive malaria parasite component that activates dendritic cells and triggers innate immune responses. J Immunol. 2010;184:4338-4348. DOI: 10.4049/jimmunol.0903824

[73] Behboudi S, Moore A, Hill AV. Splenic dendritic cell subsets prime and boost CD8 T cells and are involved in the generation of effector CD8 T cells. Cell Immunol. 2004;228:15-19. DOI: 10.1016/j.cellimm.2004.03.010

[74] Murray SA, Mohar I, Miller JL, Brempelis KJ, Vaughan AM, Kappe SH, Crispe IN. CD40 is required for protective immunity against liver stage *Plasmodium* infection. J Immunol. 2015;194:2268-2279. DOI: 10.4049/jimmunol.1401724

[75] Bruña-Romero O, Rodriguez A. Dendritic cells can initiate protective immune responses against malaria. Infect Immun. 2001;69:5173-5176. DOI: 10.1128/IAI.69.8.5173-5176.2001

[76] Vichchathorn P, Jenwithisuk R, Leelaudomlipi S, Tungpradabkul S, Hongeng S, Cui L, Sattabongkot J, Udomsangpetch R. Induction of specific immune responses against the *Plasmodium vivax* liver-stage *via in vitro* activation by dendritic cells. Parasitol Int. 2006;55:187-193. DOI: 10.1016/j.parint.2006.04.001

[77] Jung S, Unutmaz D, Wong P, Sano G, De los Santos K, Sparwasser T, Wu S, Vuthoori S, Ko K, Zavala F, Pamer EG, Littman DR, Lang RA. *In vivo* depletion of CD11c+ dendritic cells abrogates priming of CD8+ T cells by exogenous cell-associated antigens. Immunity. 2002;17:211-220. DOI: 10.1016/s1074-7613(02)00365-5

[78] Radtke AJ, Kastenmüller W, Espinosa DA, Gerner MY, Tse SW, Sinnis P, Germain RN, Zavala FP, Cockburn IA. Lymph-node resident CD8α+ dendritic cells capture antigens from migratory malaria sporozoites and induce CD8+ T cell responses. PLoS Pathog. 2015;11:e1004637. DOI: 10.1371/journal.ppat.1004637

[79] Perry JA, Rush A, Wilson RJ, Olver CS, Avery AC. Dendritic cells from malaria-infected mice are fully functional APC. J Immunol. 2004;172:475-482. DOI: 10.4049/jimmunol.172.1.475

[80] Luyendyk J, Olivas OR, Ginger LA, Avery AC. Antigen-presenting cell function during Plasmodium

yoelii infection. Infect Immun. 2002;70:2941-2949. DOI: 10.1128/iai.70.6.2941-2949.2002

[81] Seixas E, Cross C, Quin S, Langhorne J. Direct activation of dendritic cells by the malaria parasite, *Plasmodium chabaudi chabaudi*. Eur J Immunol. 2001;31:2970-2978. DOI: 10.1002/1521-4141(2001010)31:10<2970::aid-immu2970>3.0.co;2-s

[82] Leisewitz AL, Rockett KA, Gumede B, Jones M, Urban B, Kwiatkowski DP. Response of the splenic dendritic cell population to malaria infection. Infect Immun. 2004;72:4233-4239. DOI: 10.1128/IAI.72.7.4233-4239.2004

[83] Amorim KN, Chagas DC, Sulczewski FB, Boscardin SB. Dendritic Cells and Their Multiple Roles during Malaria Infection. J Immunol Res. 2016;2016:2926436. DOI: 10.1155/2016/2926436

[84] Gazzinelli-Guimaraes PH, Nutman TB. Helminth parasites and immune regulation. F1000Res. 2018;7:F1000 Faculty Rev-1685. DOI: 10.12688/f1000research.15596.1

[85] Ilic N, Gruden-Movsesijan A, Sofronic-Milosavljevic L. *Trichinella spiralis*: shaping the immune response. Immunol Res. 2012;52:111-119. DOI: 10.1007/s12026-012-8287-5

[86] Ashour DS. *Trichinella spiralis* immunomodulation: an interactive multifactorial process. Expert Rev Clin Immunol. 2013;9:669-675. DOI: 10.1586/1744666X.2013.811187.

[87] Bruschi F, Chiumiento L. Immunomodulation in trichinellosis: does *Trichinella* really escape the host immune system? Endocr Metab Immune Disord Drug Targets. 2012;12:4-15. DOI: 10.2174/187153012799279081

[88] Cieza RJ, Cao AT, Cong Y, Torres AG. Immunomodulation for gastrointestinal infections. Expert Rev Anti Infect Ther. 2012;10:391-400. DOI: 10.1586/eri.11.176

[89] Maizels RM, McSorley HJ. Regulation of the host immune system by helminth parasites. J Allergy Clin Immunol. 2016;138:666-675. DOI: 10.1016/j.jaci.2016.07.007

[90] Perrigoue JG, Marshall FA, Artis D. On the hunt for helminths: innate immune cells in the recognition and response to helminth parasites. Cell Microbiol. 2008;10:1757-1764. DOI: 10.1111/j.1462-5822.2008.01174.x

[91] Carvalho L, Sun J, Kane C, Marshall F, Krawczyk C, Pearce EJ. Review series on helminths, immune modulation and the hygiene hypothesis: mechanisms underlying helminth modulation of dendritic cell function. Immunology. 2009;126:28-34. DOI: 10.1111/j.1365-2567.2008.03008.x

[92] Muñoz-Carrillo JL, Maldonado-Tapia C, López-Luna A, Muñoz-Escobedo JJ, Flores-De La Torre JA, Moreno-García A. Current Aspects in Trichinellosis. In: G. A. Bastidas Pacheco, editor. Parasites and Parasitic Diseases. London: InTechOpen. 2018. p. 1-23. DOI: 10.5772/intechopen.80372

[93] Ilic N, Worthington JJ, Gruden-Movsesijan A, Travis MA, Sofronic-Milosavljevic L, Grencis RK. *Trichinella spiralis* antigens prime mixed Th1/Th2 response but do not induce de novo generation of Foxp3+ T cells *in vitro*. Parasite Immunol. 2011;33:572-582. DOI: 10.1111/j.1365-3024.2011.01322.x

[94] Yu YR, Deng MJ, Lu WW, Jia MZ, Wu W, Qi YF. Systemic cytokine profiles and splenic toll-like receptor expression during *Trichinella spiralis* infection. Exp Parasitol. 2013;134:92-101. DOI: 10.1016/j.exppara.2013.02.014

[95] Gruden-Movsesijan A, Ilic N, Colic M, Majstorovic I, Vasilev S, Radovic I, Sofronic-Milosavljevic Lj. The impact of *Trichinella spiralis* excretory-secretory products on dendritic cells. Comp Immunol Microbiol Infect Dis. 2011;34:429-439. DOI: 10.1016/j.cimid.2011.08.004

[96] Gentilini MV, Nuñez GG, Roux ME, Venturiello SM. *Trichinella spiralis* infection rapidly induces lung inflammatory response: the lung as the site of helminthocytotoxic activity. Immunobiology. 2011;216:1054-63. DOI: 10.1016/j.imbio.2011.02.002

[97] Muñoz-Carrillo JL, Contreras-Cordero JF, Muñoz-López JL, Maldonado-Tapia CH, Muñoz-Escobedo JJ, Moreno-García MA. Resiniferatoxin modulates the Th1 immune response and protects the host during intestinal nematode infection. Parasite Immunol. 2017;39. DOI: 10.1111/pim.12448

[98] Ilic N, Colic M, Gruden-movsesijan A, Majstorovic I, Vasilev S, Sofronic-Milosavljevic Lj. Characterization of rat bone marrow dendritic cells initially primed by *Trichinella spiralis* antigens. Parasite Immunol. 2008;30:491-495. DOI: 10.1111/j.1365-3024.2008.01049.x

[99] Ming L, Peng RY, Zhang L, Zhang CL, Lv P, Wang ZQ, Cui J, Ren HJ. Invasion by *Trichinella spiralis* infective larvae affects the levels of inflammatory cytokines in intestinal epithelial cells *in vitro*. Exp Parasitol. 2016;170:220-226. DOI: 10.1016/j.exppara.2016.10.003

[100] Muñoz-Carrillo JL, Muñoz-Escobedo JJ, Maldonado-Tapia CH, Chávez-Ruvalcaba F, Moreno-García MA. Resiniferatoxin lowers TNF-α, NO and PGE2 in the intestinal phase and the parasite burden in the muscular phase of *Trichinella spiralis* infection. Parasite Immunol. 2017;39: e12393. DOI: 10.1111/pim

[101] Langelaar M, Aranzamendi C, Franssen F, Van Der Giessen J, Rutten V, van der Ley P, Pinelli E. Suppression of dendritic cell maturation by *Trichinella spiralis* excretory/secretory products. Parasite Immunol. 2009;31:641-645. DOI: 10.1111/j.1365-3024.2009.01136.x

[102] Han C, Xu J, Liu CH, Li X, Zhai P, Hashan A, Song M. Modulation of TLR2 and TLR4 in Macrophages Following *Trichinella Spiralis* Infection. Helminthologia. 2018;55:195-203. DOI: 10.2478/helm-2018-0015

[103] Vanhoutte F, Breuilh L, Fontaine J, Zouain CS, Mallevaey T, Vasseur V, Capron M, Goriely S, Faveeuw C, Ryffel B, Trottein F. Toll-like receptor (TLR)2 and TLR3 sensing is required for dendritic cell activation, but dispensable to control Schistosoma mansoni infection and pathology. Microbes Infect. 2007;9:1606-1613. DOI: 10.1016/j.micinf.2007.09.013

[104] Aksoy E, Zouain CS, Vanhoutte F, Fontaine J, Pavelka N, Thieblemont N, Willems F, Ricciardi-Castagnoli P, Goldman M, Capron M, Ryffel B, Trottein F. Double-stranded RNAs from the helminth parasite *Schistosoma* activate TLR3 in dendritic cells. J Biol Chem. 2005;280:277-283. DOI: 10.1074/jbc.M411223200

[105] Sorobetea D, Svensson-Frej M, Grencis R. Immunity to gastrointestinal nematode infections. Mucosal Immunol. 2018;11:304-315. DOI: 10.1038/mi.2017.113

[106] Maizels RM, McSorley HJ. Regulation of the host immune system by helminth parasites. J Allergy Clin Immunol. 2016;138:666-675. DOI: 10.1016/j.jaci.2016.07.007

[107] Tussiwand R, Everts B, Grajales-Reyes GE, Kretzer NM, Iwata A, Bagaitkar J, Wu X, Wong R, Anderson DA, Murphy TL, Pearce EJ, Murphy KM. Klf4 expression in conventional dendritic cells is required for T helper 2

cell responses. Immunity. 2015;42:916-928. DOI: 10.1016/j.immuni.2015.04.017

[108] Babu S, Nutman TB, Immune responses to helminth infection. in: Robert R. Rich, Thomas A. Fleisher, William T. Shearer, Harry W. Schroeder, Anthony J. Frew, Cornelia M. Weyand, editors. Clinical Immunology. Fifth Edition. 2019. p. 437-447, 2019. DOI: 10.1016/B978-0-7020-6896-6.00031-4

[109] Muñoz-Carrillo JL, Muñoz-López JL, Muñoz-Escobedo JJ, Maldonado-Tapia C, Gutiérrez-Coronado O, Contreras-Cordero JF, Moreno-García MA. Therapeutic Effects of Resiniferatoxin Related with Immunological Responses for Intestinal Inflammation in Trichinellosis. Korean J Parasitol. 2017;55:587-599. DOI: 10.3347/kjp.2017.55.6.587

[110] Van Riet E, Everts B, Retra K, Phylipsen M, van Hellemond JJ, Tielens AG, van der Kleij D, Hartgers FC, Yazdanbakhsh M. Combined TLR2 and TLR4 ligation in the context of bacterial or helminth extracts in human monocyte derived dendritic cells: molecular correlates for Th1/Th2 polarization. BMC Immunol. 2009;10:9. DOI: 10.1186/1471-2172-10-9

[111] Schabussova I, Amer H, van Die I, Kosma P, Maizels RM. O-methylated glycans from *Toxocara* are specific targets for antibody binding in human and animal infections. Int J Parasitol. 2007;37:97-109. DOI: 10.1016/j.ijpara.2006.09.006

[112] Tawill S, Le Goff L, Ali F, Blaxter M, Allen JE. Both free-living and parasitic nematodes induce a characteristic Th2 response that is dependent on the presence of intact glycans. Infect Immun. 2004;72:398-407. DOI: 10.1128/iai.72.1.398-407.2004

[113] Everts B, Smits HH, Hokke CH, Yazdanbakhsh M. Helminths and dendritic cells: sensing and regulating *via* pattern recognition receptors, Th2 and Treg responses. Eur J Immunol. 2010;40:1525-1537. DOI: 10.1002/eji.200940109

[114] Ryffel B, Fremond C, Jacobs M, Parida S, Botha T, Schnyder B, Quesniaux V. Innate immunity to mycobacterial infection in mice: critical role for toll-like receptors. Tuberculosis (Edinb). 2005;85:395-405. DOI: 10.1016/j.tube.2005.08.021

[115] Faridgohar M, Nikoueinejad H. New findings of Toll-like receptors involved in *Mycobacterium tuberculosis* infection. Pathog Glob Health. 2017;111:256-264. DOI: 10.1080/20477724.2017.1351080

[116] Takeda K, Akira S. TLR signaling pathways. Semin Immunol. 2004;16:3-9. DOI: 10.1016/j.smim.2003.10.003

[117] Jiménez-Dalmaroni MJ, Gerswhin ME, Adamopoulos IE. The critical role of toll-like receptors--From microbial recognition to autoimmunity: A comprehensive review. Autoimmun Rev. 2016;15:1-8. DOI: 10.1016/j.autrev.2015.08.009

[118] Heo DR, Shin SJ, Kim WS, Noh KT, Park JW, Son KH, Park WS, Lee MG, Kim D, Shin YK, Jung ID, Park YM. *Mycobacterium tuberculosis* lpdC, Rv0462, induces dendritic cell maturation and Th1 polarization. Biochem Biophys Res Commun. 2011;411:642-647. DOI: 10.1016/j.bbrc.2011.07.013

[119] Basu J, Shin DM, Jo EK. Mycobacterial signaling through toll-like receptors. Front Cell Infect Microbiol. 2012;2:145. DOI: 10.3389/fcimb.2012.00145

[120] Choi S, Choi HG, Lee J, Shin KW, Kim HJ. *Mycobacterium tuberculosis* protein Rv2220 induces maturation and activation of dendritic cells.

Cell Immunol. 2018;328:70-78. DOI: 10.1016/j.cellimm.2018.03.012

[121] Bansal K, Sinha AY, Ghorpade DS, Togarsimalemath SK, Patil SA, Kaveri SV, Balaji KN, Bayry J. Src homology 3-interacting domain of Rv1917c of *Mycobacterium tuberculosis* induces selective maturation of human dendritic cells by regulating PI3K-MAPK-NF-kappaB signaling and drives Th2 immune responses. J Biol Chem. 2010;285:36511-36522. DOI: 10.1074/jbc.M110.158055

[122] Costa Franco MMS, Marim FM, Alves-Silva J, Cerqueira D, Rungue M, Tavares IP, Oliveira SC. AIM2 senses *Brucella abortus* DNA in dendritic cells to induce IL-1β secretion, pyroptosis and resistance to bacterial infection in mice. Microbes Infect. 2019;21:85-93. DOI: 10.1016/j.micinf.2018.09.001

[123] Awad F, Assrawi E, Louvrier C, Jumeau C, Georgin-Lavialle S, Grateau G, Amselem S, Giurgea I, Karabina SA. Inflammasome biology, molecular pathology and therapeutic implications. Pharmacol Ther. 2018;187:133-149. DOI: 10.1016/j.pharmthera.2018.02.011

[124] Guo H, Callaway JB, Ting JP. Inflammasomes: mechanism of action, role in disease, and therapeutics. Nat Med. 2015;21:677-687. DOI: 10.1038/nm.3893

[125] Gong T, Jiang W, Zhou R. Control of Inflammasome Activation by Phosphorylation. Trends Biochem Sci. 2018;43:685-699. DOI: 10.1016/j.tibs.2018.06.008

[126] Chavarría-Smith J, Vance RE. The NLRP1 inflammasomes. Immunol Rev. 2015;265:22-34. DOI: 10.1111/imr.12283

[127] Sharma D, Kanneganti TD. The cell biology of inflammasomes: Mechanisms of inflammasome activation and regulation. J Cell Biol. 2016;213:617-629. DOI: 10.1083/jcb.201602089

[128] Man SM, Kanneganti TD. Regulation of inflammasome activation. Immunol Rev. 2015;265:6-21. DOI: 10.1111/imr.12296

[129] He Y, Hara H, Núñez G. Mechanism and Regulation of NLRP3 Inflammasome Activation. Trends Biochem Sci. 2016;41:1012-1021. DOI: 10.1016/j.tibs.2016.09.002

[130] Miao EA, Mao DP, Yudkovsky N, Bonneau R, Lorang CG, Warren SE, Leaf IA, Aderem A. Innate immune detection of the type III secretion apparatus through the NLRC4 inflammasome. Proc Natl Acad Sci U S A. 2010;107:3076-3080. DOI: 10.1073/pnas.0913087107

[131] Duncan JA, Canna SW. The NLRC4 Inflammasome. Immunol Rev. 2018;281:115-123. DOI: 10.1111/imr.12607

[132] Marim FM, Franco MMC, Gomes MTR, Miraglia MC, Giambartolomei GH, Oliveira SC. The role of NLRP3 and AIM2 in inflammasome activation during *Brucella abortus* infection. Semin Immunopathol. 2017;39:215-223. DOI: 10.1007/s00281-016-0581-1

[133] Deng J, Yu XQ, Wang PH. Inflammasome activation and Th17 responses. Mol Immunol. 2019;107:142-164. DOI: 10.1016/j.molimm.2018.12.024

[134] Basu R, Whitley SK, Bhaumik S, Zindl CL, Schoeb TR, Benveniste EN, Pear WS, Hatton RD, Weaver CT. IL-1 signaling modulates activation of STAT transcription factors to antagonize retinoic acid signaling and control the TH17 cell-iTreg cell balance. Nat Immunol. 2015;16:286-295. DOI: 10.1038/ni.3099

[135] Peck A, Mellins ED. Precarious balance: Th17 cells in host defense. Infect Immun. 2010;78:32-38. DOI: 10.1128/IAI.00929-09

[136] Andreasen C, Powell DA, Carbonetti NH. Pertussis toxin stimulates IL-17 production in response to Bordetella pertussis infection in mice. PLoS One. 2009;4:e7079. DOI: 10.1371/journal.pone.0007079

[137] Elinav E, Strowig T, Henao-Mejia J, Flavell RA. Regulation of the antimicrobial response by NLR proteins. Immunity. 2011;34:665-679. DOI: 10.1016/j.immuni.2011.05.007

[138] Bertrand MJ, Doiron K, Labbé K, Korneluk RG, Barker PA, Saleh M. Cellular inhibitors of apoptosis cIAP1 and cIAP2 are required for innate immunity signaling by the pattern recognition receptors NOD1 and NOD2. Immunity. 2009;30:789-801. DOI: 10.1016/j.immuni.2009.04.011

[139] Chamaillard M, Hashimoto M, Horie Y, Masumoto J, Qiu S, Saab L, Ogura Y, Kawasaki A, Fukase K, Kusumoto S, Valvano MA, Foster SJ, Mak TW, Nuñez G, Inohara N. An essential role for NOD1 in host recognition of bacterial peptidoglycan containing diaminopimelic acid. Nat Immunol. 2003;4:702-707. DOI: 10.1038/ni945

[140] Stafford CA, Lawlor KE, Heim VJ, Bankovacki A, Bernardini JP, Silke J, Nachbur U. IAPs Regulate Distinct Innate Immune Pathways to Co-ordinate the Response to Bacterial Peptidoglycans. Cell Rep. 2018;22:1496-1508. DOI: 10.1016/j.celrep.2018.01.024

[141] Velloso FJ, Trombetta-Lima M, Anschau V, Sogayar MC, Correa RG. NOD-like receptors: major players (and targets) in the interface between innate immunity and cancer. Biosci Rep. 2019;39:BSR20181709. DOI: 10.1042/BSR20181709

[142] Kanneganti TD, Lamkanfi M, Núñez G. Intracellular NOD-like receptors in host defense and disease. Immunity. 2007;27:549-559. DOI: 10.1016/j.immuni.2007.10.002

[143] Park JH, Kim YG, McDonald C, Kanneganti TD, Hasegawa M, Body-Malapel M, Inohara N, Núñez G. RICK/RIP2 mediates innate immune responses induced through Nod1 and Nod2 but not TLRs. J Immunol. 2007;178:2380-2386. DOI: 10.4049/jimmunol.178.4.2380

[144] Bozza S, Montagnoli C, Gaziano R, Rossi G, Nkwanyuo G, Bellocchio S, Romani L. Dendritic cell-based vaccination against opportunistic fungi. Vaccine. 2004;22:857-864. DOI: 10.1016/j.vaccine.2003.11.031

[145] Romani L. Immunity to fungal infections. Nat Rev Immunol. 2011;11:275-288. DOI: 10.1038/nri2939

[146] Romani L. Immunity to fungal infections. Nat Rev Immunol. 2004;4:1-23. DOI: 10.1038/nri1255

[147] Muranski P, Restifo NP. Essentials of Th17 cell commitment and plasticity. Blood. 2013;121:2402-2414. DOI: 10.1182/blood-2012-09-378653

[148] de Castro LF, Longhi LNA, Paião MR, Justo-Júnior ADS, de Jesus MB, Blotta MHSL, Mamoni RL. NLRP3 inflammasome is involved in the recognition of *Paracoccidioides brasiliensis* by human dendritic cells and in the induction of Th17 cells. J Infect. 2018;77:137-144. DOI: 10.1016/j.jinf.2018.03.004

[149] Espinosa V, Rivera A. Cytokines and the regulation of fungus-specific CD4 T cell differentiation. Cytokine. 2012;58:100-106. DOI: 10.1016/j.cyto.2011.11.005

[150] Shi D, Li D, Wang Q, Kong X, Mei H, Shen Y, Liu W. Silencing SOCS1 in dendritic cells promote survival of mice with systemic *Candida*

albicans infection *via* inducing Th1-cell differentiation. Immunol Lett. 2018;197:53-62. DOI: 10.1016/j.imlet.2018.03.009

[151] Hasebe A, Saeki A, Yoshida Y, Shibata KI. Differences in interleukin-1β release-inducing activity of *Candida albicans* toward dendritic cells and macrophages. Arch Oral Biol. 2018;93:115-125. DOI: 10.1016/j.archoralbio.2018.06.004

Chapter 9

Innate Immunity Modulation during Zika Virus Infection on Pregnancy: What We Still Need to Know for Medical Sciences Breakthrough

Tamiris Azamor, Amanda Torrentes-Carvalho, Zilton Vasconcelos, Ana Paula Dinis Ano Bom and Juliana Gil Melgaço

Abstract

Zika virus (ZIKV), an arthropod-borne flavivirus, was classified as reemerging infectious disease and included as neglected tropical disease. During the recent ZIKV outbreak in South America, it has been demonstrated that ZIKV infection during pregnancy is strongly associated with fetal loss, malformations and neurological disorders in newborns. Despite the first line of host immune defense is related to innate immunity activation, the immunological homeostasis is essential for pregnancy success. Although the dynamic changes in maternal-fetal immunity is not completely understood and poorly investigated, the knowledge of immune responses during gestation is very important for infectious disease prevention and control, as ZIKV. Here, we put together more and new information about the innate immunity during gestation, highlighting three parts probably involved with clinical outcome and/or not well explored in literature: 1) type III interferon; 2) innate regulatory cells; and 3) cell death pathways modulation. Additionally, we will be focused on discussing how the dynamic responses of innate immune system during pregnancy and its effects in newborns, could be modulated by ZIKV, as well as how efforts on development of new/old drugs and vaccines could be effective for ZIKV prevention and control to provide a successful pregnancy.

Keywords: innate immunity, pregnancy, zika, technological development

1. Introduction

Zika virus (ZIKV) is an arthropod-borne flavivirus, considered a reemerging infectious disease as well as a neglected tropical disease [1]. Moreover, ZIKV was also classified as sexually transmitted disease (STD), since viral RNA and infectious particles were detectable in reproductive organs and others described some cases related to sexual transmission [2, 3]. Although the major concern about ZIKV infection is the intrauterine transmission [4–6].

Innate immunity during pregnancy still needs attention when some infection compromises pregnancy success. Recently, the world testified a huge public health problem during Zika virus (ZIKV) outbreak in Latin American countries [7–9], in which poor outcomes were observed firstly in Brazilian newborns from mothers infected on early pregnancy phase (1st -2nd trimester) [7, 8]. Consequences of viral infections on newborns are irreversible and public health and social costs are immensurable [10], making World Health Organization consider Zika infection a public health emergency in 2016 February [11].

Due to its neurotropic features, the infection caused by ZIKV has been evidenced [12–14], which show a correlation between clinical manifestations based on its tropism by brain neuronal cells of fetuses and neonates born from infected pregnant women, with a strong association to neurological damage, including microcephaly and other fetal neurological disorders, collectively named as Congenital Zika Syndrome (CZS) or Zika Associated with Birth Defect (ZABD) [15–18].

The immune system is composed of a set of flexible mechanisms that are fundamental to maintain homeostasis, allowing many interactions and coexistence between different populations of microorganisms and the host. The imbalance of homeostasis can be caused by a microorganism because of its pathogenic behavior. With the establishment of an active infection and consequent immune response, inflammatory mediators, produced initially, collaborate to activate cellular populations of the innate immunity, promoting antiviral and cytotoxic responses, for example. At first, these effector responses would influence the viremia resolution with the re-establishment of homeostasis. However, the loss or dysfunction of this immune response can generate a harmful environment that triggers an uncontrolled damage inflammation and consequent cell death due to a direct cytopathic effect caused by the microorganism [19].

Some studies were conducted to understand the mechanisms involved in vertical transmission. During pregnancy, the transfer of ZIKV to the placenta occurs after an infection of decidua, the placenta maternal region, since studies have shown that decidua cells are permissive to ZIKV infection and remain permissive throughout pregnancy [20, 21]. From the infection of the decidua, there are two routes by which ZIKV reaches the fetus: infection of syncytiotrophoblasts (SBTs) through capillaries containing maternal blood or infection of Extravilous Trophoblast (EVTs) by cell-to-cell propagation [4]. In vitro studies have shown that ZIKV can infect first-trimester cytotrophoblasts CTBs and EVTs [4, 20, 21]. On the other hand, STBs are high producers of type III interferon and remain relatively resistant to viral infection throughout pregnancy, therefore, the main route hypothesis for transplacental transmission of ZIKV is that of the spread of decidua to EVTs [21, 22]. Additionally, infection of placental macrophages, the Hofbauer cells by ZIKV may contribute to both intrauterine transmission and immunomodulation [23, 24]. Further, transplacental transfer of ZIKV is more likely to occur in the pro-inflammatory environment and tolerant to placental immunity in the first trimester.

Histopathological and immunological studies in placentas have shown that infections by ZIKV lead to an increase in important inflammation markers such as TNF, CCL5, and altered vascular permeability such as metalloproteinases [25]. In addition, in vitro experiments demonstrate that trophoblastic cells become progressively more resistant to infection by ZIKV during pregnancy, partly through the secretion of IFNs [26]. In this context, a lot of efforts were raised to provide funds to deeply investigate how to avoid another spread of Zika virus infection, as well as drugs tests and vaccine development based on viral proteins, DNA vaccines, Virus Like Particles (VLP), chimeric viruses, among other strategies [27–30]. Therefore, there are few studies to investigate the pregnancy immunity and how the immune interface mother-to-child could contribute to infection spread with drastic consequences to fetus [21, 31–34]. To our knowledge, the imbalance of normal pregnancy immunity is already cause of metabolic disorders and the poor outcome is related to abortion [35–37]. Then, a viral infection can make this picture worst and tragic [8, 13, 15, 38, 39].

Like other Flaviviruses, ZIKV life cycle modulates machinery and functions of target immune host cells, making essential virus-cells interactions for pathogenesis development. Moreover, while several human and animal models' studies have argued and proved ZIKV neurotropism, there are still many answers regarding viral pathogenesis in mother and its influence the fetal neural system and persistence, and clinical outcome. In this chapter we will put together the information about innate immunity during gestation, highlighting three parts probably involved with clinical outcome: 1) interferon type III; 2) innate regulatory cells; and 3) cell death pathways modulation. Additionally, we will focus on discussing how the dynamic responses of innate immune system during pregnancy and its effects in newborns, could be modulated by ZIKV, as well as how efforts on development of new/old drugs and vaccines could be effective to help pregnancy success.

2. Type III interferon

The success of pregnancy is dependent on a coordinated balance between the "invading" fetal trophoblast and a receptive maternal decidua in the placenta, maintaining a dynamic and responsive immune system. The longest period of the pregnancy, fetal growth, demands a symbiotic and tolerogenic environment, but congenital viral infections can disrupt this equilibrium. In order to avoid infection severity placenta actively modulates the immunologic profile of the maternal-fetal interface [40, 41]. In this context, recent studies demonstrated that placenta responds to ZIKV infection by production of the newest interferon group type III interferons [21, 42, 43].

Type III interferon (IFN-λ 1–4) comprising a group of cytokines with action pathways under strengthen discovery [44–46], basically acting with shared inflammatory regulation and antiviral properties [47]. IFN-λs receptor was identified as a complex composed of two subunits: IFN-λR1 and IL-10R2, which is also a receptor subunit of the regulatory cytokines IL10, IL22, and IL26 [48]. In contrast with the classical pro-inflammatory type I interferons which receptors are expressed in almost all cell types, the IFNLR1/IL10RB complex is expressed primarily in cells of epithelial origin and few immune cells conferring selective IFN-λ responsiveness to them: neutrophils [49], myeloid dendritic cells (DCs) [50, 51] and plasmacytoid dendritic cells (pDC) [52]. Because of the restricted cell types producing IFN-λs, this cytokine acts locally as an immunologic barrier in organs with suppressing innate pro-inflammatory responses and limiting host damaging effects associated with inflammation [53]. Moreover, IFN-λs utilize mechanisms to suppress viral infections which induce a strong antiviral state following receptor binding with non-translational and translational processes [49, 54].

Between the different inflammatory regulation actions already described for IFN-λs, the suppression of neutrophil gains prominence because they are the immune cells that present higher expression.

of IFN-λR1 at the steady-state [55–57]. Neutrophils contribute to various stages of the reproductive process since conception and implantation, ensuring fetal wellbeing during pregnancy and finally contributing to parturition and postpartum maternal health. On the other hand, aberrant neutrophil activity is associated with severe pregnancy-related disorders such as pre-eclampsia, recurrent fetal loss or gestational diabetes mellitus [58–60]. In murine models, it was demonstrated that neutrophil exposed to IFN-λ can induce antiviral interferon-stimulated genes (ISGs); and IFN-λ (but not IFN-β) specifically activated a translation-independent signaling pathway that diminished the production of reactive oxygen species and degranulation in neutrophils, which might permit a controlled development of the inflammatory process [49].

Studies utilizing a cellular model of collagen-induced arthritis demonstrated that IFN-λ2 was protective and could stop the progression of the disease, diminishing infiltration of neutrophils to the inflamed joints as well as the production of IL-1β upon treatment with pegylated recombinant IFN-λ2 [57]. *Ex vivo* experiments with cardiopathic patients` cells demonstrated that IFN-λ inhibits Neutrophil Extracellular Traps (NETs) [61]. NETosis has been appointed as critical agents during pregnancy, particularly involved an auto-inflammatory process involving the release of placental micro-debris in preeclampsia and recurrent fetal loss [62]. In collagen-induced arthritis murine models, it was demonstrated that IFN-λ exerts its anti-inflammatory effect by restricting recruitment of IL-1β–expressing neutrophils, which are important for amplification of inflammation, and reducing IL-17–producing Th17 and γδ T cells in the joints and inguinal lymph nodes, without affecting T cell proliferative responses [57].

IFN-λ is strongly associated with DCs activity inducing an effector adaptive immunity response [63, 64]. Studies with a mice model of influenza A virus infection demonstrated that IFN-λ directed acts in the migration and function of CD103(+) dendritic cells, also regulating DC IL-10 network [65]. Migratory CD103(+) DCs derived from skin, lung, and intestine, efficiently present exogenous antigens in their corresponding draining lymph nodes to specific CD8(+) T cells through a mechanism known as cross-presentation, demonstrating the IFN-λ importance for the development of specific CD8+ T cell responses [65, 66]. Moreover, IFN-λ contributes to the formation of tolerogenic DCs cell, contributing to control inflammatory responses and homeostasis by fostering the conversion of naive T cells into induced Foxp3(+) regulatory T cells [66]. In vitro studies demonstrated that IFN- λ directs DCs to a regulatory phenotype with diminished capacity to stimulate T cell proliferation in a PD-1/PD-L1 dependent manner with contribution from the imbalanced cytokine milieu, such as low IL-12 and IL-2 and/or high IL-10 production [50]. Another study using mixed lymphocyte cultures demonstrated that IFN- λ -treated DCs specifically induced IL-2-dependent proliferation of a CD4(+) CD25(+) Foxp3(+) T-cell subset with contact-dependent suppressive activity on T-cell proliferation initiated by fully mature DCs [51].

Plasmacytoid dendritic cells (pDC) are rare cells found in peripheral blood and lymphoid tissues, considered to be "professional" type I IFN-producing cells and produce 10- to 100-fold more IFN-α than other cell types in response to enveloped viruses. However, in vitro IFN-λ treatment of pDC resulted in increased virus-induced expression of both IFN-α and IFN-λ, indicating that pDC are high producers of IFN-λ1 and -λ2 in response to viral stimulation and the consequences of this high IFN-λ production by pDC should be further explored [52].

In human congenital ZIKV infections, it was demonstrated that ZIKV infection leads to a typical inflammatory response in the placenta, including the expression of anti-viral Type I interferon genes (*IFIT5*, *IFNA1*, and *IFNB*), type II interferon (*IFI16*), cytokine signaling (*IL22RA* and *IP10*), and interferon regulatory factors (*IRF7* and *IRF9*). Furthermore, the CZS cases present a gene expression profile with impaired *IFNL2* response, accompanied by an exacerbated type I IFN response; with an increased expression of *IFIT5*, parallel to a decrease in *ISG15* mRNA [67], which was already identified as negative modulator of type I IFN and protective against ZIKV ocular manifestations [68]. These results are corroborated by *in vitro*

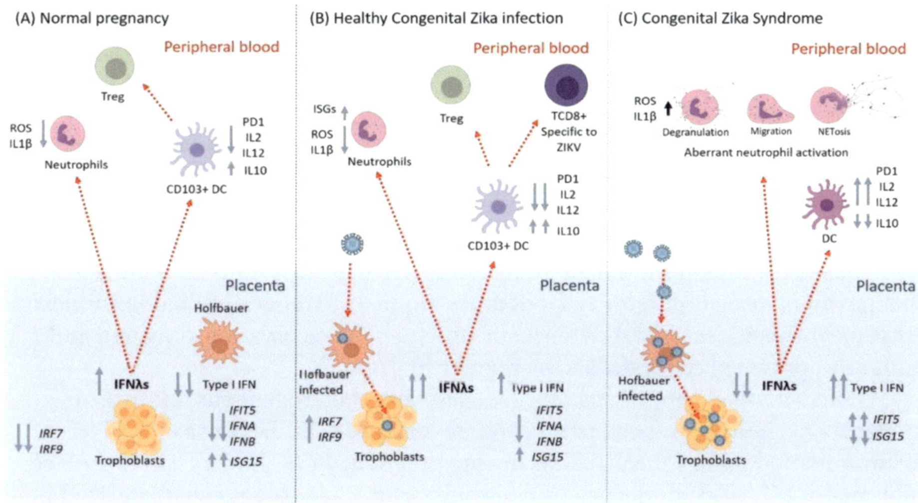

Figure 1.
Summary of Interferon lambda (IFN-λ) function during normal pregnancy (A), Healthy Congenital Zika infection (B), and Zika-Associated Birth Defects (C). (A) In normal pregnancy, trophoblasts exhibit a constitutive IFN-λ production, contributing to the general tolerogenic environment demanded by pregnancy (A1); Considering the peripheral blood tissue IFN-λ Interact with: (A2) neutrophils leading to a decrease in ROS and IL1β, and (A3) migratory CD103+Dendritic cells (DC) that present low levels of PD1, IL2 and IL12 together with high IL10. These CD103+DC foster the conversion of naive T cells into induced Foxp3(+) regulatory T cells (Treg) (A4). In the placenta, the constitutive IFN-λ is accompanied by decreased type I IFN pathway: low expression of IFIT5, IFNA, and IFNB, and high expression of type I IFN the negative regulator ISG15 (A5). In the lack of viral infection, the interferon regulatory factors IRF7 and IRF9 present low expression levels (A7). (B) In healthy congenital Zika infections, the placenta expresses high levels of IFN-λ to protect the fetus from congenital defects (B1). In this low damage antiviral response, high levels of IFN-λ elicits the production of ISGs and the decrease of ROS and IL1β by circulating neutrophils (B2), meanwhile the CD103+ DC presents an accented regulatory profile (B3), with induction of high specific anti-ZIKV response by Treg (B4) and TCD8+ cells (B5). In the placental level type, I interferon pathway shows a slight increase, together with the enhance of IRF7 and IRF9, forming a balanced antiviral response. (C) In Congenital Zika Syndrome (CZS) the lack of IFN-λ contributes to a damaging outcome (C1). Diminished levels of IFN-λ could not control the neutrophil activity, culminating in augmented ROS and IL1β (C2), and presence of aberrant activation forms as well as degranulation, migration, and NETosis (C3). Without IFN-λ the Dendritic Cells (DC) present a pró-inflammatory profile, with augmented PD1, IL2, and IL12 and diminished IL10 (C4). The placenta shows an exacerbated type I interferon response, which together with low IFN-λ levels (C5), leads to an imbalanced damaging antiviral response. Grey arrows represent the production or expression levels (up = high, down = low). Double arrows represent a high magnitude of production or expression. Red dashed arrows represent the direction of function/induction events that have been known and those suggested. Figure created using Biorender software (https://www.biorender.com).

studies that showed induction of *IFNL1* expression by susceptible placental cells after ZIKV infection, acting as an antiviral agent [43], reinforcing that IFN-λs are protective factors in ZIKV congenital infections. Studies with *ex vivo* placental 3D cultures from a different trimester of healthy pregnant volunteers showed that IFN-λs are expressed mostly by deciduous (the maternal portion of the placenta), already indicating that mothers are the agents on the immunoregulation of CZS outcome (**Figure 1**) [21].

3. Innate regulatory cells - myeloid-derived suppressor cells (MDSC)

Immunity during pregnancy is very important to be explored since successful pregnancy requires that immunoregulatory mechanisms are triggered to suppress activated fetal-specific T cells lymphocytes [36, 37]. Maternal immune cells can recognize paternal antigens on fetus. Thus, it has been very well described that dysfunction of immune cells during pregnancy can lead to immunologic fetal rejection by mother, in which the consequences are related to abortion, preterm delivery, or other severe complications [35–37].

Then, maternal-fetal tolerance involves the regulation of mother's immune system to tolerate the semi allogeneic fetus expressing paternal antigens without immune rejection. Even though, some studies showed that regulatory T cells are the main cells which plays an important role in suppressing activated T cells during gestation; since then innate immunity system is poorly investigated [69–71].

Considering infections during pregnancy, it is also important to know that changes on maternal immune responses are required to induce limited immunosuppression without loss of host defense, in which a balance between activated and immunosuppressed cells needs to be regular [35].

Myeloid-derived suppressor cells (MDSC) are a heterogeneous mixture of immature myeloid cells, been part of innate immune cells, having a crucial role in immunomodulatory mechanisms during pregnancy [36, 72, 73]. There are two subtypes of MDSC, a monocytic and granulocytic. Phenotype is characterized by expression of CD33 and CD11b in humans, CD14 by monocytic MDSC and CD15 by granulocytic MDSC cells but lacks the maturation marker HLA-DR. But both subtypes share the characteristic of immune-suppressive function inhibiting activated NK and T cell expansion [73, 74].

Normally, immature myeloid cells as MDSC are scarcely found in peripheral blood, and their maturation includes macrophages, dendritic cells, and granulocytes formation. Nevertheless, the MDSC are also recognized by their role in some pathological conditions, like cancer, sepsis, stress, autoimmune disorders and infectious diseases [38, 75, 76].

Several studies have been reported that a decrease of MDSC during pregnancy may lead to poor outcomes, as miscarriage [77]. Also, it has been shown that progesterone levels increase MDSC during pregnancy in mice, as well as high levels of TNF and IL-1β, pro-inflammatory cytokines [38, 78].

In murine models, it was demonstrated that MDSC can produce TGF-β and IL-10, as immunosuppressive cytokines, similarly to regulatory T cells. Adding to that, MDSC can suppress T cell activation and function by arginase-1 (Arg-1) secretion, as well as nitric oxide synthase and indoleamine 2,3 dioxygenase aimed to deplete nutrients for T cell proliferation, as I-arginine (I-Arg). According to Ismail 2018, arginine is also involved in replication, and virulence of several agents, as viruses and bacteria. Then, it is suggested that an accumulation of MDSC in placenta could influence an increase of arginase activity, and it would serve for a dual purpose, inhibiting the adaptive immune system whilst also providing potential protection against infection by arginine auxotrophic pathogens [79].

Nitric oxide (NO) has been related to embryo successful implantation during early pregnancy, but excessive NO production by decidual macrophages seems to be harmful and was linked with early pregnancy loss [37, 80, 81]. Another study suggests that in early pregnancy in decidua CD33+ cells express nitric oxide synthase, playing an important role to maintained pregnancy during this phase, while in later pregnancy CD33+ cells lose the expression of this enzyme [35, 37].

Kostlin-Gille *et al* 2019 showed that hypoxia condition is important to normal placenta development and its driven by a hypoxia-inducible factor 1 (HIF-1), a key regulator responsible for initiate transcription of several genes. The subunit HIF-1α is highly expressed in placenta during early gestation period, characterized by low oxygen pressure conditions. This study used myeloid HIF-1 knockout mice to evaluate the role of HIF-1α on myeloid-derived suppressor cell function, showing that HIF-1α deficiency in myeloid cells leads to diminished suppressive activity of MDSC in uterus from pregnant mice, but the expression of chemokine receptor or integrins was not altered. Despite MDSC recruitment to uterus was not altered, it was observed a lower MDSC accumulation as well as an increase of MDSC apoptosis, contributing to an elevated abortion rate in knockout mice [73].

Regarding Zika virus, there are few studies showing the presence of MDSC on women blood and during pregnancy, and considering the facts, it will be very important to know any relationship of their presence with congenital syndrome, as observed in 2016, Brazil [82, 83]. A study with 10 non-pregnant women with Zika infection showed that frequencies of circulating MDSC did not change over time [84]. Another study with pregnant monkeys infected with Zika virus showed that an imbalance on blood frequencies of MDSC and activated CD8 T cells in the acute phase may lead to poor outcome to the fetus. Adding to that, the high frequency of MDSC on placenta from pregnant monkeys showed a positive effect on pregnancy outcome, even more if a drug antiviral treatment was used [85].

Furthermore, it is worth to note that immune signature, sometimes is the key factor to explain some diseases progressions. Despite Dengue viruses is more related to signals and symptoms with Zika virus infection [86, 87], some similarities with hepatitis C virus (HCV) were also noted, and mechanisms of immune evasion have been described, as inhibition of interferon pathway, allowing virus life cycle for a long-term period, up to 100 days [88, 89]. To note, ZIKV infection is also classified as an immune-mediated viral disease, like Dengue and other viruses [86, 87, 90]. Disease progression in HCV patients to chronic infection has been associated to an increase of MDSC phenotype in peripheral blood mediated by viral proteins [38]. Wang et al., 2017 examined Japanese encephalitis virus (JEV) infection leading acute encephalopathy depending on suppression of adaptive immune response, especially T follicular helper cells, mediated by enhanced MDSC populations, such as an involvement of MDSC on splenic B cells reduction, and in lower levels of total IgM JEV-specific neutralizing antibodies in mice models [39]. Burrack et al., also suggests that MDSC has an important suppressive T cells activity and may contribute to reduce the immune-mediated disease during Chikungunya infection [90].

Otherwise, the immunosuppressive activity triggered by RNA viruses, MDSC has been associated with metabolic regulation of immunopathology induced by DNA viruses, like hepatitis B virus (HBV) [91]. Pallett et al., 2015 showed that frequencies of MDSC on liver from HBV patients without liver damage, monitored by levels of liver transaminase enzymes, were higher in comparison with patients with liver damage, showing a protective effect for patients with immune-mediated viral disease, as hepatitis B [91].

In the new coronavirus pandemic (COVID-19), the MDSC have been reported to play an important role in the early phase of symptoms, increasing their frequency on blood in the first days of signals and symptoms, and it was related to poor outcome in severe acute respiratory syndrome in hospitalized patients. Pregnancy is a risk factor for COVID-19 severity, given the Brazilian high mortality rate of 12.7% in June 2020 withing pregnant, which may be associated with the change of the immunity [92–94].

Although few studies involving MDSC frequencies on blood during Zika infection were published yet, those cell type needs to be investigated, even though in animal models for medical science breakthroughs. The technique to characterize this cell phenotype is simpler than to characterize regulatory T cells, once the procedure does not require intracellular staining [95].

If those MDSC are crucial to maintaining a healthy pregnancy, any adverse effects, as Zika virus infection could trigger an imbalance between MDSC and T cells. This dysfunction may induce a deactivation of functional MDSC on blood and placenta with failure to attempt to eliminate viral infection. In addition, T cell function during ZIKV infection is known to be delayed throughout interferences

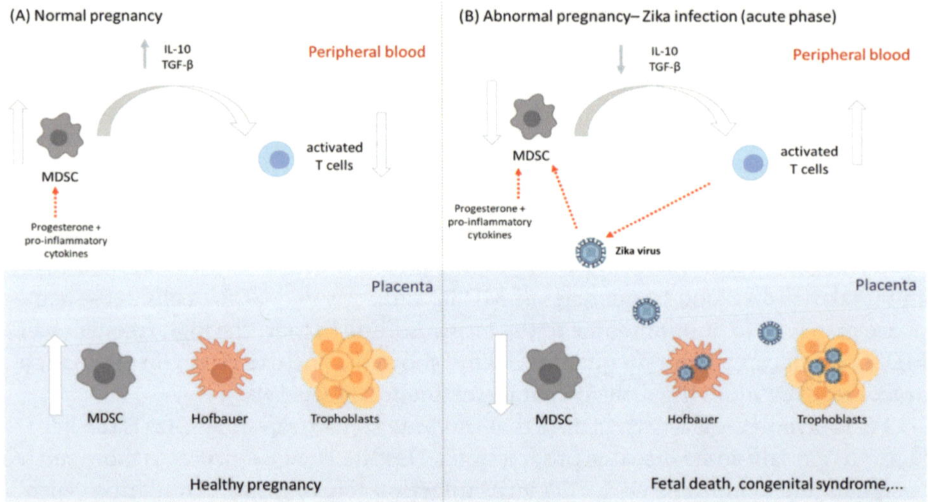

Figure 2.
Myeloid-derived suppressor cell (MDSC) activation and regulation triggered by normal pregnancy and by Zika virus infection. Summary of MDSC functionality during normal pregnancy (A) and during acute phase of Zika virus infection (B) as suggested by others into an innate immunity dysregulation observed in abnormal pregnancies on monkeys [35, 37, 38, 73, 77–81, 85]. Hormone and cytokines produced in normal pregnancy induce an equilibrium in peripheral blood maintaining frequency of MDSC elevated (1.A), as well as levels of IL-10 and TGF-beta. Meanwhile, circulating levels of T cell frequencies are reduced and controlled. In placenta, Hofbauer cells (macrophages) are responsible for immune surveillance also intermediating the cross-talking between fetus-maternal interface, with equilibrium of MDSC and T cells to maintain a healthy pregnancy. In abnormal pregnancy, also suggestive for Zika virus infection during pregnancy of non-human primates, the equilibrium is broken. Once ZIKV is circulating, there is a reduction of MDSC frequency (B), compromising pregnancy immunosuppression, with elevation of activated T cells, attempting to virus elimination. In the placental parenchyma, MDSC has a reduction in their frequency. This scenario also suggests an immune dysfunction in fetus-maternal environment, diminishing functional macrophages (Hofbauer cells), which are infected by virus. All events together can induce several poor outcomes (abortion, neurological disorders). Black arrows filled with white color represent the frequency of cells (up = high, down = low). Grey arrows represent levels of cytokines (up = high, down = low). Red dashed arrows represent the direction of function/induction events that have been known during Zika infection during pregnancy. Figure was created using Biorender software (https://www.biorender.com).

on interferon pathway, as described above. Then, this scenario may contribute to immune evasion of ZIKV, in which viral replication on maternal-fetal environment is unavoidable, inducing poor outcomes during pregnancy: fetal death, congenital syndrome, abortion, neurological disorders, etc. (**Figure 2**).

4. Programmed cell death: A host innate immune protection or a virus evasion strategy

It has been described that a protective response by innate immune cells to viruses is triggered by several distinct mechanisms including apoptosis, necrosis, paraptosis, pyroptosis, autophagy cell death, and others. Each one is depending on several aspects of infection, including where the microorganism was detected, susceptible target-cells, through signaling systems discharging the death signal, and its intensity. During the innate immune response to infections, programmed cell death may occur as a direct pathogenic mechanism of viral spread and escape from the immune system or represents an appropriate host response to limit pathogen replication. Apoptosis of lymphocytes and monocytes also plays an important role in the control of inflammatory responses, as well as in the development of maternal-fetal tolerance [96–99].

Type 1 programmed cell death, also known as apoptosis, is defined by inter-nucleosomal DNA fragmentation, marked irreversible apoptotic characteristic indicating chromatin condensation, degradation of cytoskeleton and nuclear proteins, protein crosslinking, apoptotic bodies' formation baring ligands for receptors of phagocytic cells and, finally, the uptake by these phagocytes [97–99]. Type 2, or autophagic cell death, presents unique characteristics organelles formation including autophagosomes and autophagolysosomes in the dying cell, sources of self-degradation, and recycling [100].

Two pathways can regulate the apoptosis program in different aspects: extrinsic and intrinsic. Extrinsic pathway is activated by a transduction signal through death receptors, in which TNF, Fas ligand, or TRAIL bind to their respective receptors, such as TNF receptor family: TNFR1, Fas (CD95/APO-1) and TRAIL-R1/2. A complex signal mediated by this binding leads to an enzymatic cascade of cell degradation, and at this point caspase-3 is activated promoting DNA damage [101]. Intrinsic pathway involves intracellular mitochondria, which its membrane is the local for many Bcl-2 family members and their activity in inducing / inhibiting the mitochondrial apoptosis program implies in those proteins lead to membrane collapse as well as a transition from mitochondrial permeability promoting apoptosis process [96, 101–105].

Taking together, type 2, or autophagic cell death, consists of a conserved catabolic process that contributes to degradation and recycling of many intracellular substances, through lysosome activity. In this sense, many studies have shown its importance in immune responses, including degradation of microbes, direct viral peptides MHC class I presentation [106] and even altering T-cell signaling and tolerance [107, 108]. At first, autophagy is necessary to keep the cell alive under stress conditions that precede their demise. Such kind of cell death could be achieved by several mechanisms, including prolonged hypoxia or digestion of vital factors, regulatory molecules or essential organelles. In a stress situation, caused by virus, an infected cell can induce intracellular signals of autophagy, inhibiting cell proliferation, arresting cell cycle and eventually leading to cell death [106–111].

In the acute ZIKV infection during pregnancy, macrophages and dendritic cells are involved in inflammatory cytokines production, in which CARD9 expression, an important regulator of caspase activity playing an important role in cell apoptosis regulation, is elevated allowing that pattern recognition receptors (PRR) induce pro-inflammatory cytokines cascade, as the first step on CZS, as suggested [67]. According to Quicke et al., Hofbauer cells infected with ZIKV in placenta induces IFN type I activation, reactive oxygen species production, as well as pro-inflammatory cytokines, but with minimal cell death, showing a scape of innate immune response [23]. Recently, Cao et al., showed that ZIKV could activate and increase an autophagic process in pregnant mice, suggesting an imbalance of trophoblastic cells in placenta, and relation with fetal loss [112]. Corroborating, Ribeiro et al. using a human model of placenta explants for in vitro infection demonstrated tissue injury as consequence of the association between fetal pro-inflammatory responses mediated by IL-1β, IL-6 and TNF and extrinsic caspase 3 dependent apoptosis (TNF-TNFR pathway). Together data suggest that ZIKV infection corroborates to placenta innate immune and hormonal dysfunction, increasing loss barrier integrity [42] Thus, this inflammatory status could trigger cell death and barrier loss, allowing ZIKV cross placenta and infect fetuses' neural stem cells (**Figure 3**) [23, 113–115]. Interesting, autophagosomes are present in neural stem cells and it could facilitate ZIKV replication [116], although inflammation generated as well as the cytopathic effect itself culminate in extensive caspase-dependent neuronal cell death.

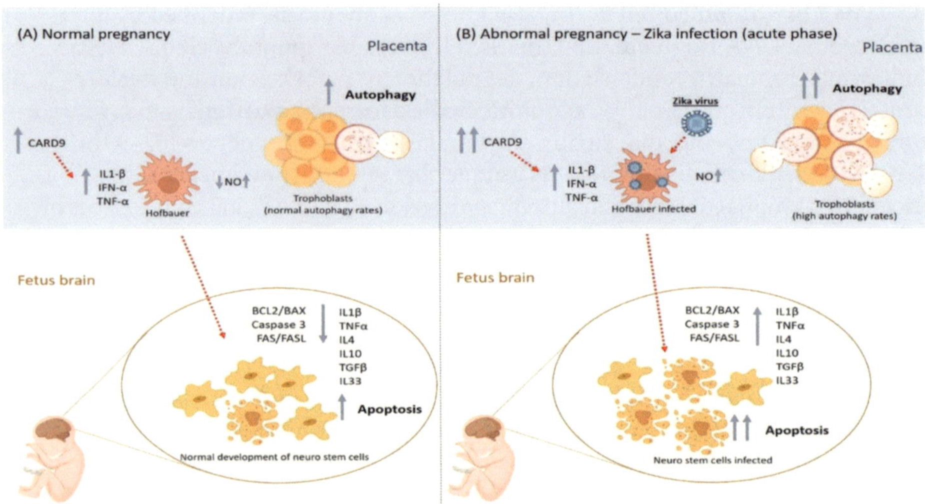

Figure 3.
Programmed cell death activation during normal pregnancy and abnormal pregnancy induced by Zika virus. Normal pregnancy equilibrium is driven by regulation of number of innate immune cells in placenta leading by programmed cell death. In this situation, caspase activity starts on CARD9 expression with cytokines production by Hofbauer cells (1.A), which oxide nitric (NO) regulates trophoblasts autophagy (2.A, 3.A). Products of Hofbauer cells activity in the surveillance in placental parenchyma contributing to extrinsic (Fas/Fas-L) and intrinsic pathway (BCL2/BAX) activation in fetus brain with low expression of pro-inflammatory cytokines, regulating number of neural stem cells and microglia by apoptosis (4.A), maintaining the healthy pregnancy. Acute ZIKV infection during pregnancy suggests that macrophages and DCs are involved in pro-inflammatory cytokines production, in which CARD9 is upregulated, increasing caspase activity, allowing pro-inflammatory cytokines and reactive species cascade (1.B, 2.B), exacerbating autophagy in placenta (3.B). Taking together this innate immune dysfunction, fetus brain is affected by high activation of apoptosis pathway (4.B), provoking a cascade of cell death with an abrupt reduction of neural cells, causing severe damage [113–115]. Grey arrows represent the production or expression levels (up = high, down = low). Double arrows represent a high magnitude of production or expression. Red dashed arrows represent the direction of function/induction events that have been known and those suggested. Figure created using Biorender software (https://www.biorender.com).

Corroborating, Lum et al. has shown that ZIKV mainly infects fetal microglia and induces high levels of pro-inflammatory cytokines that could be harmful to the fetus [117]. In addition, the analysis of in vitro culture, fetal brain histology and *ex vivo* studies with children presenting evidence of congenital infections demonstrated that, in fact, ZIKV promotes microglial activation, suggesting viral disseminating, neuronal death and an abnormal increase of astrocytes due to neurons destruction [117].

Thus, once in fetus central nervous system, ZIKV may contribute to extrinsic (Fas/Fas-L) and intrinsic (Bcl-2) pathways activation for programmed cell death, reducing number of neuronal cells. Thus, the risk of congenital syndrome is eminent, mainly in the first trimester, as well documented (**Figure 3**) [67, 118–123]. Some studies with fetuses' autopsies and infants with microcephaly have been demonstrated a broad spectrum of microscopic neuropathological abnormalities and brain damage, with direct virus cytopathic effects in neural glial cells. In this way, these data support the strong association with apoptotic cell death and microcalcifications [13, 23, 124].

5. Prevention and control of ZIKV infection: Potential candidates in pregnant women

In general, pregnancy is a challenge for prevention and control infectious diseases regard to a safe drug or vaccine development to do not disturb the innate/adaptive immunity homeostasis, however, there were no drugs approved for ZIKV infection treatment [28–30]. Here, drugs and vaccines candidates tested in animal models or in newborns will be described with details (**Table 1**).

5.1 Type III interferon: Potential efficacy and safety for immunotherapy

Type III interferon has been emerging as an efficient and low damaging therapeutic agent not only directed for the virus but also for fungal and bacterial infections, as well as cancer, autoimmune, and vascular diseases [54]. The more restricted expression of IFNLR1 likely contributes to the improved safety profile of IFN-λl in clinical studies compared to type I IFN. Pegylated IFN-λ1 have already been tested in phase 2b clinical trial to chronic hepatitis C treatment and hepatitis B, associated with improved rates of virologic response with fewer extrahepatic adverse events compared to pegylated IFN-α [125]. Even though it was deemed less effective than alternative treatments for these infections, pegylated- IFN- λ can be potential candidate ready for deployment if new indications are identified [126]. There are other viral targets for IFN- λ therapy been tested in murine models: norovirus [127], and influenza virus [128], and west nile virus – last one is another member of Flaviviridae family. It is noteworthy the effect of IFN-λ on infection with west nile virus, an encephalitic flavivirus: Treatment of IFNLR1 knockout mice with pegylated IFN-λ2 resulted in decreased blood–brain barrier permeability, reducing west nile virus infection in the brain without affecting viremia, and improved survival against lethal virus challenge [129].

The effectiveness and low damage treatments for other correlated viral infections, combined with the protagonist of IFN-λs as immunoregulatory and antiviral agent in ZIKV raise the idea of IFN-λs as ZIKV therapy, and some groups already achieve exciting good results. Concerning ZIKV infections, Jagger, et al., (2017) suggest that IFN-λ2 treatment could be a safe solution to minimize Congenital Zika Syndrome severe outcomes. Using a type III interferon-deficient mouse model, authors showed that these animals had an increase of ZIKV replication in the placenta under ZIKV infection, and treatment of pregnant mice with IFN-λ2 reduced ZIKV viremia [26]. Considering the vaginal epithelium as the first line of defense against sexually transmitted ZIKV, treatment of primary human vaginal and cervical epithelial cells lineages with IFN-λ induces host defense transcriptional signatures with augmented expression of ISGs (IFI44L, OASL, OAS1, and MX1) and inhibition of ZIKV replication. Female mice submitted to treatment with IFN-λ and intravaginal ZIKV transmission showed low levels of virus replication in the female reproductive tract with a hormonal stage-dependent role [130].

5.2 Direct-acting antiviral therapy based on RNA-dependent RNA polymerase inhibitors

Some studies were driving to evaluate effects of independent direct-acting antiviral drugs on Zika virus infection (**Table 1**), as sofosbuvir, an FDA-approved nucleotide analog inhibitor of the hepatitis C (HCV) RNA-dependent RNA polymerase (RdRp) [131, 132]. In vitro and *in vivo* studies have been demonstrated effectiveness of sofosbuvir as antiviral drugs to treat Zika and Dengue virus infection [133–135]. Mesci et al., 2018 reported that sofosbuvir was promisor to block vertical transmission of Zika virus in pregnancy using mice models [136]. Again, sofosbuvir shows to play a role in virus replication inhibition. Another flaviviral inhibitor NITD008, an adenosine analog inhibiting the RNA-dependent RNA polymerase activity through chain-termination [137], has been shown to reduce the

Therapy	classification	Mechanism of action	Immune effect	Pregnancy safety	References
Peg Interferon-λ2	Not approved	Antiviral immunobiological	Enhance IFNL-λ pathway activity	Yes/Mice models	Jagger et al., 2017 [26]
Sofosbuvir	Category B/Approved for hepatitis C treatment	Direct-acting antiviral drugs	Not explored	Yes/Mice models	Mesci et al., 2018 [136]
NITD008	Not approved	Direct-acting antiviral drugs	Not explored	Yes/Mice models	Watanabe et al., 2019 [27]
Hydroxycloroquine	Category C/Approved for malaria and autoimmune diseases therapy	Cell membrane interaction to induce cell death	Reduction of autophagy activity	Yes/Pregnant women	Cao et al., 2017 [112]
rVSV vaccine	Not approved	Recombinant viral vector vaccine	Increases in CD8+/CD44high/IFN-γ + T cell populations on spleen	Yes/Mice models	Betancourt et al., 2017 [147]
VRC5283	Clinical trial phase II (VRC-ZKADNA090–00-VP)	DNA plasmid vaccine	Induce antigen-specific antibody production/ induce of CD8+ T cells response	Yes/Mice models	Richner et al.,2017 [155]
mRNA-LNP vaccine	Clinical trial phase I (NCT03014089)	mRNA vaccine	Induce antigen-specific antibody production/ induce of CD8+ T cells response/Minimizes ADE	Yes/Mice models	Richner et al.,2017 [156]

Table 1.
Therapeutic agents or vaccine candidates targeting virus or immunity with promisor potential to use during ZIKV infection in pregnant women.

Zika virus replication in placenta, and fetal infection, thus minimizing the risk of maternal-fetal transmission of ZIKV [27].

There are few studies investigating innate immunity during antiviral therapy, especially when its concern to Flaviviridae family [38, 135, 138, 139]. Scarce literature revealed knowledge about antiviral therapy immune effects only during hepatitis C infection [138, 139]. Antiviral drugs, as pegylated interferon (PEG-IFN), ribavirin, and direct-acting antiviral agents (DAA) have been related with a reduction of innate regulatory cells, as MDSC, in peripheral blood from hepatitis C chronic patients, in which T cells were increased and immune function was reestablished [138, 139]. Nevertheless, all those drugs are aimed to interrupt viral replication and any dysregulation of immune cells during pregnancy is not safe, then those drugs are not recommended to be used during gestational period [140]. Besides no immune response evaluation was related to DAA therapy, it has been known that small molecules with specific activity should not induce any immune alterations in maternal-fetal immunity [140].

Safety and effectiveness of sofosbuvir on Zika virus infection should be addressed to immune response evaluation, which is poorly explored, even more in pregnant animal models. More studies and investments are needed for non-clinical and clinical studies, to get safety therapeutic protocols aimed to pregnant women with Zika virus or other flavivirus infection.

5.3 Cell death modulation during antiviral therapy

Genetic manipulation has been proven to be a promising tool for vaccine and therapy development. Considering the type 2 of programmed death, autophagy is activated by ZIKV in placental parenchyma and is involved in poor outcome during pregnancy, this cell death pathway has been a target for therapies [112, 141–143].

Recently, a study showed the role of an autophagy gene (Atg16I1) during ZIKV infection in pregnant mice model, in which inducing a deficiency in this gene limited ZIKV vertical transmission, as well fetal damage, improving placental and fetal outcomes [112]. In addition, an antiviral compound approved to be used by pregnant women for malaria and autoimmune diseases [141], hydroxychloroquine (HCQ), has been used to dampen autophagic activity *in vivo* [142]. Thus, Cao et al., showed that HCQ administered with a dose of 40 mg/kg/day has *in vivo* inhibitory effects on autophagy sustained lower levels of ZIKV RNA compared with saline buffer treatment [112].

Based on the knowledge of ZIKV infection that can trigger a caspase-3 activation contributing to cell death of neural progenitor cells during pregnancy, it is an extremely relevant approaches targeting cell death pathways for antiviral treatments even though for therapeutic vaccines.

5.4 Recombinant viral vectors as vaccine candidate

Recombinant viral vectors have been highlighted as therapeutic alternatives to prevent and treat infectious disease [144, 145], considering its specificity and the adverse effects of antiviral drugs and some vaccines [140, 146]. Betancourt et al., 2017 showed that a recombinant viral vector from vesicular stomatitis virus (rVSV) anti-ZIKV vaccine increased IFN-γ production by splenic CD8+ T cells as well as high neutralizing anti-ZIKV antibody titers from pregnant mice. This study also demonstrates that neonatal mouse from vaccinated dams was partially protected against neurological manifestations of ZIKV infection following wild-type virus challenge [147]. This rVSV using pre membrane and envelope region together obtained from a ZIKV strain as reference had the potential to protect from ZIKV infection during prenatal and neonatal development, likely through the transmission of maternal IgG. Despite rVSV vaccine induces IFN-γ production in pregnant mice, this vaccine needs to be evaluated for other types of interferon, mainly its effects on placental tissues .

5.5 Potential DNA and mRNA vaccines

mRNA vaccines as well as DNA-based vaccines represent a versatile vaccine platform and an alternative to conventional vaccine approaches because of their high potency, capacity for rapid development and potential for low-cost manufacture and safe administration [148]. Recent technological advances have allowed mRNA vaccines to demonstrate encouraging results in both animal and human models. Regarding prophylactic mRNA vaccines, a number of reports have demonstrated the potency and versatility of mRNA to elicited protective immunity against a variety of infectious agents in animal models against, including influenza virus, Ebola virus, Zika virus, Human Immunodeficiency virus 1 (HIV-1), herpes simplex virus, cytomegalovirus, hepatitis C and respiratory syncytial virus [149–151]. It has been noted that approximately ten mRNA vaccines programs have entered clinical trials [152].

The importance of mRNA-based vaccines and therapies is emphasized when mRNA-based biopharmaceuticals are entering the market with guidance of new biopharmaceutical companies. Modern Therapeutics, an mRNA therapy company evaluated various mRNA vaccine technologies to identify immunogenic and scalable candidates. The pipeline of this company shows different investigative stages mRNA vaccines of the following vaccines Respiratory Syncytial virus (RSV), Cytomegalovirus (CMV), human metapneumovirus (hMPV) + Parainfluenza virus Type 3 (PIV3), Influenza A subtypes H10N8, and H7N9, Zika, and Chikungunya. Curevac is the first biopharmaceutical company that developed the first prophylactic mRNA vaccine in the clinics, recently they showed that RNActive® vaccines induced long-lived and protective immunity to influenza A virus infections in various animal models [153].

Thus, big pharmaceutic companies, such as Merck & Co., have been invested in Modern Therapeutics aiming to expand the field of mRNA vaccine (https://www.modernatx.com/). Indeed, nucleic acid vaccine platform has been presented to combat the emergence of acute viral diseases, mainly to rapidly contain emerging outbreaks before they spread out of control. In this context, two vaccines were developed to combat the ZIKV outbreak (1) DNA plasmid vaccine encoding the prM-E genes of ZIKV and (VRC5283) (2) mRNA vaccine (mRNA-LNP), both vaccines mediate protection from ZIKV infection in mouse models. The DNA plasmid vaccine is in phase 2 human clinical trials (VRC-ZKADNA090–00-VP) and vaccine mRNA-LNP is in phase 1 clinical trial (NCT03014089) [154–156].

Considering that vaccine trials might not be performed in pregnant women and have not yet tested vaccines against ZIKV vertical transmission, there is a need for establishing the efficacy of ZIKV vaccines against mother-to-child transmission in animal models. In order to address those questions, it has been shown that vaccination with DNA plasmid encoding Zika virus prM-E and a lipid-encapsulated mRNA vaccine-elicited antigen-specific antibody and CD8+ T cell responses in mice, being able to generate a high level of protection against vertical transmission. Moreover, the mRNA-LNP vaccine not only inhibited vertical transmission but also ensured that fetuses are protected therefore, reinforcing its potential as promising vaccine for pregnant women [155]. Since there are few studies in the field of ZIKV vaccine candidates that evaluated vertical transmission, intrinsic maternal factors as well as fetal health, nucleic acid vaccines are pointed as a great opportunity to contain ZIKV infection.

DOI: http://dx.doi.org/10.5772/intechopen.94861

6. Conclusion

Considering the normal pregnancy, the innate immunity balance is conduct by downregulation of effector T cells and NK cells leading by innate regulatory cells (MDSC) and upregulation of pro-inflammatory cytokines. This innate immune modulation that occurs mainly at the placenta, includes interferon pathway and cell death modulation as shown in **Figure 4A**. Gestation has its own difficulties to successful outcomes regarding maternal immune tolerance. Zika virus infection becomes classified as disease-causing birth defects, developing an abnormal pregnancy, as consequence of immune dysregulation (**Figure 4B**). Thus, antiviral therapy is the key to control this immune imbalance showing positive effects in innate immunity on pregnant mice models. It has been known that efforts through vaccines development targeting pregnant women will be the solution for ZIKV prevention, as well as for other arboviral infections, to maintain immune homeostasis and generate healthy babies. Finally, this chapter brings some new thoughts that help for targeted improvements in medical science considering Zika infection

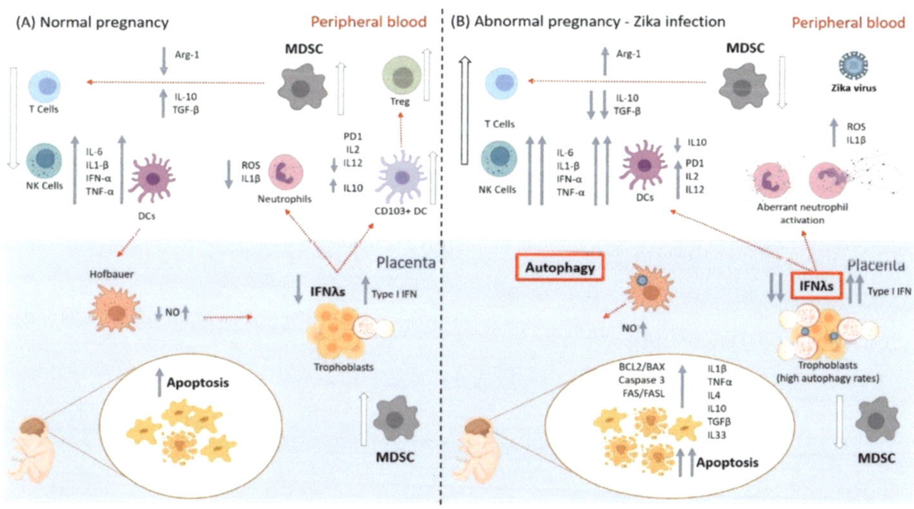

Figure 4.
Summary of innate immunity functionality during normal pregnancy and in Zika virus infection focus on interferon III, myeloid-derived suppressor cells, and programmed cell death activities. During pregnancy, initial signal is dependent on nidation process and placenta formation leading by trophoblasts expansion and activation. Following this process, innate cells, such as neutrophils, DCs, and cytokines are activated (1.A, 2.A) with IL10 and TGF-beta production in periphery, allowing immunosuppressive functionality triggered by regulatory cells (MDSC and Treg) (3.A). This condition facilitates suppression of effector cells (NK and lymphocytes) in peripheral blood and in placenta triggered by MDSC (4.A), whereas Hofbauer cells maintain reactive species (NO) balanced (5.A) as well as the IFN-λ downregulation, IFN type I upregulation, and trophoblast autophagy (6.A), contributing to the cross-linking in the fetus-maternal interface. Adding to that, programmed cell death contributes to control the accelerated growth of neural cells in fetus brain (7.A), corroborating with a successful pregnancy. Zika virus has been related to abnormal pregnancy, leading to massive innate immune alteration, causing severe brain damage to fetus. Given that, when the virus is in the blood, there is a gross activation of innate cells, elevation of cytokines and chemokines (1.B, 2.B), and suppressive activity by regulatory cells is compromised (3.B), generating early activation of NK and T cells in blood (4.B) and macrophages in placenta (5.B). Virus invasion in placenta through Hofbauer and trophoblast cells results in high autophagy activity with interferon type I gene highly expressed combined with super downregulation of interferon type III (6.B). This imbalance also contributes to fetal brain damage, orchestra by high activation of apoptosis pathway, avoiding neural cells growing progress. Thus, Zika provides severe damage to fetus, in which drugs, vaccines and immunotherapies have been designed suggesting a modulation of three important keys of innate immunity to control virus replication and spread into fetus-maternal interface: interferon type III expression, MDSC frequency, and autophagy process (highlighted with red rectangles) to avoid severe fetus brain damage, allowing a healthy pregnancy. This figure was made based on the information from ***Figures 1–3****. Figure created using Biorender software (https://www.biorender.com).*

on pregnancy, and innate immune system linked to therapies previewing the prevention and control.

Acknowledgements

The authors would like to thank Directory of Technological Development from Immunobiological Technology Institute, Biomanguinhos, Oswaldo Cruz Foundation, Rio de Janeiro, Brazil for founding support.

Conflict of interest

Authors to declare no conflicts of interest.

Author details

Tamiris Azamor[1*], Amanda Torrentes-Carvalho[2], Zilton Vasconcelos[3], Ana Paula Dinis Ano Bom[1] and Juliana Gil Melgaço[1]

1 Immunobiological Technology Institute, Biomanguinhos, Oswaldo Cruz Foundation, Rio de Janeiro, Brazil

2 Biology Institute, Fluminense Federal University, Niterói, Brazil

3 Fernandes Figueira Institute, Oswaldo Cruz Foundation, Rio de Janeiro, Brazil

*Address all correspondence to: tamiris.azamor@bio.fiocruz.br

References

[1] Petersen LR, Jamieson DJ, Powers AM, Honein MA. Zika Virus. N Engl J Med 2016;374:1552-63. https://doi.org/10.1056/NEJMra1602113.

[2] D'Ortenzio E, Matheron S, de Lamballerie X, Hubert B, Piorkowski G, Maquart M, et al. Evidence of Sexual Transmission of Zika Virus. N Engl J Med 2016;374:2195-8. https://doi.org/10.1056/NEJMc1604449.

[3] Counotte MJ, Kim CR, Wang J, Bernstein K, Deal CD, Broutet NJN, et al. Sexual transmission of Zika virus and other flaviviruses: A living systematic review. PLoS Med 2018;15:e1002611. https://doi.org/10.1371/journal.pmed.1002611.

[4] Tabata T, Petitt M, Puerta-Guardo H, Michlmayr D, Wang C, Fang-Hoover J, et al. Zika Virus Targets Different Primary Human Placental Cells, Suggesting Two Routes for Vertical Transmission. Cell Host & Microbe 2016;20:155-66. https://doi.org/10.1016/j.chom.2016.07.002.

[5] Pereira L. Congenital Viral Infection: Traversing the Uterine-Placental Interface. Annu Rev Virol 2018;5:273-99. https://doi.org/10.1146/annurev-virology-092917-043236.

[6] Gregory CJ, Oduyebo T, Brault AC, Brooks JT, Chung K-W, Hills S, et al. Modes of Transmission of Zika Virus. J Infect Dis 2017;216:S875-83. https://doi.org/10.1093/infdis/jix396.

[7] Brasil P, Pereira JP, Moreira ME, Ribeiro Nogueira RM, Damasceno L, Wakimoto M, et al. Zika Virus Infection in Pregnant Women in Rio de Janeiro. New England Journal of Medicine 2016;375:2321-34. https://doi.org/10.1056/NEJMoa1602412.

[8] Jaenisch T, Rosenberger KD, Brito C, Brady O, Brasil P, Marques ET. Risk of microcephaly after Zika virus infection in Brazil, 2015 to 2016. Bull World Health Organ 2017;95:191-8. https://doi.org/10.2471/BLT.16.178608.

[9] Campos GS, Bandeira AC, Sardi SI. Zika Virus Outbreak, Bahia, Brazil. Emerg Infect Dis 2015;21:1885-6. https://doi.org/10.3201/eid2110.150847.

[10] Freitas P de SS, Soares GB, Mocelin HJS, Lacerda LCX, Prado TN do, Sales CMM, et al. Síndrome congênita do vírus Zika: perfil sociodemográfico das mães. Revista Panamericana de Salud Pública 2018;43:1. https://doi.org/10.26633/RPSP.2019.24.

[11] WHO. WHO statement on the first meeting of the International Health Regulations (2005) (IHR 2005) Emergency Committee on Zika virus and observed increase in neurological disorders and neonatal malformations 2016.

[12] Watrin L, Ghawché F, Larre P, Neau J-P, Mathis S, Fournier E. Guillain–Barré Syndrome (42 Cases) Occurring During a Zika Virus Outbreak in French Polynesia: Medicine 2016;95:e3257. https://doi.org/10.1097/MD.0000000000003257.

[13] Martines RB, Bhatnagar J, de Oliveira Ramos AM, Davi HPF, Iglezias SD, Kanamura CT, et al. Pathology of congenital Zika syndrome in Brazil: a case series. The Lancet 2016;388:898-904. https://doi.org/10.1016/S0140-6736(16)30883-2.

[14] Sarno M, Aquino M, Pimentel K, Cabral R, Costa G, Bastos F, et al. Progressive lesions of central nervous system in microcephalic fetuses with suspected congenital Zika virus syndrome. Ultrasound Obstet Gynecol 2017;50:717-22. https://doi.org/10.1002/uog.17303.

[15] de Araújo TVB, Ximenes RA de A, Miranda-Filho D de B, Souza WV, Montarroyos UR, de Melo APL, et al. Association between microcephaly, Zika virus infection, and other risk factors in Brazil: final report of a case-control study. Lancet Infect Dis 2018;18:328-36. https://doi.org/10.1016/S1473-3099(17)30727-2.

[16] Roma JHF, Alves RC, Silva VS da, Ferreira MJ, Araújo C de, Pavoni JHC. Descriptive study of suspected congenital Zika syndrome cases during the 2015-2016 epidemic in Brazil. Rev Soc Bras Med Trop 2019;52:e20190105. https://doi.org/10.1590/0037-8682-0105-2019.

[17] Mlakar J, Korva M, Tul N, Popović M, Poljšak-Prijatelj M, Mraz J, et al. Zika Virus Associated with Microcephaly. N Engl J Med 2016;374:951-8. https://doi.org/10.1056/NEJMoa1600651.

[18] Pool K-L, Adachi K, Karnezis S, Salamon N, Romero T, Nielsen-Saines K, et al. Association Between Neonatal Neuroimaging and Clinical Outcomes in Zika-Exposed Infants From Rio de Janeiro, Brazil. JAMA Netw Open 2019;2:e198124. https://doi.org/10.1001/jamanetworkopen.2019.8124.

[19] Kollmann TR, Kampmann B, Mazmanian SK, Marchant A, Levy O. Protecting the Newborn and Young Infant from Infectious Diseases: Lessons from Immune Ontogeny. Immunity 2017;46:350-63. https://doi.org/10.1016/j.immuni.2017.03.009.

[20] El Costa H, Gouilly J, Mansuy J-M, Chen Q, Levy C, Cartron G, et al. ZIKA virus reveals broad tissue and cell tropism during the first trimester of pregnancy. Sci Rep 2016;6:35296. https://doi.org/10.1038/srep35296.

[21] Weisblum Y, Oiknine-Djian E, Vorontsov OM, Haimov-Kochman R, Zakay-Rones Z, Meir K, et al. Zika Virus Infects Early- and Midgestation Human Maternal Decidual Tissues, Inducing Distinct Innate Tissue Responses in the Maternal-Fetal Interface. J Virol 2017;91:e01905-16, /jvi/91/4/e01905-16.atom. https://doi.org/10.1128/JVI.01905-16.

[22] Yockey LJ, Iwasaki A. Interferons and Proinflammatory Cytokines in Pregnancy and Fetal Development. Immunity 2018;49:397-412. https://doi.org/10.1016/j.immuni.2018.07.017.

[23] Quicke KM, Bowen JR, Johnson EL, McDonald CE, Ma H, O'Neal JT, et al. Zika Virus Infects Human Placental Macrophages. Cell Host & Microbe 2016;20:83-90. https://doi.org/10.1016/j.chom.2016.05.015.

[24] Simoni MK, Jurado KA, Abrahams VM, Fikrig E, Guller S. Zika virus infection of Hofbauer cells. Am J Reprod Immunol 2017;77:e12613. https://doi.org/10.1111/aji.12613.

[25] Rabelo K, de Souza LJ, Salomão NG, Machado LN, Pereira PG, Portari EA, et al. Zika Induces Human Placental Damage and Inflammation. Front Immunol 2020;11:2146. https://doi.org/10.3389/fimmu.2020.02146.

[26] Jagger BW, Miner JJ, Cao B, Arora N, Smith AM, Kovacs A, et al. Gestational Stage and IFN-λ Signaling Regulate ZIKV Infection In Utero. Cell Host Microbe 2017;22:366-376.e3. https://doi.org/10.1016/j.chom.2017.08.012.

[27] Watanabe S, Tan NWW, Chan KWK, Vasudevan SG. Assessing the utility of antivirals for preventing maternal-fetal transmission of zika virus in pregnant mice. Antiviral Res 2019;167:104-9. https://doi.org/10.1016/j.antiviral.2019.04.013.

[28] McArthur MA. Zika Virus: Recent Advances towards the Development of Vaccines and Therapeutics. Viruses

2017;9. https://doi.org/10.3390/v9060143.

[29] Bernatchez JA, Tran LT, Li J, Luan Y, Siqueira-Neto JL, Li R. Drugs for the Treatment of Zika Virus Infection. J Med Chem 2020;63:470-89. https://doi.org/10.1021/acs.jmedchem.9b00775.

[30] Fontes-Garfias CR, Baker CK, Shi P-Y. Reverse genetic approaches for the development of Zika vaccines and therapeutics. Current Opinion in Virology 2020;44:7-15. https://doi.org/10.1016/j.coviro.2020.05.002.

[31] Khaiboullina SF, Lopes P, de Carvalho TG, Real ALCV, Souza DG, Costa VV, et al. Host Immune Response to ZIKV in an Immunocompetent Embryonic Mouse Model of Intravaginal Infection. Viruses 2019;11. https://doi.org/10.3390/v11060558.

[32] Heffron AS, Mohr EL, Baker D, Haj AK, Buechler CR, Bailey A, et al. Antibody responses to Zika virus proteins in pregnant and non-pregnant macaques. PLoS Negl Trop Dis 2018;12:e0006903. https://doi.org/10.1371/journal.pntd.0006903.

[33] Ornelas AMM, Pezzuto P, Silveira PP, Melo FO, Ferreira TA, Oliveira-Szejnfeld PS, et al. Immune activation in amniotic fluid from Zika virus-associated microcephaly. Ann Neurol 2017;81:152-6. https://doi.org/10.1002/ana.24839.

[34] Rau C-S, Wu S-C, Chen Y-C, Chien P-C, Hsieh H-Y, Kuo P-J, et al. Stress-Induced Hyperglycemia, but Not Diabetic Hyperglycemia, Is Associated with Higher Mortality in Patients with Isolated Moderate and Severe Traumatic Brain Injury: Analysis of a Propensity Score-Matched Population. Int J Environ Res Public Health 2017;14. https://doi.org/10.3390/ijerph14111340.

[35] Köstlin N, Kugel H, Spring B, Leiber A, Marmé A, Henes M, et al. Granulocytic myeloid derived suppressor cells expand in human pregnancy and modulate T-cell responses. Eur J Immunol 2014;44:2582-91. https://doi.org/10.1002/eji.201344200.

[36] Köstlin N, Ostermeir A-L, Spring B, Schwarz J, Marmé A, Walter CB, et al. HLA-G promotes myeloid-derived suppressor cell accumulation and suppressive activity during human pregnancy through engagement of the receptor ILT4. Eur J Immunol 2017;47:374-84. https://doi.org/10.1002/eji.201646564.

[37] Ghaebi M, Nouri M, Ghasemzadeh A, Farzadi L, Jadidi-Niaragh F, Ahmadi M, et al. Immune regulatory network in successful pregnancy and reproductive failures. Biomed Pharmacother 2017;88:61-73. https://doi.org/10.1016/j.biopha.2017.01.016.

[38] Goh C, Narayanan S, Hahn YS. Myeloid-derived suppressor cells: the dark knight or the joker in viral infections? Immunol Rev 2013;255:210-21. https://doi.org/10.1111/imr.12084.

[39] Wang C, Zhang N, Qi L, Yuan J, Wang K, Wang K, et al. Myeloid-Derived Suppressor Cells Inhibit T Follicular Helper Cell Immune Response in Japanese Encephalitis Virus Infection. J Immunol 2017;199:3094-105. https://doi.org/10.4049/jimmunol.1700671.

[40] Mor G, Aldo P, Alvero AB. The unique immunological and microbial aspects of pregnancy. Nat Rev Immunol 2017;17:469-82. https://doi.org/10.1038/nri.2017.64.

[41] Aluvihare VR, Kallikourdis M, Betz AG. Regulatory T cells mediate maternal tolerance to the fetus. Nat Immunol 2004;5:266-71. https://doi.org/10.1038/ni1037.

[42] Ribeiro MR, Moreli JB, Marques RE, Papa MP, Meuren LM, Rahal P, et al.

Zika-virus-infected human full-term placental explants display pro-inflammatory responses and undergo apoptosis. Arch Virol 2018;163:2687-99. https://doi.org/10.1007/s00705-018-3911-x.

[43] Bayer A, Lennemann NJ, Ouyang Y, Bramley JC, Morosky S, Marques ETDA, et al. Type III Interferons Produced by Human Placental Trophoblasts Confer Protection against Zika Virus Infection. Cell Host & Microbe 2016;19:705-12. https://doi.org/10.1016/j.chom.2016.03.008.

[44] Vilcek J. Novel interferons. Nat Immunol 2003;4:8-9. https://doi.org/10.1038/ni0103-8.

[45] Kotenko SV, Gallagher G, Baurin VV, Lewis-Antes A, Shen M, Shah NK, et al. IFN-λs mediate antiviral protection through a distinct class II cytokine receptor complex. Nat Immunol 2003;4:69-77. https://doi.org/10.1038/ni875.

[46] Sheppard P, Kindsvogel W, Xu W, Henderson K, Schlutsmeyer S, Whitmore TE, et al. IL-28, IL-29 and their class II cytokine receptor IL-28R. Nat Immunol 2003;4:63-8. https://doi.org/10.1038/ni873.

[47] Andreakos E, Zanoni I, Galani IE. Lambda interferons come to light: dual function cytokines mediating antiviral immunity and damage control. Curr Opin Immunol 2019;56:67-75. https://doi.org/10.1016/j.coi.2018.10.007.

[48] Kotenko SV, Langer JA. Full house: 12 receptors for 27 cytokines. Int Immunopharmacol 2004;4:593-608. https://doi.org/10.1016/j.intimp.2004.01.003.

[49] Broggi A, Tan Y, Granucci F, Zanoni I. IFN-λ suppresses intestinal inflammation by non-translational regulation of neutrophil function. Nat Immunol 2017;18:1084-93. https://doi.org/10.1038/ni.3821.

[50] Dolganiuc A, Kodys K, Marshall C, Saha B, Zhang S, Bala S, et al. Type III Interferons, IL-28 and IL-29, Are Increased in Chronic HCV Infection and Induce Myeloid Dendritic Cell-Mediated FoxP3+ Regulatory T Cells. PLoS ONE 2012;7:e44915. https://doi.org/10.1371/journal.pone.0044915.

[51] Mennechet FJD, Uzé G. Interferon-λ–treated dendritic cells specifically induce proliferation of FOXP3-expressing suppressor T cells. Blood 2006;107:4417-23. https://doi.org/10.1182/blood-2005-10-4129.

[52] Yin Z, Dai J, Deng J, Sheikh F, Natalia M, Shih T, et al. Type III IFNs Are Produced by and Stimulate Human Plasmacytoid Dendritic Cells. JI 2012;189:2735-45. https://doi.org/10.4049/jimmunol.1102038.

[53] Kotenko SV, Durbin JE. Contribution of type III interferons to antiviral immunity: location, location, location. Journal of Biological Chemistry 2017;292:7295-303. https://doi.org/10.1074/jbc.R117.777102.

[54] Wells AI, Coyne CB. Type III Interferons in Antiviral Defenses at Barrier Surfaces. Trends in Immunology 2018;39:848-58. https://doi.org/10.1016/j.it.2018.08.008.

[55] Ericson JA, Duffau P, Yasuda K, Ortiz-Lopez A, Rothamel K, Rifkin IR, et al. Gene Expression during the Generation and Activation of Mouse Neutrophils: Implication of Novel Functional and Regulatory Pathways. PLoS ONE 2014;9:e108553. https://doi.org/10.1371/journal.pone.0108553.

[56] Rivera A. Interferon Lambda's New Role as Regulator of Neutrophil Function. J Interferon Cytokine Res 2019. https://doi.org/10.1089/jir.2019.0036.

[57] Blazek K, Eames HL, Weiss M, Byrne AJ, Perocheau D, Pease JE, et al.

IFN-λ resolves inflammation via suppression of neutrophil infiltration and IL-1β production. J Exp Med 2015;212:845-53. https://doi.org/10.1084/jem.20140995.

[58] Giaglis S, Stoikou M, Grimolizzi F, Subramanian BY, van Breda SV, Hoesli I, et al. Neutrophil migration into the placenta: Good, bad or deadly? Cell Adhesion & Migration 2016;10:208-25. https://doi.org/10.1080/19336918.2016.1148866.

[59] Hahn S, Hasler P, Vokalova L, van Breda SV, Lapaire O, Than NG, et al. The role of neutrophil activation in determining the outcome of pregnancy and modulation by hormones and/or cytokines. Clin Exp Immunol 2019;198:24-36. https://doi.org/10.1111/cei.13278.

[60] Stoikou M, Grimolizzi F, Giaglis S, Schäfer G, van Breda SV, Hoesli IM, et al. Gestational Diabetes Mellitus Is Associated with Altered Neutrophil Activity. Front Immunol 2017;8:702. https://doi.org/10.3389/fimmu.2017.00702.

[61] Chrysanthopoulou A, Kambas K, Stakos D, Mitroulis I, Mitsios A, Vidali V, et al. Interferon lambda1/IL-29 and inorganic polyphosphate are novel regulators of neutrophil-driven thromboinflammation. J Pathol 2017;243:111-22. https://doi.org/10.1002/path.4935.

[62] Hahn S, Giaglis S, Hoesli I, Hasler P. Neutrophil NETs in reproduction: from infertility to preeclampsia and the possibility of fetal loss. Front Immun 2012;3. https://doi.org/10.3389/fimmu.2012.00362.

[63] Ye L, Schnepf D, Staeheli P. Interferon-λ orchestrates innate and adaptive mucosal immune responses. Nature Reviews Immunology 2019:1. https://doi.org/10.1038/s41577-019-0182-z.

[64] Finotti G, Tamassia N, Cassatella MA. Interferon-λs and Plasmacytoid Dendritic Cells: A Close Relationship. Front Immunol 2017;8:1015. https://doi.org/10.3389/fimmu.2017.01015.

[65] Hemann EA, Green R, Turnbull JB, Langlois RA, Savan R, Gale M. Interferon-λ modulates dendritic cells to facilitate T cell immunity during infection with influenza A virus. Nat Immunol 2019;20:1035-45. https://doi.org/10.1038/s41590-019-0408-z.

[66] del Rio M-L, Bernhardt G, Rodriguez-Barbosa J-I, Förster R. Development and functional specialization of CD103 + dendritic cells. Immunological Reviews 2010;234:268-81. https://doi.org/10.1111/j.0105-2896.2009.00874.x.

[67] Azamor T, Cunha DP, da Silva AMV, de Lima Bezerra OC, Ribeiro-Alves M, Calvo TL, et al. Congenital Zika Syndrome is associated with maternal genetic background. Genetics; 2019. https://doi.org/10.1101/715862.

[68] Singh PK, Guest J-M, Kanwar M, Boss J, Gao N, Juzych MS, et al. Zika virus infects cells lining the blood-retinal barrier and causes chorioretinal atrophy in mouse eyes. JCI Insight 2017;2:e92340. https://doi.org/10.1172/jci.insight.92340.

[69] Salvany-Celades M, van der Zwan A, Benner M, Setrajcic-Dragos V, Bougleux Gomes HA, Iyer V, et al. Three Types of Functional Regulatory T Cells Control T Cell Responses at the Human Maternal-Fetal Interface. Cell Rep 2019;27:2537-2547.e5. https://doi.org/10.1016/j.celrep.2019.04.109.

[70] Robertson SA, Green ES, Care AS, Moldenhauer LM, Prins JR, Hull ML,

et al. Therapeutic Potential of Regulatory T Cells in Preeclampsia-Opportunities and Challenges. Front Immunol 2019;10:478. https://doi.org/10.3389/fimmu.2019.00478.

[71] Tsuda S, Nakashima A, Shima T, Saito S. New Paradigm in the Role of Regulatory T Cells During Pregnancy. Front Immunol 2019;10:573. https://doi.org/10.3389/fimmu.2019.00573.

[72] Zhao A-M, Xu H-J, Kang X-M, Zhao A-M, Lu L-M. New insights into myeloid-derived suppressor cells and their roles in feto-maternal immune cross-talk. J Reprod Immunol 2016;113:35-41. https://doi.org/10.1016/j.jri.2015.11.001.

[73] Köstlin-Gille N, Dietz S, Schwarz J, Spring B, Pauluschke-Fröhlich J, Poets CF, et al. HIF-1α-Deficiency in Myeloid Cells Leads to a Disturbed Accumulation of Myeloid Derived Suppressor Cells (MDSC) During Pregnancy and to an Increased Abortion Rate in Mice. Front Immunol 2019;10:161. https://doi.org/10.3389/fimmu.2019.00161.

[74] Zhang Y, Qu D, Sun J, Zhao L, Wang Q, Shao Q, et al. Human trophoblast cells induced MDSCs from peripheral blood CD14(+) myelomonocytic cells via elevated levels of CCL2. Cell Mol Immunol 2016;13:615-27. https://doi.org/10.1038/cmi.2015.41.

[75] Marigo I, Dolcetti L, Serafini P, Zanovello P, Bronte V. Tumor-induced tolerance and immune suppression by myeloid derived suppressor cells. Immunol Rev 2008;222:162-79. https://doi.org/10.1111/j.1600-065X.2008.00602.x.

[76] Makarenkova VP, Bansal V, Matta BM, Perez LA, Ochoa JB. CD11b+/Gr-1+ myeloid suppressor cells cause T cell dysfunction after traumatic stress. J Immunol 2006;176:2085-94. https://doi.org/10.4049/jimmunol.176.4.2085.

[77] Nair RR, Sinha P, Khanna A, Singh K. Reduced Myeloid-derived Suppressor Cells in the Blood and Endometrium is Associated with Early Miscarriage. Am J Reprod Immunol 2015;73:479-86. https://doi.org/10.1111/aji.12351.

[78] Spallanzani RG, Dalotto-Moreno T, Raffo Iraolagoitia XL, Ziblat A, Domaica CI, Avila DE, et al. Expansion of CD11b(+)Ly6G (+)Ly6C (int) cells driven by medroxyprogesterone acetate in mice bearing breast tumors restrains NK cell effector functions. Cancer Immunol Immunother 2013;62:1781-95. https://doi.org/10.1007/s00262-013-1483-x.

[79] Ismail AQT. Does placental MDSC-mediated modulation of arginine levels help protect the foetus from auxotrophic pathogens? J Matern Fetal Neonatal Med 2018;31:1667-9. https://doi.org/10.1080/14767058.2017.1319935.

[80] Sengupta J, Dhawan L, Lalitkumar PGL, Ghosh D. Nitric oxide in blastocyst implantation in the rhesus monkey. Reproduction 2005;130:321-32. https://doi.org/10.1530/rep.1.00535.

[81] Kang X, Zhang X, Liu Z, Xu H, Wang T, He L, et al. CXCR2-Mediated Granulocytic Myeloid-Derived Suppressor Cells' Functional Characterization and Their Role in Maternal Fetal Interface. DNA Cell Biol 2016;35:358-65. https://doi.org/10.1089/dna.2015.2962.

[82] Sousa AQ, Cavalcante DIM, Franco LM, Araújo FMC, Sousa ET, Valença-Junior JT, et al. Postmortem Findings for 7 Neonates with Congenital Zika Virus Infection. Emerging Infect Dis 2017;23:1164-7. https://doi.org/10.3201/eid2307.162019.

[83] Brady OJ, Osgood-Zimmerman A, Kassebaum NJ, Ray SE, de Araújo VEM, da Nóbrega AA, et al. The association between Zika virus infection and microcephaly in Brazil 2015-2017: An observational analysis of over 4 million births. PLoS Med 2019;16:e1002755. https://doi.org/10.1371/journal.pmed.1002755.

[84] Tonnerre P, Melgaço JG, Torres-Cornejo A, Pinto MA, Yue C, Blümel J, et al. Evolution of the innate and adaptive immune response in women with acute Zika virus infection. Nat Microbiol 2020;5:76-83. https://doi.org/10.1038/s41564-019-0618-z.

[85] Gardinali NR, Marchevsky RS, Oliveira JM, Pelajo-Machado M, Kugelmeier T, Castro MP, et al. Sofosbuvir shows a protective effect against vertical transmission of Zika virus and the associated congenital syndrome in rhesus monkeys. Antiviral Research 2020;182:104859. https://doi.org/10.1016/j.antiviral.2020.104859.

[86] Noorbakhsh F, Abdolmohammadi K, Fatahi Y, Dalili H, Rasoolinejad M, Rezaei F, et al. Zika Virus Infection, Basic and Clinical Aspects: A Review Article. Iran J Public Health 2019;48:20-31.

[87] Mehta R, Soares CN, Medialdea-Carrera R, Ellul M, da Silva MTT, Rosala-Hallas A, et al. The spectrum of neurological disease associated with Zika and chikungunya viruses in adults in Rio de Janeiro, Brazil: A case series. PLoS Negl Trop Dis 2018;12:e0006212. https://doi.org/10.1371/journal.pntd.0006212.

[88] Paz-Bailey G, Rosenberg ES, Doyle K, Munoz-Jordan J, Santiago GA, Klein L, et al. Persistence of Zika Virus in Body Fluids - Final Report. N Engl J Med 2018;379:1234-43. https://doi.org/10.1056/NEJMoa1613108.

[89] Huits R, De Smet B, Ariën KK, Van Esbroeck M, Bottieau E, Cnops L. Zika virus in semen: a prospective cohort study of symptomatic travellers returning to Belgium. Bull World Health Organ 2017;95:802-9. https://doi.org/10.2471/BLT.17.181370.

[90] Burrack KS, Tan JJL, McCarthy MK, Her Z, Berger JN, Ng LFP, et al. Myeloid Cell Arg1 Inhibits Control of Arthritogenic Alphavirus Infection by Suppressing Antiviral T Cells. PLoS Pathog 2015;11:e1005191. https://doi.org/10.1371/journal.ppat.1005191.

[91] Pallett LJ, Gill US, Quaglia A, Sinclair LV, Jover-Cobos M, Schurich A, et al. Metabolic regulation of hepatitis B immunopathology by myeloid-derived suppressor cells. Nat Med 2015;21:591-600. https://doi.org/10.1038/nm.3856.

[92] Agrati C, Sacchi A, Bordoni V, Cimini E, Notari S, Grassi G, et al. Expansion of myeloid-derived suppressor cells in patients with severe coronavirus disease (COVID-19). Cell Death Differ 2020. https://doi.org/10.1038/s41418-020-0572-6.

[93] Reis HLB dos, Boldrini NAT, Caldas JVJ, Paz APC da, Ferrugini CLP, Miranda AE. Severe coronavirus infection in pregnancy: challenging cases report. Rev Inst Med Trop S Paulo 2020;62:e49. https://doi.org/10.1590/s1678-9946202062049.

[94] Takemoto MLS, Menezes M de O, Andreucci CB, Nakamura-Pereira M, Amorim MMR, Katz L, et al. The tragedy of COVID-19 in Brazil: 124 maternal deaths and counting. Int J Gynecol Obstet 2020:ijgo.13300. https://doi.org/10.1002/ijgo.13300.

[95] Rodríguez-Perea AL, Arcia ED, Rueda CM, Velilla PA. Phenotypical characterization of regulatory T cells in humans and rodents. Clin Exp Immunol 2016;185:281-91. https://doi.org/10.1111/cei.12804.

[96] Czabotar PE, Lessene G, Strasser A, Adams JM. Control of apoptosis by the BCL-2 protein family: implications for physiology and therapy. Nat Rev Mol Cell Biol 2014;15:49-63. https://doi.org/10.1038/nrm3722.

[97] Vaux DL, Haecker G, Strasser A. An evolutionary perspective on apoptosis. Cell 1994;76:777-9. https://doi.org/10.1016/0092-8674(94)90350-6.

[98] Cohen GM. Caspases: the executioners of apoptosis. Biochemical Journal 1997;326:1-16. https://doi.org/10.1042/bj3260001.

[99] Rai NK, Tripathi K, Sharma D, Shukla VK. Apoptosis: A Basic Physiologic Process in Wound Healing. The International Journal of Lower Extremity Wounds 2005;4:138-44. https://doi.org/10.1177/1534734605280018.

[100] Klionsky DJ. Autophagy as a Regulated Pathway of Cellular Degradation. Science 2000;290:1717-21. https://doi.org/10.1126/science.290.5497.1717.

[101] Feig C, Peter ME. How apoptosis got the immune system in shape. Eur J Immunol 2007;37:S61-70. https://doi.org/10.1002/eji.200737462.

[102] Chinnaiyan AM. The Apoptosome: Heart and Soul of the Cell Death Machine. Neoplasia 1999;1:5-15. https://doi.org/10.1038/sj.neo.7900003.

[103] Hill MM, Adrain C, Duriez PJ, Creagh EM, Martin SJ. Analysis of the composition, assembly kinetics and activity of native Apaf-1 apoptosomes. EMBO J 2004;23:2134-45. https://doi.org/10.1038/sj.emboj.7600210.

[104] van Loo G, Saelens X, van Gurp M, MacFarlane M, Martin SJ, Vandenabeele P. The role of mitochondrial factors in apoptosis: a Russian roulette with more than one bullet. Cell Death Differ 2002;9:1031-42. https://doi.org/10.1038/sj.cdd.4401088.

[105] Li LY, Luo X, Wang X. Endonuclease G is an apoptotic DNase when released from mitochondria. Nature 2001;412:95-9. https://doi.org/10.1038/35083620.

[106] English L, Chemali M, Duron J, Rondeau C, Laplante A, Gingras D, et al. Autophagy enhances the presentation of endogenous viral antigens on MHC class I molecules during HSV-1 infection. Nat Immunol 2009;10:480-7. https://doi.org/10.1038/ni.1720.

[107] Nedjic J, Aichinger M, Emmerich J, Mizushima N, Klein L. Autophagy in thymic epithelium shapes the T-cell repertoire and is essential for tolerance. Nature 2008;455:396-400. https://doi.org/10.1038/nature07208.

[108] Shibutani ST, Saitoh T, Nowag H, Münz C, Yoshimori T. Autophagy and autophagy-related proteins in the immune system. Nat Immunol 2015;16:1014-24. https://doi.org/10.1038/ni.3273.

[109] Baehrecke EH. Autophagy: dual roles in life and death? Nat Rev Mol Cell Biol 2005;6:505-10. https://doi.org/10.1038/nrm1666.

[110] Yorimitsu T, Klionsky DJ. Autophagy: molecular machinery for self-eating. Cell Death Differ 2005;12:1542-52. https://doi.org/10.1038/sj.cdd.4401765.

[111] He C, Klionsky DJ. Regulation Mechanisms and Signaling Pathways of Autophagy. Annu Rev Genet 2009;43:67-93. https://doi.org/10.1146/annurev-genet-102808-114910.

[112] Cao B, Parnell LA, Diamond MS, Mysorekar IU. Inhibition of autophagy limits vertical transmission of Zika virus in pregnant mice. Journal of

Experimental Medicine 2017;214:2303-13. https://doi.org/10.1084/jem.20170957.

[113] Torrentes-Carvalho A, Azeredo EL, Reis SR, Miranda AS, Gandini M, Barbosa LS, et al. Dengue-2 infection and the induction of apoptosis in human primary monocytes. Mem Inst Oswaldo Cruz 2009;104:1091-9. https://doi.org/10.1590/S0074-02762009000800005.

[114] Azeredo EL, Neves-Souza PC, Alvarenga AR, Reis SRNI, Torrentes-Carvalho A, Zagne S-MO, et al. Differential regulation of toll-like receptor-2, toll-like receptor-4, CD16 and human leucocyte antigen-DR on peripheral blood monocytes during mild and severe dengue fever: Monocyte activation after dengue infection. Immunology 2010;130:202-16. https://doi.org/10.1111/j.1365-2567.2009.03224.x.

[115] Limonta D, Torrentes-Carvalho A, Marinho CF, de Azeredo EL, de Souza LJ, Motta-Castro ARC, et al. Apoptotic mediators in patients with severe and non-severe dengue from Brazil: Apoptotic Mediators in Dengue. J Med Virol 2014;86:1437-47. https://doi.org/10.1002/jmv.23832.

[116] Liang Q, Luo Z, Zeng J, Chen W, Foo S-S, Lee S-A, et al. Zika Virus NS4A and NS4B Proteins Deregulate Akt-mTOR Signaling in Human Fetal Neural Stem Cells to Inhibit Neurogenesis and Induce Autophagy. Cell Stem Cell 2016;19:663-71. https://doi.org/10.1016/j.stem.2016.07.019.

[117] Lum F-M, Low DKS, Fan Y, Tan JJL, Lee B, Chan JKY, et al. Zika Virus Infects Human Fetal Brain Microglia and Induces Inflammation. Clin Infect Dis 2017;64:914-20. https://doi.org/10.1093/cid/ciw878.

[118] Garcez PP, Loiola EC, Madeiro da Costa R, Higa LM, Trindade P, Delvecchio R, et al. Zika virus impairs growth in human neurospheres and brain organoids. Science 2016;352:816-8. https://doi.org/10.1126/science.aaf6116.

[119] Calvet G, Aguiar RS, Melo ASO, Sampaio SA, de Filippis I, Fabri A, et al. Detection and sequencing of Zika virus from amniotic fluid of fetuses with microcephaly in Brazil: a case study. The Lancet Infectious Diseases 2016;16:653-60. https://doi.org/10.1016/S1473-3099(16)00095-5.

[120] Martines RB, Bhatnagar J, Keating MK, Silva-Flannery L, Muehlenbachs A, Gary J, et al. Notes from the Field : Evidence of Zika Virus Infection in Brain and Placental Tissues from Two Congenitally Infected Newborns and Two Fetal Losses — Brazil, 2015. MMWR Morb Mortal Wkly Rep 2016;65:159-60. https://doi.org/10.15585/mmwr.mm6506e1.

[121] Clinical features and neuroimaging (CT and MRI) findings in presumed Zika virus related congenital infection and microcephaly: retrospective case series study. BMJ 2016:i3182. https://doi.org/10.1136/bmj.i3182.

[122] Oliveira Melo AS, Malinger G, Ximenes R, Szejnfeld PO, Alves Sampaio S, Bispo de Filippis AM. Zika virus intrauterine infection causes fetal brain abnormality and microcephaly: tip of the iceberg?: Physician Alert. Ultrasound Obstet Gynecol 2016;47:6-7. https://doi.org/10.1002/uog.15831.

[123] de Sousa JR, Azevedo RSS, Martins Filho AJ, Araujo MTF, Moutinho ERC, Baldez Vasconcelos BC, et al. Correlation between Apoptosis and in Situ Immune Response in Fatal Cases of Microcephaly Caused by Zika Virus. The American Journal of Pathology 2018;188:2644-52. https://doi.org/10.1016/j.ajpath.2018.07.009.

[124] Thompson MR, Kaminski JJ, Kurt-Jones EA, Fitzgerald KA. Pattern

Recognition Receptors and the Innate Immune Response to Viral Infection. Viruses 2011;3:920-40. https://doi.org/10.3390/v3060920.

[125] Muir AJ, Arora S, Everson G, Flisiak R, George J, Ghalib R, et al. A randomized phase 2b study of peginterferon lambda-1a for the treatment of chronic HCV infection. Journal of Hepatology 2014;61:1238-46. https://doi.org/10.1016/j.jhep.2014.07.022.

[126] O'Brien TR, Young HA, Donnelly RP, Prokunina-Olsson L. Meeting Overview: Interferon Lambda-Disease Impact and Therapeutic Potential. J Interferon Cytokine Res 2019. https://doi.org/10.1089/jir.2019.0018.

[127] Baldridge MT, Nice TJ, McCune BT, Yokoyama CC, Kambal A, Wheadon M, et al. Commensal microbes and interferon- determine persistence of enteric murine norovirus infection. Science 2015;347:266-9. https://doi.org/10.1126/science.1258025.

[128] Galani IE, Triantafyllia V, Eleminiadou E-E, Koltsida O, Stavropoulos A, Manioudaki M, et al. Interferon-λ Mediates Non-redundant Front-Line Antiviral Protection against Influenza Virus Infection without Compromising Host Fitness. Immunity 2017;46:875-890.e6. https://doi.org/10.1016/j.immuni.2017.04.025.

[129] Lazear HM, Daniels BP, Pinto AK, Huang AC, Vick SC, Doyle SE, et al. Interferon-λ restricts West Nile virus neuroinvasion by tightening the blood-brain barrier. Sci Transl Med 2015;7:284ra59-284ra59. https://doi.org/10.1126/scitranslmed.aaa4304.

[130] Caine EA, Scheaffer SM, Arora N, Zaitsev K, Artyomov MN, Coyne CB, et al. Interferon lambda protects the female reproductive tract against Zika virus infection. Nat Commun 2019;10:280. https://doi.org/10.1038/s41467-018-07993-2.

[131] Barrows NJ, Campos RK, Powell ST, Prasanth KR, Schott-Lerner G, Soto-Acosta R, et al. A Screen of FDA-Approved Drugs for Inhibitors of Zika Virus Infection. Cell Host Microbe 2016;20:259-70. https://doi.org/10.1016/j.chom.2016.07.004.

[132] Bullard-Feibelman KM, Govero J, Zhu Z, Salazar V, Veselinovic M, Diamond MS, et al. The FDA-approved drug sofosbuvir inhibits Zika virus infection. Antiviral Res 2017;137:134-40. https://doi.org/10.1016/j.antiviral.2016.11.023.

[133] Sacramento CQ, de Melo GR, de Freitas CS, Rocha N, Hoelz LVB, Miranda M, et al. The clinically approved antiviral drug sofosbuvir inhibits Zika virus replication. Sci Rep 2017;7:40920. https://doi.org/10.1038/srep40920.

[134] Ferreira AC, Zaverucha-do-Valle C, Reis PA, Barbosa-Lima G, Vieira YR, Mattos M, et al. Sofosbuvir protects Zika virus-infected mice from mortality, preventing short- and long-term sequelae. Sci Rep 2017;7:9409. https://doi.org/10.1038/s41598-017-09797-8.

[135] Gan CS, Lim SK, Chee CF, Yusof R, Heh CH. Sofosbuvir as treatment against dengue? Chem Biol Drug Des 2018;91:448-55. https://doi.org/10.1111/cbdd.13091.

[136] Mesci P, Macia A, Moore SM, Shiryaev SA, Pinto A, Huang C-T, et al. Blocking Zika virus vertical transmission. Sci Rep 2018;8:1218. https://doi.org/10.1038/s41598-018-19526-4.

[137] Yin Z, Chen Y-L, Schul W, Wang Q-Y, Gu F, Duraiswamy J, et al. An adenosine nucleoside inhibitor of dengue virus. Proc Natl Acad Sci

USA 2009;106:20435-9. https://doi.org/10.1073/pnas.0907010106.

[138] Li Y, Zeng Y, Zeng G, Li J, Zhang X, Cai Q, et al. The effects of direct-acting antiviral agents on the frequency of myeloid-derived suppressor cells and natural killer cells in patients with chronic hepatitis C. J Med Virol 2019;91:278-86. https://doi.org/10.1002/jmv.25302.

[139] Liu Y, She L-H, Wang X-Y, Zhang G-L, Yan Y, Lin C-S, et al. Expansion of myeloid-derived suppressor cells from peripheral blood decreases after 4-week antiviral treatment in patients with chronic hepatitis C. Int J Clin Exp Med 2014;7:998-1004.

[140] Xie X, Zou J, Shan C, Shi P-Y. Small Molecules and Antibodies for Zika Therapy. J Infect Dis 2017;216:S945-50. https://doi.org/10.1093/infdis/jix406.

[141] Kaplan YC, Ozsarfati J, Nickel C, Koren G. Reproductive outcomes following hydroxychloroquine use for autoimmune diseases: a systematic review and meta-analysis. Br J Clin Pharmacol 2016;81:835-48. https://doi.org/10.1111/bcp.12872.

[142] Rosenfeldt MT, O'Prey J, Morton JP, Nixon C, MacKay G, Mrowinska A, et al. p53 status determines the role of autophagy in pancreatic tumour development. Nature 2013;504:296-300. https://doi.org/10.1038/nature12865.

[143] Xu M, Lee EM, Wen Z, Cheng Y, Huang W-K, Qian X, et al. Identification of small-molecule inhibitors of Zika virus infection and induced neural cell death via a drug repurposing screen. Nat Med 2016;22:1101-7. https://doi.org/10.1038/nm.4184.

[144] Domenger C, Grimm D. Next-generation AAV vectors-do not judge a virus (only) by its cover. Hum Mol Genet 2019;28:R3-14. https://doi.org/10.1093/hmg/ddz148.

[145] Rodriguez SE, Cross RW, Fenton KA, Bente DA, Mire CE, Geisbert TW. Vesicular Stomatitis Virus-Based Vaccine Protects Mice against Crimean-Congo Hemorrhagic Fever. Sci Rep 2019;9:7755. https://doi.org/10.1038/s41598-019-44210-6.

[146] Swedish Council on Health Technology Assessment. Vaccines to Children: Protective Effect and Adverse Events: A Systematic Review. Stockholm: Swedish Council on Health Technology Assessment (SBU); 2009.

[147] Betancourt D, de Queiroz NMGP, Xia T, Ahn J, Barber GN. Cutting Edge: Innate Immune Augmenting Vesicular Stomatitis Virus Expressing Zika Virus Proteins Confers Protective Immunity. JI 2017;198:3023-8. https://doi.org/10.4049/jimmunol.1602180.

[148] Pardi N, Hogan MJ, Porter FW, Weissman D. mRNA vaccines - a new era in vaccinology. Nat Rev Drug Discov 2018;17:261-79. https://doi.org/10.1038/nrd.2017.243.

[149] Brito LA, Chan M, Shaw CA, Hekele A, Carsillo T, Schaefer M, et al. A cationic nanoemulsion for the delivery of next-generation RNA vaccines. Mol Ther 2014;22:2118-29. https://doi.org/10.1038/mt.2014.133.

[150] Brito LA, Kommareddy S, Maione D, Uematsu Y, Giovani C, Berlanda Scorza F, et al. Self-amplifying mRNA vaccines. Adv Genet 2015;89:179-233. https://doi.org/10.1016/bs.adgen.2014.10.005.

[151] Bogers WM, Oostermeijer H, Mooij P, Koopman G, Verschoor EJ, Davis D, et al. Potent immune responses in rhesus macaques induced by nonviral delivery of a self-amplifying RNA vaccine expressing HIV type 1 envelope with a cationic nanoemulsion. J Infect Dis 2015;211:947-55. https://doi.org/10.1093/infdis/jiu522.

[152] U. S. National Library of Medicine. https://clinicaltrials.gov 2020.

[153] Petsch B, Schnee M, Vogel AB, Lange E, Hoffmann B, Voss D, et al. Protective efficacy of in vitro synthesized, specific mRNA vaccines against influenza A virus infection. Nat Biotechnol 2012;30:1210-6. https://doi.org/10.1038/nbt.2436.

[154] Makhluf H, Shresta S. Development of Zika Virus Vaccines. Vaccines (Basel) 2018;6. https://doi.org/10.3390/vaccines6010007.

[155] Richner JM, Jagger BW, Shan C, Fontes CR, Dowd KA, Cao B, et al. Vaccine Mediated Protection Against Zika Virus-Induced Congenital Disease. Cell 2017;170:273-283.e12. https://doi.org/10.1016/j.cell.2017.06.040.

[156] Richner JM, Himansu S, Dowd KA, Butler SL, Salazar V, Fox JM, et al. Modified mRNA Vaccines Protect against Zika Virus Infection. Cell 2017;169:176. https://doi.org/10.1016/j.cell.2017.03.016.

Chapter 10

Role of Toll-Like Receptor (TLR)-Signaling in Cancer Progression and Treatment

Shyam Babu Prasad and Rahul Kumar

Abstract

Toll-like receptors (TLRs) are the most essential pattern recognition receptors in mediating the effects of innate immunity. It plays a pivotal role in inducing immune response against a number of pathogens, various diseases conditions including pathogenesis of cancer. Inflammation is often associated with the development and progression of most of cancer, where TLRs interplay very crucial roles. Moreover, TLRs activation can impact the initiation, progression and treatment of cancer by modulating the inflammatory microenvironment. Rapidly growing number of evidences related to TLRs function and expression in cancer cells, suggests its critical association with chemoresistance and tumourigenesis. The current chapter describes the development of various agonist and antagosist for TLRs and their application in cancer therapeutics. The aim of this book chapter is to highlights basic features of TLRs, and its role in cancer progression. It also addresses, how a defect in the TLRs signaling pathway can contributes towards carcinogenesis and recent development of cancer therapeutics that target TLR signaling pathways.

Keywords: toll-like receptors (TLRs), cancer progression, TLR agonists, inflammation, signaling

1. Introduction to the toll-like receptors (TLRs)

TLRs are trans-membrane proteins receptors that trigger the signal transduction cascades upon binding with specific pathogen-associated molecular patterns (PAMPs) ligands, and earlier have been thought to be restricted to immune cells. TLRs play a key role in the innate immune system as well as subsequent induction of adaptive immune responses against microbial infection or tissue injury [1]. TLRs receptors triggers immune response against various invading pathogens by recognizing receptor specific to PAMPs, which is highly conserved and derived from potential pathogenic microorganism such as bacteria, viruses, fungi and parasites [2, 3]. The very well-known one such PAMPs is lipopolysaccharides (LPS) acts as ligands for TLRs, which is found on outer cell wall of gram negative bacteria [4]. Moreover, TLR receptors also recognize endogenous damage-associated molecular patterns (DAMPs), derived from injured host cells including necrotic cancer cells, dead or dying cells, or products released from cells in response to signals such as hypoxia and epithelial cells [5]. These PAMPs and DAMPs together help in discriminating both self and non-self-danger signals [1, 2]. Specific TLR receptors recognizes distinct microbial ligands i.e. lipopeptides, lipoteichoic acids, LPS, peptidoglycans, flagellins, viral and bacterial nucleic acids etc. [6]. These ligands binds to specific TLR receptors, initiate cascade pathway which plays important role in maintenance of cellular homeostasis, cell proliferation or apoptosis, cell differentiation, as well as induction of inflammatory cytokines like interferons (IFNs), interleukins (IL2, IL6, IL8, IL12, IL16), and TNF-α to get rid of pathogens [3, 7].

Cancer develops when uncontrolled growth of abnormal cells occurs anywhere in a body and further metastasized to distant part of the body. In order to deepen our understanding of cancer biology, it is very important to address the factors that are involved in the tissue repair process, such as cytokines, chemokines, growth factors and TLR signaling, which are the key determinants of cancer progression [8, 9]. TLR signaling is known to activate nuclear factor-κB (NFκB) and mitogen-activated protein kinase (MAPK) pathways [10]. NF-κB in turn, regulates the expression of anti-apoptotic genes, and activation of the complement pathway depending upon type of ligands it sensed [11, 12]. Furthermore, TLRs are expressed not only on the surface of immune cells but also on cancerous cells [13]. In humans, TLRs (TLR1-TLR10) play very important role in diseases progression and the TLRs signaling have been well studied in various diseases including cancer [14]. The TLRs and their intracellular signaling components play very important role in the onset of inflammatory diseases [4]. Recent studies have revealed that chronic inflammation can increases the risk of cancer development and also promote its progression [14]. TLRs signaling also plays a crucial role in the development of chemo-resistance; *Michael et al., (2006)* shows TLR4/MYD88 signaling promotes tumor growth and contributes to chemo-resistance against paclitaxel in ovarian cancer [9]. Moreover, a recent study delineates that high TLR7 and TLR8 expression promotes chemo-resistance, leading to increases increased tumor cell proliferation in human pancreatic cancer [15]. However, the role of TLR signaling is still not completely understood in cancer progression; some studies suggest it has both pro-tumor as well as anti-tumor effects. To date, TLRs are documented to play supportive role for initiation, progression and metastatic potential of cancer [16, 17]. One the other hand, they are capable of maintaining antitumor environment by eliciting activation of anti-tumor mediators such as type I interferon [18]. This book chapter highlights the current understanding of role of TLRs and addresses a crucial link between carcinogenesis and immune cells, TLRs signaling and antagonist.

2. TLRs genetics and regulation of signaling

TLRs were first described in Drosophila in 1984, and were later discovered in vertebrates including humans [1, 3]. Till date, 13 TLRs are discovered in mammals, and 10 are functional in humans [2]. Genes encoding human TLRs are located on chromosomes 1 (TLR5), 3 (TLR9), 4 (TLR1, TLR2, TLR3, TLR6 and TLR10), 9 (TLR4) and X (TLR7 and TLR8) [1, 5]. TLR1–9 is conserved in both human and mice; however, mouse TLR10 is not functional because of a retrovirus insertion, and TLR11–13 has been lost from the human genome [8, 19, 20].

A number of genetic changes like single nucleotide polymorphisms (SNPs) within the TLR genes has been reported in humans which can influence receptor binding and function, that ultimately influences the risk for the inflammatory diseases as well as cancers [21]. Although there have been numerous studies reporting the impact of polymorphisms on TLR function and disease development, there is still a lot of contradiction in terms of outcomes [22].

DOI: http://dx.doi.org/10.5772/intechopen.94423

A recent report has shown that functional TLRs are expressed not only on immune cells, but also on cancer cells, thus implicating a role of TLRs in cancer biology. Overwhelming evidence supports that TLR signaling provides a micro-environment that is necessary for tumor cells to proliferate and evade the immune response for further growth and migration [23]. The TLR family can be largely divided into 2 subgroups, extracellular and intracellular, depending on their cellular localization. TLR1, TLR2, TLR4, TLR5, TLR6, and TLR11 are located on the cell surface, while TLR3, TLR7, TLR8, and TLR9 are localized to the endosomal/ lyso-somal compartment [10]. The subcellular localization of TLR4 is unique because it is localized to both plasma membrane as well as endosomal vesicles [24]. TLRs are type I transmembrane proteins that consist of three major domains: (1) a leucine rich extracellular domain, (2) a transmembrane domain, (3) A cytoplasmic TIR (Toll/Interleukin-1 Receptor) domain. The recognition of ligand by TLRs is mediated by the extracellular domain that harbor a leucine rich repeat (LRR) composed of 19–25 tandem copies of the "xLxxLxLxx" motif [25]. TLR signaling was extensively studied in the recent years. There are two important TLR pathways: one is dependent on myeloid differentiation factor 88 (MYD88) adaptor proteins and the other is independent of MYD88.

All TLRs except TLR3, which exclusively uses the TIR-domain-containing adapter-inducing interferon-β (TRIF) pathway, use MYD88 as the downstream adapter protein that activate the classical/canonical inflammatory signaling pathway [26–29]. After activation with their specific ligands, TLRs recruit MYD88, leading to subsequent activation of three main transcription factors: interferon-regulatory factors (IRF3, IRF5 and IRF7), NF-kB, MAPK and AP1 [21–25, 27–32]. Subsequently, it promote the transcription of cytokines such as TNF-α, IL-6 and IL-1, chemokines and interferons which are key mediators of inflammation [30].

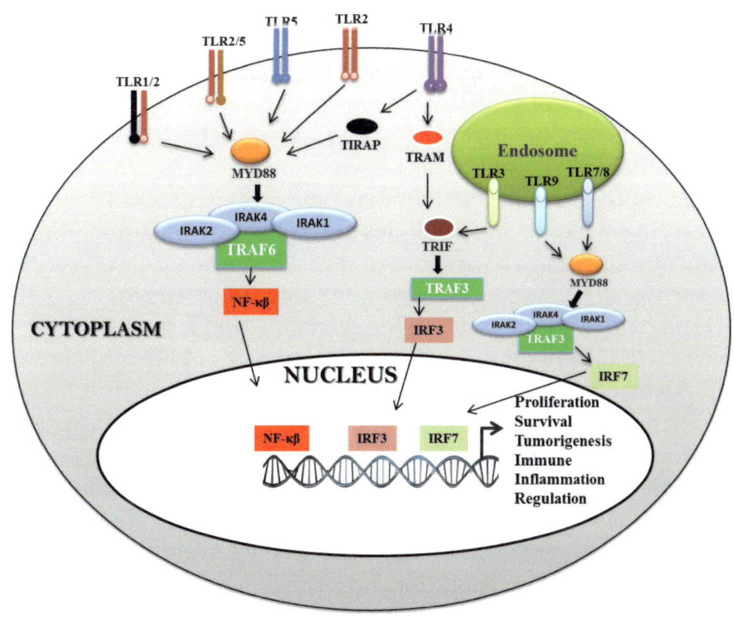

Figure 1.
Toll-like receptors (TLRs) signaling pathway:Toll-like receptors (TLRs) recognize different ligands and triggered innate immune responses. The activation of the TLR signaling pathway originates from the cytoplasmic TIR domain that associates with an adaptor, MYD88. IRAK is activated by phosphorylation and associates with TRAF6, leading to activation NF-κB. Activation of MYD88- independent pathways occurs via TRIF and TRAF activates interferon--regulatory factor (IRF). Then they promote the transcription of inflammation mediators: Cytokines, chemokines and interferons.

Expression of cytokines also leads to maturation of dendritic cells and activation of B-cells and T-cells, which underlies the involvement of TLR in adaptive immunity [23]. TLR2 and TLR4 upon binding with their respective ligands form dimeric complexes, followed by recruitment of 5 specific adapters, including 1) MYD88, 2) TIR domain containing adaptor protein (TIRAP)/MYD88 adaptor like (Mal), 3) TRIF, 4) TRIF-related adaptor molecule (TRAM), and 5) sterile α and armadillo motif-containing protein (SARM) [19, 33]. This response elicits the downstream responses like proliferation, invasion, inflammation and tumorigenesis etc. The schematic representation of the role of various TLRs signaling pathways is shown in **Figure 1**. This alternative/non-canonical pathway culminates in the activation of TRAF3 and interferon regulatory factor 3 (IRF3), which results in the secretion of type I IFNs, which are required for an effective antiviral response [31].

3. TLRs biology in the pathogenesis of cancer

In the host cell, TLRs are expressed either on cell membrane or in intracellular compartments (i.e. endosomes) [10]. TLRs belong to a family of pattern recognition receptors (PRRs) that are best-known for their role in host defense mechanism against a number of pathogens. Infection with potential microbial pathogens (bacteria, viruses, protozoa, and fungi) provokes innate and adaptive immune system [26]. In vertebrates, interactions between innate and adaptive immunity leads to highly efficient recognition and clearance of pathogens. Innate immune response elicits nonspecific activation of immune cells (neutrophils, monocytes, macrophages, dendritic cells (DCs), natural killer (NK) cells) and complements system [33, 34]. Inflammation is the immune system's response to protect our body against any harmful stimuli like pathogens, cell damage and harmful/toxic compound. However, uncontrolled acute inflammation may become chronic; contributing to a variety of diseases including cancer [19]. In 1858, Rudolf Virchow noticed that the site of chronic inflammation is highly susceptible to cancer development [35]. He also hypothesized that chronic inflammation could promote the proliferation of cells and thus, the development of cancer. An association between the inflammation and development of cancer has long been appreciated [33]. In 2000, Hanahan and Weinberg proposed a model to define six hallmarks of cancer progression [36]. However, emerging evidence also reiterates the role of inflammation in cancer development. Various studies have shown a close link between chronic inflammation and cancer, such as long standing *H.pylori* infection and gastric cancer [37], chronic pancreatitis and pancreatic cancer, chronic bronchitis and lung cancer, human papillomavirus (HPVs) infection mediated cervical cancer [38], and chronic cholecystitis with gall bladder cancer [33]. Besides inflammatory response, TLR signaling has been shown to regulate apoptosis through the expression of anti-apoptotic proteins or inhibitors of apoptosis [39]. TLRs regulate variety of cellular responses which include the anti-apoptotic effect of NF-κB, a transcription factor commonly engaged in inflammatory conditions [12, 14]. Although this response can be initiated by several types of PRRs, and TLRs are the best-characterized key players. TLRs also regulate cell proliferation, apoptosis, invasion, and survival by recruiting more immune cells to enhance inflammation in the tumor microenvironment [40]. These tumor cells further release proangiogenic factors and growth factors, which enhance their resistance to cytotoxic lymphocyte attack, thereby leading to immune evasion. As mentioned earlier, TLRs function as double-edged swords, with both pro- and antitumor consequences. However, up-regulation of TLRs in tumor cells may directly or indirectly contribute to carcinogenesis in different organs. Engagement of TLRs on the surface of tumor cells with their ligands can activate subsequent signaling cascades involving cytokine and chemokine production. Subsequently, these factors can promote tumor invasion, tumor cell survival (apoptosis resistance), chemo-resistance, tumor progression and metastasis.

4. Molecular mechanism of TLRs in cancer progression

Overall, as discussed earlier, the activation of TLRs can promote as well as inhibit tumor growth and cancer progression, but the actual underlying molecular mechanism still remains elusive. TLRs are also involved in controlling many important cellular processes like cell proliferation, survival, apoptosis, cell migration, metastasis and angiogenesis [16]. TLR signaling has been implicated in various autoimmune, chronic inflammation and inflammatory diseases. This situation creates a microenvironment that is rich in growth and survival factors, which leads to the development of various types of cancer [41]. High TLR expression has been reported in several cancer types including cancerous cell lines. It was known that TLR4 and TLR5 are over expressed in gastric epithelium infected with *H. pylori* as well as in precursor lesions [37, 42]. It is considered that TLRs enable cells to interact with *H. pylori* which can induce the expression of tumorigenic factors and may promote cancer development. TLR over–expression has also been found in colon cancer, hepatocellular carcinoma, ovarian and cervical cancers, breast and prostate cancers, lung cancer, melanoma and neuroblastoma [43]. TLR expression in cancer cells has been linked with cancer progression, evasion of immune surveillance, apoptosis and survival. Recent studies have shown high expression of TLR4 in lung cancer cells, which is linked with expression of immunosuppressive cytokines (TGFb), angiogenic factors VEGF and IL-8, and increased resistance to apoptosis [16, 42]. Cell proliferation and production of pro-inflammatory cytokines IL-6 and IL-8 can be significantly decreased by silencing of TLR4 expression in breast cancer cell line (MDA-MB-231) [44]. Other studies in ovarian cancer and cell lines has shown that TLR4 and NF-kB activation by LPS and paclitaxel respectively promotes production of IL-8, IL-6, VEGF and MCP-1 while TLR4 silencing lead to loss of resistance to Paclitaxel [45]. TLR2 mRNA expression was significantly higher in sporadic colorectal cancer cells than in noncancerous cells [45]. On the basis of above mentioned facts, we can deduce that various TLRs might trigger different signaling pathways in cancer initiation and progression [46].

A recent report found that activation of TLRs may induce cancerous cells to secrete a number of soluble factors, which play distinct roles in cancer development. The role of TLRs in cancer progression needs to be further investigated, and in depth precise underlying mechanism must be elucidated for further development of TLR agonists as therapeutic agents.

5. TLRs modulation in cancer treatments

TLR agonists play an important role in activation of immune system, both innate and adaptive. In *in vivo* models, TLR antagonism have been shown to reduce tumor growth in treatment group, receiving combination of therapeutic agents, such chemotherapy drugs, monoclonal antibodies (mAb), subunit or DNA vaccines [47, 48]. The selection of TLR agonists has been premised on their ability to activate antigen-presenting cells (APCs), particularly DCs. The involvement of specific TLRs on cancer cells may impact tumor growth by various mechanisms, such as inducing apoptosis and potentiating the effects of chemotherapy [45]. Inhibition of TLRs can be achieved either by 1) Preventing the interaction between TLR and respective binding partner and 2) By blocking the interaction between TLRs and respective adaptor protein. TLR inhibitors can be broadly classified into (i) Small molecule inhibitors, (ii) Antibodies, (iii) Oligonucleotides, and (iv) Lipid-A analogs. The following section illustrates the anticancer effects of inhibiting TLR signaling pathways on tumor growth and developments.

Small molecule inhibitors (SMI): These are synthetic or naturally derived small molecules with the ability to cross plasma membranes due to their small size and amphipathic nature. Interestingly, one of the most commonly used anti-malarial drug chloroquine has been shown to possess inhibitory effects against endosomal TLR7/8/9 [49]. Inhibition of TLR7 and 9 by chloroquine inhibits the growth of hepatocellular carcinoma in both cellular *in vitro* models and mouse xenograft models via down regulation of p-Akt [50]. However due to their non specific mode of actions continuous efforts have been made to develop more efficient and specific derivatives of chloroquine as anti-cancer agents by targeting TLRs. One such derivative, CpG-52,363 has immunosuppressor functions but its role in cancer therapeutics is yet to be discovered [51]. SM934, a derivative of another anti-malarial drug artemensin can inhibit the proliferation and metastasis in breast cancer probably via inhibition of TLR signaling [52]. TAK-242 specifically inhibits TLR-4 by binding to cysteine 747 in the intracellular domain and consequently suppresses the progression of breast cancer [53]. Therapeutic role of SM934, another artemensin derivative has been well documented in inflammatory disease [54], but its role in cancer prevention is still need to be explored.

Antibodies: Various antibodies with therapeutic potential have been raised against TLRs to treat a wide spectrum of inflammatory diseases and cancer. Therapeutic role of OPN-305, the first fully humanized IgG against TLR2, against Myelodysplastic Syndromes (MDS) has been reported in different clinical trials [55]. Several antibodies have been developed beside OPN-305, like NI-0101and T2.5, but their role in cancer has not been determined yet [56].

Oligonucleotides: Specific nucleotide sequences are known to inhibit the function of endosomal TLRs by blocking their binding with respective ligands. These includes immunoregulatory DNA sequence (IRS) 661 (TLR7 specific), IRS-869 (TLR9 specific) and IRS-954 (both TLR7 and 9 specific). Recent report suggest that TLR antagonism using immune modulator oligonucleotide-3100 (IMO-3100) can serve as a potential therapeutic for the management of pancreatic cancer associated cachexia [57].

Lipid A analogs: Eritoran, a syhnthetic analogue of lipid A from *Rhodobacter sphaeroides,* is known to inhibit TLR4 by binding to MD2 pocket and thereby preventing the interaction between TLR4 and lipid A. Bacterial LPS induced colon cancer can be prevented by the administration of Eritoran by mechanism involving inhibition TLR4 and induction of CD14/Src/PKCζ-mediated apoptosis [58].

It is important to mention here that TLR acts as double edged sword and its agonism can also prevents the progression of cancer by activating the immune response against cancer cells. The following section describes TLR agonists which had shown the potential to prevent cancer progression.

Calmette–Guerin strain (BCG) a live-attenuated *Mycobacterium bovis* can activate TLR2, TLR4, and TLR9. The activation of TLR in urothelial cell carcinomas with BCG induced cell death and decreased proliferation as well as metaststais. The anti-cancer effects of BCG have been associated with increased production of cytotoxic NO in cell lines, as well as in patients [46]. These studies also emphasize the development of vaccination strategies that incorporate TLR ligands to stimulate immune responses and make cancerous cells specific targets for immune system mediated death. In human colon cancer cells, TLR3 activation with Polyriboinosinic-polyribocytidylic acid (Poly I:C) can induce apoptosis alone or when used in synergy with 5-fluorouracil or IFN-α [16]. Poly I:C is a synthetic analogue of viral dsRNA. The expression of TLR5 on cancer cells has been shown to revoke cell growth in certain types of cancer [16]. For instance, in breast cancer, when TLR5 is over–expression with flagellin inhibits tumor cell proliferation and downregulates expression of cyclin B1, cyclin D1 and cyclin E2 in a murine model [59].

Irradiation along with activation of TLR9 signaling pathway in human glioma cell line can decrease cell proliferation by arresting cell-cycle, which is mediated by NF-κB and nitric oxide (NO) [60]. This therapeutic effect could be used to sensitize the cancerous cells to the toxic effects of radiation treatment [61]. Also, CpG-island mediated activation of TLR9 in neuroblastoma cell has been revealed to decrease cell proliferation and increase caspase-dependent apoptosis and leads to an increased survival in tumor-bearing mice. Several TLR agonists have been approved by the food and drug administration (FDA) for use in the treatment of cancer patients like BCG (which activate TLR2, TLR3, TLR4, and TLR9), MPL (TLR4 agonist) and imiquimod (TLR7 agonist) [62]. TLR agonists should be used in combination with other agents to synergistically increase their immune stimulatory response. An important TLR modulators are summarized in **Table 1** which having anticancer activity.

Future direction of TLRs based treatment of cancer:

In this book chapter, we summarized the role of TLRs signaling in inflammation, cell proliferation, apoptosis and chemo-resistance, which are the major attributes of cancerous cells. Beside these, several TLRs agonists and antagonists have been developed and/or are in clinical trials as cancer therapeutics. TLRs play a critical role in imparting immunity against tumor, and their antitumor effects are noticeable as depicted from previous studies. It is quite interesting to note that activation of same TLR in one tumor type might induce cell death, and in a different tumor could exert pro-tumor effects. Using TLR agonists or antagonist as cancer therapeutics must be decided on the basis of TLR expression profile of tumor cells and resulting response within a specific cancer type [19]. The prospective approach for future cancer treatment will be the combination of specific TLR agonists or antagonists with traditional cancer treatments to improve treatment outcome. The role of TLRs in both promoting and inhibiting tumor growth and metastasis has been confirmed in various studies. However, the specific mechanism of action is still unclear as cancer is a multifactorial\disease, and the research of TLRs on tumor immunity is still in the nascent phase. Further in depth studies will help us to develop better

Name	Targets	Antagonist/ agonist	Ref.
Chloroquine	TLR 7 and 9	Antagonist	[50]
TAK-242	TLR 4	Antagonist	[53]
IMO-3100	TLR 7, 8 and 9	Antagonist	[57]
Eritoran	TLR 4	Antagonist	[58]
Polyriboinosinic-polyribocytidylic acid (Poly I:C)	TLR 3	Agonist	[16]
Calmette–Guerin strain (BCG)	TLR 2, 4 and 9	Agonist	[46]

Table 1.
Different TLR modulators having anticancer activity.

understanding of TLRs role in tumorigenesis, tumor immunity, and tumor metastasis which in turn can provide new strategies and prospects for more effective cancer management. We anticipate that future studies on the role of TLRs in cancer progression and development will provide us a better insight into the mechanisms underplaying. Therefore, understanding the roles of TLRs in tumor biology may pave the way for the discovery of novel therapeutic targets in cancer therapy.

Author details

Shyam Babu Prasad[1*] and Rahul Kumar[2]

1 Department of Zoology, School of Life sciences, Mahatma Gandhi Central University Bihar (MGCUB), Motihari, India

2 Department of Biotechnology, GITAM Institute of Sciences, GITAM University, Vishakhapatnam, India

*Address all correspondence to: shyambabuprasad@mgcub.ac.in

References

[1] Andersen, J.M., D. Al-Khairy, and R.R. Ingalls, *Innate immunity at the mucosal surface: role of toll-like receptor 3 and toll-like receptor 9 in cervical epithelial cell responses to microbial pathogens.* Biol Reprod, 2006. **74**(5): p. 824-31.

[2] Takeda, K., T. Kaisho, and S. Akira, *Toll-like receptors.* Annu Rev Immunol, 2003. **21**: p. 335-76.

[3] Heil, F., et al., *Species-specific recognition of single-stranded RNA via toll-like receptor 7 and 8.* Science, 2004. **303**(5663): p. 1526-9.

[4] Mogensen, T.H., *Pathogen recognition and inflammatory signaling in innate immune defenses.* Clin Microbiol Rev, 2009. **22**(2): p. 240-73, Table of Contents.

[5] Geddes, K., J.G. Magalhaes, and S.E. Girardin, *Unleashing the therapeutic potential of NOD-like receptors.* Nat Rev Drug Discov, 2009. **8**(6): p. 465-79.

[6] Bell, J.K., et al., *Leucine-rich repeats and pathogen recognition in Toll-like receptors.* Trends Immunol, 2003. **24**(10): p. 528-33.

[7] Verma, S., et al., *Release of cytokines by brain endothelial cells: A polarized response to lipopolysaccharide.* Brain Behav Immun, 2006. **20**(5): p. 449-55.

[8] Balkwill, F. and L.M. Coussens, *Cancer: an inflammatory link.* Nature, 2004. **431**(7007): p. 405-6.

[9] Kelly, M.G., et al., *TLR-4 signaling promotes tumor growth and paclitaxel chemoresistance in ovarian cancer.* Cancer Res, 2006. **66**(7): p. 3859-68.

[10] Kawasaki, T. and T. Kawai, *Toll-like receptor signaling pathways.* Front Immunol, 2014. **5**: p. 461.

[11] Singh, V.K. and H.B. Pollard, *Patents for Toll-like receptor ligands as radiation countermeasures for acute radiation syndrome.* Expert Opin Ther Pat, 2015. **25**(10): p. 1085-92.

[12] Kawai, T. and S. Akira, *TLR signaling.* Cell Death Differ, 2006. **13**(5): p. 816-25.

[13] Goto, Y., et al., *Activation of Toll-like receptors 2, 3, and 4 on human melanoma cells induces inflammatory factors.* Mol Cancer Ther, 2008. **7**(11): p. 3642-53.

[14] Mantovani, A., et al., *Cancer-related inflammation.* Nature, 2008. **454**(7203): p. 436-44.

[15] Grimmig, T., et al., *TLR7 and TLR8 expression increases tumor cell proliferation and promotes chemoresistance in human pancreatic cancer.* Int J Oncol, 2015. **47**(3): p. 857-66.

[16] Kaczanowska, S., A.M. Joseph, and E. Davila, *TLR agonists: our best frenemy in cancer immunotherapy.* J Leukoc Biol, 2013. **93**(6): p. 847-63.

[17] Karin, M. and F.R. Greten, *NF-kappaB: linking inflammation and immunity to cancer development and progression.* Nat Rev Immunol, 2005. **5**(10): p. 749-59.

[18] Rakoff-Nahoum, S. and R. Medzhitov, *Toll-like receptors and cancer.* Nat Rev Cancer, 2009. **9**(1): p. 57-63.

[19] Coussens, L.M. and Z. Werb, *Inflammation and cancer.* Nature, 2002. **420**(6917): p. 860-7.

[20] Medvedev, A.E., *Toll-like receptor polymorphisms, inflammatory and infectious diseases, allergies, and cancer.* J Interferon Cytokine Res, 2013. **33**(9): p. 467-84.

[21] Tipping, P.G., *Toll-like receptors: the interface between innate and adaptive*

immunity. J Am Soc Nephrol, 2006. **17**(7): p. 1769-71.

[22] Pandey, N.O., et al., *Association of TLR4 and TLR9 polymorphisms and haplotypes with cervical cancer susceptibility.* Sci Rep, 2019. **9**(1): p. 9729.

[23] Kagan, J.C., et al., *TRAM couples endocytosis of Toll-like receptor 4 to the induction of interferon-beta.* Nat Immunol, 2008. **9**(4): p. 361-8.

[24] Jin, M.S. and J.O. Lee, *Structures of the toll-like receptor family and its ligand complexes.* Immunity, 2008. **29**(2): p. 182-91.

[25] Kawai, T. and S. Akira, *Pathogen recognition with Toll-like receptors.* Curr Opin Immunol, 2005. **17**(4): p. 338-44.

[26] Akira, S., S. Uematsu, and O. Takeuchi, *Pathogen recognition and innate immunity.* Cell, 2006. **124**(4): p. 783-801.

[27] Takeda, K. and S. Akira, *TLR signaling pathways.* Semin Immunol, 2004. **16**(1): p. 3-9.

[28] O'Neill, L.A., *How Toll-like receptors signal: what we know and what we don't know.* Curr Opin Immunol, 2006. **18**(1): p. 3-9.

[29] McDermott, E.P. and L.A. O'Neill, *Ras participates in the activation of p38 MAPK by interleukin-1 by associating with IRAK, IRAK2, TRAF6, and TAK-1.* J Biol Chem, 2002. **277**(10): p. 7808-15.

[30] Pandey, S. and D.K. Agrawal, *Immunobiology of Toll-like receptors: emerging trends.* Immunol Cell Biol, 2006. **84**(4): p. 333-41.

[31] Marodi, L., *Neonatal innate immunity to infectious agents.* Infect Immun, 2006. **74**(4): p. 1999-2006.

[32] Grivennikov, S.I., F.R. Greten, and M. Karin, *Immunity, inflammation, and cancer.* Cell, 2010. **140**(6): p. 883-99.

[33] Balkwill, F. and A. Mantovani, *Inflammation and cancer: back to Virchow?* Lancet, 2001. **357**(9255): p. 539-45.

[34] Lee, M.S. and Y.J. Kim, *Signaling pathways downstream of pattern-recognition receptors and their cross talk.* Annu Rev Biochem, 2007. **76**: p. 447-80.

[35] Korniluk, A., et al., *From inflammation to cancer.* Ir J Med Sci, 2017. **186**(1): p. 57-62.

[36] Hanahan, D. and R.A. Weinberg, *The hallmarks of cancer.* Cell, 2000. **100**(1): p. 57-70.

[37] Ernst, P.B., H. Takaishi, and S.E. Crowe, *Helicobacter pylori infection as a model for gastrointestinal immunity and chronic inflammatory diseases.* Dig Dis, 2001. **19**(2): p. 104-11.

[38] Kim, W.Y., et al., *Increased expression of Toll-like receptor 5 during progression of cervical neoplasia.* Int J Gynecol Cancer, 2008. **18**(2): p. 300-5.

[39] Pikarsky, E., et al., *NF-kappaB functions as a tumour promoter in inflammation-associated cancer.* Nature, 2004. **431**(7007): p. 461-6.

[40] Kawai, T. and S. Akira, *The role of pattern-recognition receptors in innate immunity: update on Toll-like receptors.* Nat Immunol, 2010. **11**(5): p. 373-84.

[41] Schmausser, B., et al., *Toll-like receptors TLR4, TLR5 and TLR9 on gastric carcinoma cells: an implication for interaction with Helicobacter pylori.* Int J Med Microbiol, 2005. **295**(3): p. 179-85.

[42] Yu, L. and S. Chen, *Toll-like receptors expressed in tumor cells: targets for therapy.* Cancer Immunol Immunother, 2008. **57**(9): p. 1271-8.

[43] Yang, H., et al., *Reduced expression of Toll-like receptor 4 inhibits human breast cancer cells proliferation and*

inflammatory cytokines secretion. J Exp Clin Cancer Res, 2010. **29**: p. 92.

[44] Szajnik, M., et al., *TLR4 signaling induced by lipopolysaccharide or paclitaxel regulates tumor survival and chemoresistance in ovarian cancer.* Oncogene, 2009. **28**(49): p. 4353-63.

[45] Nihon-Yanagi, Y., et al., *Tissue expression of Toll-like receptors 2 and 4 in sporadic human colorectal cancer.* Cancer Immunol Immunother, 2011. **61**(1): p. 71-7.

[46] Jego, G., et al., *Pathogen-associated molecular patterns are growth and survival factors for human myeloma cells through Toll-like receptors.* Leukemia, 2006. **20**(6): p. 1130-7.

[47] Triozzi, P.L., W. Aldrich, and S. Ponnazhagan, *Inhibition and promotion of tumor growth with adeno-associated virus carcinoembryonic antigen vaccine and Toll-like receptor agonists.* Cancer Gene Ther, 2011. **18**(12): p. 850-8.

[48] Davis, M.B., et al., *Intratumoral administration of TLR4 agonist absorbed into a cellular vector improves antitumor responses.* Clin Cancer Res, 2011. **17**(12): p. 3984-92.

[49] Kuznik, A., et al., *Mechanism of endosomal TLR inhibition by antimalarial drugs and imidazoquinolines.* J Immunol, 2011. **186**(8): p. 4794-804.

[50] Mohamed, F.E., et al., *Effect of toll-like receptor 7 and 9 targeted therapy to prevent the development of hepatocellular carcinoma.* Liver Int, 2014. **35**(3): p. 1063-76.

[51] Anwar, M.A., et al., *Recent clinical trends in Toll-like receptor targeting therapeutics.* Med Res Rev, 2018. **39**(3): p. 1053-1090.

[52] Gu, X., et al., *A novel derivative of artemisinin inhibits cell proliferation and metastasis via down-regulation of cathepsin K in breast cancer.* Eur J Pharmacol, 2019. **858**: p. 172382.

[53] Zandi, Z., et al., *The anticancer effect of the TLR4 inhibition using TAK-242 (resatorvid) either as a single agent or in combination with chemotherapy: A novel therapeutic potential for breast cancer.* J Cell Biochem, 2019. **121**(2): p. 1623-1634.

[54] Hou, L.F., et al., *Oral administration of artemisinin analog SM934 ameliorates lupus syndromes in MRL/lpr mice by inhibiting Th1 and Th17 cell responses.* Arthritis Rheum, 2011. **63**(8): p. 2445-55.

[55] Reilly, M., et al., *Randomized, double-blind, placebo-controlled, dose-escalating phase I, healthy subjects study of intravenous OPN-305, a humanized anti-TLR2 antibody.* Clin Pharmacol Ther, 2013. **94**(5): p. 593-600.

[56] Gao, W., et al., *Inhibition of Toll-Like Receptor Signaling as a Promising Therapy for Inflammatory Diseases: A Journey from Molecular to Nano Therapeutics.* Front Physiol, 2017. **8**: p. 508.

[57] Calore, F., et al., *The TLR7/8/9 Antagonist IMO-8503 Inhibits Cancer-Induced Cachexia.* Cancer Res, 2018. **78**(23): p. 6680-6690.

[58] Yesudhas, D., et al., *Multiple roles of toll-like receptor 4 in colorectal cancer.* Front Immunol, 2014. **5**: p. 334.

[59] Cai, Z., et al., *Activation of Toll-like receptor 5 on breast cancer cells by flagellin suppresses cell proliferation and tumor growth.* Cancer Res, 2011. **71**(7): p. 2466-75.

[60] Li, X., S. Jiang, and R.I. Tapping, *Toll-like receptor signaling in cell proliferation and survival.* Cytokine, 2010. **49**(1): p. 1-9.

[61] Li, X., et al., *CpG ODN107 potentiates radiosensitivity of human glioma cells via TLR9-mediated NF-kappaB activation and NO production.* Tumour Biol, 2012. **33**(5): p. 1607-18.

[62] Patinote, C., et al., *Agonist and antagonist ligands of toll-like receptors 7 and 8: Ingenious tools for therapeutic purposes.* Eur J Med Chem, 2020. **193**: p. 112238.